D0207991

THE DEFECT CHEMISTRY
OF METAL OXIDES

D. M. Smyth

Lehigh University

New York   Oxford
OXFORD UNIVERSITY PRESS
2000

Oxford University Press

Oxford   New York
Athens   Auckland   Bangkok   Bogotá   Buenos Aires   Calcutta
Cape Town   Chennai   Dar es Salaam   Delhi   Florence   Hong Kong   Istanbul
Karachi   Kuala Lumpur   Madrid   Melbourne   Mexico City   Mumbai
Nairobi   Paris   São Paulo   Singapore   Taipei   Tokyo   Toronto   Warsaw

and associated companies in
Berlin   Ibadan

Published by Oxford University Press, Inc.,
198 Madison Avenue, New York, New York, 10016
http://www.oup-usa.org

Oxford is a registered trademark of Oxford University Press

Library of Congress Cataloging-in-Publication Data
Smyth, Donald Morgan, 1930–
    The defect chemistry of metal oxides / D. M. Smyth.
        p.   cm.
    Includes bibliographical references and index.
    ISBN 0-19-511014-5 (cloth)
    1. Metallic oxides.   2. Crystals—Defects.   I. Title.
QD181.01 S55   2000
548'.8—dc21   99-087179

Printing (last digit): 9 8 7 6 5 4 3 2 1

Printed in the United States of America
on acid-free paper

This book is dedicated to my wife,
Elisabeth Luce Smyth,
my partner in life.

Contents

Preface

This book is designed to help those with little or no background in the field of defect chemistry to apply its principles and to interpret the related behavior of materials. It is the product of a course for advanced undergraduates and graduate students that was taught by the author at Lehigh University for over twenty years. The course is highly interdisciplinary and has been attended by students from the departments of chemical engineering, chemistry, electrical engineering and computer science, geology, materials science and engineering, and physics. The only prerequisite is introductory chemistry, although some very basic thermodynamics will be helpful. The book is intended for use either as a text for a similar course or as a reference work that covers the major principles of defect chemistry. While it starts from a very basic level, it proceeds to fairly advanced interpretations and should also be of use to those with some acquaintance with the field. Unfortunately, only rarely is an entire course devoted to defect chemistry, a field commonly covered in two to four weeks in courses on diffusion, electrical properties, or more general physical properties. That brief exposure almost always proves to be inadequate for any useful application of the material. It is to be hoped that the availability of a book designed to be an introductory text will encourage more thorough coverage.

The subject is developed in a very systematic and sequential way, and an attempt has been made to define terms and symbols clearly and consistently. Hopefully this will help to bring some order to a somewhat chaotic state of terminology and notation. The defect notation used throughout is that proposed forty years ago by Kröger and Vink. It is clear and self-consistent, and, in the opinion of the author, no deviation from it should be tolerated in any publication.

For those who have some background in defect chemistry, it will not be necessary to plow dutifully through the book from beginning to end, or even through complete chapters, since many sections have been designed to be reasonably complete by themselves. A few problems have been included. It is the experience of the author that while readers may feel that they understand the material, there may still be a problem in applying it to real or hypothetical cases. Attempts to do the problems often lead to a much better understanding.

Much of the treatment of defect chemistry can be shortened by recognizing that once an adequate number of simultaneous equations have been established, they can be fed into a computer that will tabulate or plot out the results. However, this does not lead to an understanding of the chemical interactions of the system. For that reason, this book takes the more laborious and traditional approach of breaking the system down into its component parts so that the chemistry is fully exposed to view.

Special acknowledgment must be given to two books that have proved to be particularly helpful in the development of both the course and this book. *Introduction to Solid State Physics*, by Charles Kittel (3rd edition, John Wiley & Sons, New York, 1966) has an unusually lucid treatment of the electronic properties of semiconducting solids, and chapters by A. D. Franklin, R. G. Fuller, and A. S. Nowick in *Point Defects in Solids*, Volume 1, *General and Ionic Crystals* (edited by J. H. Crawford Jr. and L. M. Slifkin, Plenum Press, New York, 1972) have been particularly useful as sources of experimental data and illustrative examples.

The book has benefited greatly from the patience of the students who have been subjected to the evolution of understanding and presentation that inevitably grows out of the teaching interaction. It has also been a special privilege to share the learning process with so many dedicated graduate students and postdoctoral associates who have worked in the author's laboratory at Lehigh. Also most helpful have been discussions and friendly arguments with many colleagues, including Harlan Anderson, Nick Eror, George Shirn, and Terry Tripp at the now defunct Research and Development Laboratories of the Sprague Electric Company, and later, at Lehigh University, especially Sid Butler, Helen Chan, Martin Harmer, and the late Frank Feigl. In addition, a substantial portion of this book was written while the author was on leave at the Sandia National Laboratories in Albuquerque, New Mexico, and the opportunity to collaborate with the research activities there, especially with Duane Dimos, Bill Hammetter, Bruce Tuttle, and Bill Warren, is deeply appreciated. I am also grateful to Ms. Sharon Balogh for the beautiful line drawings. Finally, and most importantly, none of this could have been accomplished without the support and patience of my wife, Betty, who has suffered through years of trying to communicate while my brain was wandering through some crystalline lattice looking for vacancies, interstitials, and stray electrons.

Introduction

1

The orderly array of atoms in a crystalline solid is one of the many beauties of the natural world. This was first appreciated in the macroscopic form of naturally occurring crystals, such as quartz or feldspar. Subsequently, x-ray and other diffraction techniques have given detailed information on the geometric arrangement of the atoms. More recently, it has been possible to image the individual planes of atoms, or even the atoms themselves, by high resolution transmission electron microscopy or by scanning tunneling microscopy. The most striking feature of such crystalline materials is their structural orderliness.

Defect chemistry is the study of deviations from perfect order in crystalline inorganic compounds, and the effects of such disorder on their properties. The ideally perfect crystal does not exist in the real world, given the practical limitations on purification and the impossibility of achieving thermodynamic equilibrium at 0 K, but in some cases the amount of disorder may be low enough to be without effect on observable properties. Such nearly perfect crystals may be useful for optical or dielectric applications that depend on bulk properties such as the refractive index or dielectric permittivity, or they may have decorative value as gemstones, but they will be nearly inert in terms of the transport of mass or charge. Thus ionic conductivity, electronic conductivity, and diffusion all depend on the presence of local deviations from perfect crystalline order. This was first appreciated by Frenkel (1926) when he realized that an atomic species in a crystalline solid cannot diffuse from one site to another without becoming a defect, a temporary deviation from the perfect crystal. That observation was the beginning of defect chemistry and has been commemorated by the use of the term "Frenkel defects" to designate one of the most important types of lattice disorder.

A knowledge of defect chemistry has contributed to an understanding of the formation and development of the latent image in the photographic process. It allows a prediction of the rate at which an oxide scale will grow on a metal exposed to air at high temperatures. It suggests how to design mixed ionic–electronic conductors for use as electrodes in high temperature fuel cells. It explains why some oxides must be processed in an oxidizing atmosphere, such as air, in order to be electrically insulating, while other oxides must be processed in highly reducing atmospheres to obtain similar properties. It contributes to an understanding of the sintering process whereby powders are thermally consolidated into the dense solid products needed for structural and electrical applications. It gives guidance for the formulation and processing of metal oxides for electronic applications. These are only a few examples that illustrate the contribution of defect chemistry to real problems and applications.

Defects are defined as any deviation from the ideally perfect crystal, which therefore serves as a thermodynamic standard state. For the most part, we will deal with point defects, those that involve only a single atomic species or lattice site. More complex types of disorder such as dislocations and grain boundaries are also defects, but they arise for quite different reasons and contribute to a different set of properties. They are not generally part of the equilibrium state of a crystal. Point defects include normally occupied lattice sites that are empty, called vacancies, and atoms that occupy the wrong site, whether a normally unoccupied interstice in the lattice, hence the name interstitial, or substituted for another normal component of the crystal. In the latter case, the species located where it is not expected may also be an impurity atom. Any electron that is not in its lowest available energy state is considered to be a defect, as is a missing electron, which is called a hole.

Defects can be present in all kinds of crystals, inorganic, organic, polymeric, metallic, and so on. Because of their practical importance, we emphasize defects in inorganic compounds, primarily metal oxides, but with some discussion of halides, since they are particularly useful for demonstrating various defect-related behaviors. The elemental semiconductors such as Si and Ge, and their closely related neighbors the III-V compounds (e.g., GaAs and InSb), are used only to introduce electronic disorder. There is a strong emphasis on binary compounds such as NaCl, NiO, and TiO_2, but a more complex compound, $BaTiO_3$, also is described because of the increasing practical importance of more complicated materials. It is assumed that an ionic model of the compounds is an acceptable approximation to the true mix of ionic and covalent bonding. The complications that result from explicitly including the covalent contributions largely cancel out in defect chemistry, and our concentration on metal oxides and halides ensures that we will be dealing mostly with materials having a very substantial amount of ionic character.

Defects must be considered in the context of the crystal structure in which they are found. A type of defect that may be quite important in one structure may be very unfavorable in another. For this reason, we start by describing a few of the most important structures for the compounds of interest, such as the NaCl, fluorite, rutile, and perovskite structures. The number of structures described is very limited, and the description is in the terms most useful for defect chemistry. This means that local coordination numbers and symmetry, and the relative sizes of ions and lattice sites are emphasized, rather than more mathematical, diffraction-related descriptions. The discussion of crystal structures is followed by a chapter establishing the validity and usefulness of the mass-action approach based on a dilute solution view of the equilibrium state. Defects are then introduced in the form of intrinsic ionic defects, defects that can be part of the thermodynamic equilibrium state of the crystal. There follows a discussion of extrinsic ionic defects, those that are related to the presence of an impurity. A discussion of defect interactions then sets the stage for an analysis of ionic conduction and diffusion processes in ionic solids. Intrinsic electronic disorder requires a rudimentary version of the electronic band model. Extrinsic electronic defects then introduce the very important concept of nonstoichiometry, the deviation from a specific, defined composition designated as the stoichiometric composition. The application of defect chemistry to some real systems is described, and the book concludes with a brief discussion of systems in which large defect concentrations are eliminated by local adjustments in the crystal structure. Examples of practical implications are scattered throughout the text.

Defects can be treated as identifiable species dissolved in an otherwise perfect crystal. This is directly analogous to aqueous salt solutions, whose ions are viewed as identifiable species dissolved in the otherwise pure water. It is seldom necessary, in either case, to explicitly include the matrix material, whether water or the defect-free parts of a crystal, in the description. We are primarily concerned with the deviations from those standard states. The same types of thermodynamic relationship can be used in both systems, such as the equilibrium chemical reactions and mass-action expressions that are so familiar in aqueous chemistry. The approximations of dilute solution thermodynamics can be applied to surprisingly high defect concentrations in the solid state.

A major theme of this book, and one of its main differences from other books in the field, is the repeated use of the known chemical properties of the constituent ionic species in a predictive mode. This includes not only the relative sizes and charges of the ions, but their chemistry in the sense of whether they have single, fixed oxidation states, or can be oxidized to a higher state or reduced to a lower state, and, if so, how easily. A little bit of chemical knowledge of this type can be enormously helpful in predicting the behavior and properties of specific systems.

Finally, this is not a book replete with mathematical theories, but one in which the emphasis is on visualizing physical models. The concepts of defect chemistry are not difficult, but are mainly based on a few rather obvious rules. If those rules are applied rigorously and sensibly, all goes well. However, there are pitfalls in the seeming simplicity, as can be seen by even a cursory glance at the published literature in the field.

REFERENCE

Frenkel, I. Z. *Phys.* 35:652, 1926.

A Few Useful Crystal Structures

2

INTRODUCTION

To be able to make judgments about which types of lattice defect are most likely in various compounds, and to estimate their relative mobilities, it is necessary to consider the defects in the context of each specific structure. The amount of space available and the electrostatic environment of various sites are important considerations in interpreting defect-related behavior. It is particularly significant that the ideal crystal structures represent thermodynamic standard states for the treatment of defect equilibria and are the reference structures from which defects are defined. Therefore, it is essential to be able to understand a few key structures in sufficient detail to permit a self-consistent discussion of defect chemistry. We will suffer from the usual limitation of trying to describe three-dimensional objects on a two-dimensional page. People differ greatly in their ability to visualize the additional dimension, and the reader is encouraged to use three-dimensional crystal models to supplement the following discussion. It is not necessary to absorb all of the contents of this chapter before plunging into the subsequent discussion of defect chemistry. An understanding of the NaCl and fluorite structures will suffice for most of the early treatment.

A purely ionic model will prove to be adequate for the defect chemistry of metal oxides and halides. Thus the constituents of a crystal can be treated as classical ions with integral charges. After all, if the actual charge on the cation is less than the classical charge, then charge neutrality demands that the charge on the anion be similarly reduced. The differences balance out because defects must always occur in electrically neutral combinations. Only in a few specific circumstances will it be helpful to consider explicitly the covalent contribution to bonding.

There are several different ways to describe crystals, and we shall use those that are most useful for each specific case. However, a few basic principles are usually followed. In an ionic crystal, the lattice energy will be optimized by surrounding each cation with as much anionic charge as possible, and vice versa. Likewise, cations and anions should be located as far away from their own kind as possible. This desire for large coordination numbers must be tempered by the relative sizes of the cations and anions. Thus it may be electrostatically more favorable to have six anions closely packed around a central cation than to have eight anions that crowd each other out to a more distant spacing. In any case, the relative coordination numbers of the cations and anions must precisely reflect the stoichiometry of the compound, as must the ratio of cation to anion sites. In many cases, the optimization of electrostatic energy results in an efficient filling of crystal space; such cases are called close-packed structures.

Cations are usually smaller than anions, and this affects the way crystals are constructed. When electrons are removed from an atom to make a cation, the remaining electrons are more strongly attracted on the average by the positive nuclear charge, and the electron cloud shrinks. Therefore, cations are smaller than the atoms from which they are derived, and become increasingly so with increasing charge. By an analogous argument, anions are bigger than the atoms from which they are formed. This generalization is clearly demonstrated in a compound such as MgF_2. Both the Mg^{2+} and the F^- ions have the same electronic structure: $1s^2/2s^22p^6$, the stable configuration of the rare gas neon. In Mg^{2+}, these 10 electrons are attracted by 12 positive charges in the nucleus, and the ionic radius is 0.072 nm (or 0.72 Å, or 72 pm, depending on the preferred units). However, in F^- the 10 electrons are attracted

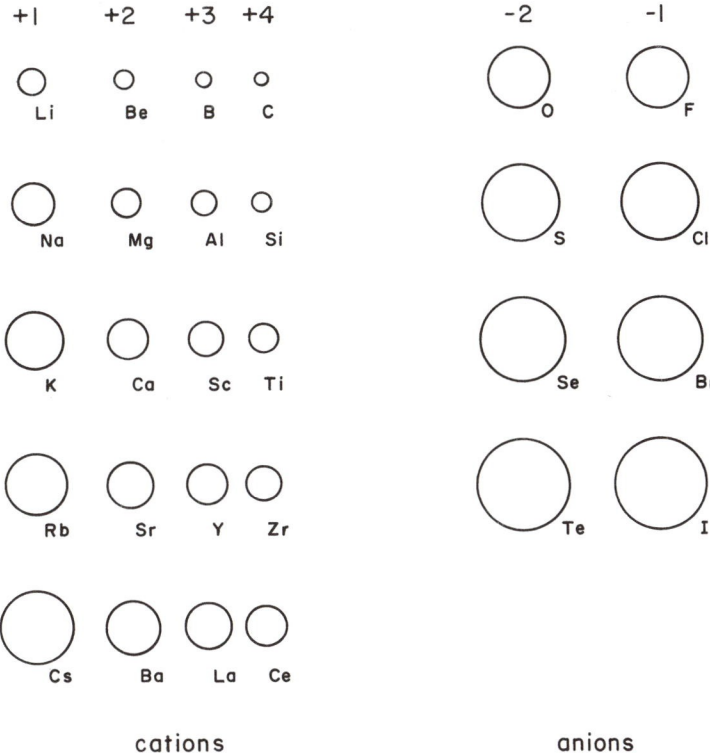

Figure 2.1 The relative sizes of selected ions, demonstrating that cations are generally much smaller than anions. The usual oxidation states of the ions are listed across the top.

by only 9 positive nuclear charges, and the ionic radius is 0.133 nm, almost twice the size of the isoelectronic Mg^{2+}. The relative sizes of selected cations and anions are depicted schematically in Fig. 2.1: only in the cases of the largest cations and the smallest anions are the sizes comparable. As a result it is often convenient to visualize a crystal as being made up of the anions packed as efficiently as possible (i.e., in a true or approximately close-packed structure) and then to see how the cations fit into the remaining space.

CLOSE-PACKED STRUCTURES

Proceeding as just described, let us first consider how to fill space most efficiently with anions treated as rigid spheres of equal size. The closest possible packing in a single layer is shown in Fig. 2.2a. Two sets of three adjacent anions are cross-hatched so that they can be identified as we add successive layers. A second layer of the same kind can then be placed on top such that its anions nestle efficiently into the depressions between the anions of the first layer. It is seen that there are two alternative sets of depressions into which the second layer can be placed; in the case shown in Fig. 2.2b, the set marked by dots in Fig 2.2a have been used. For only two layers, the alternative orientations are equivalent, but when a third layer is added, the two choices result in two different geometries. In one option, the third layer can be located directly over the first layer; this results in a layer sequence of A-B-A-B- which has hexagonal symmetry

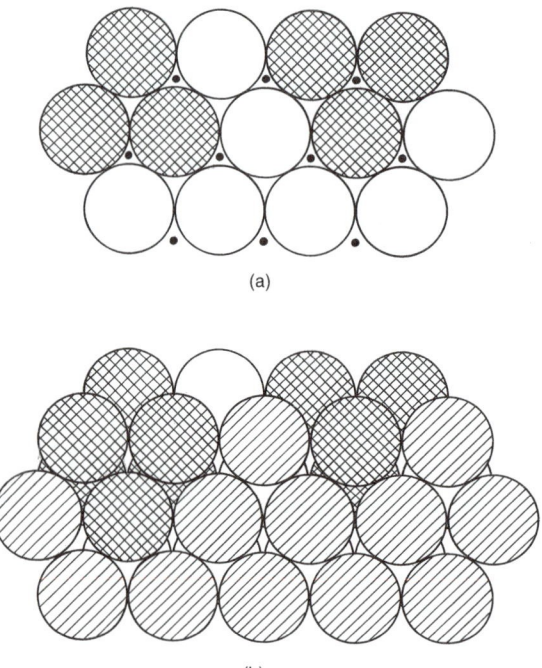

(a)

(b)

Figure 2.2 (a) A single close-packed layer of equal-sized spheres. The dots indicate the location of the spheres in the second layer shown in Fig. 2.2b. The cross-hatched spheres represent part of an octahedral (left) and a tetrahedral (right) array of anions around central cations as shown in Fig. 2.2b. (b) A second layer of close-packed spheres has been placed over the dots shown in Fig. 2.2a. Note that there is an alternate location for the upper layer. The cross-hatched spheres represent completed octahedral (left) and tetrahedral (right) arrays of anions around central cation positions.

as shown in Fig. 2.3a. This is called the hexagonal-close-packed, or hcp, arrangement. In the other option, the third layer is not over the first, but the fourth layer is; this gives a layer sequence of A-B-C-A-B-C-, which has cubic symmetry as shown in Fig. 2.3b. This is the cubic-close-packed, or ccp, arrangement. The alternative view of the ccp arrangement in Fig. 2.3c shows that it is a special case of the face-centered-cubic, or fcc, structure. In Fig. 2.3c close-packed planes are shown as perpendicular to a line drawn from the lower-left-front corner to the upper-back-right corner. They are the 111 planes. The hcp and ccp arrangements are equally efficient in filling space, with the spheres occupying 74% of the total available volume.

Both the hcp and the ccp arrangements have the same two types of interstices between the close-packed layers of anions, into which smaller cations can be placed. These are indicated schematically by the cross-hatched spheres in Fig. 2.2b. In the set

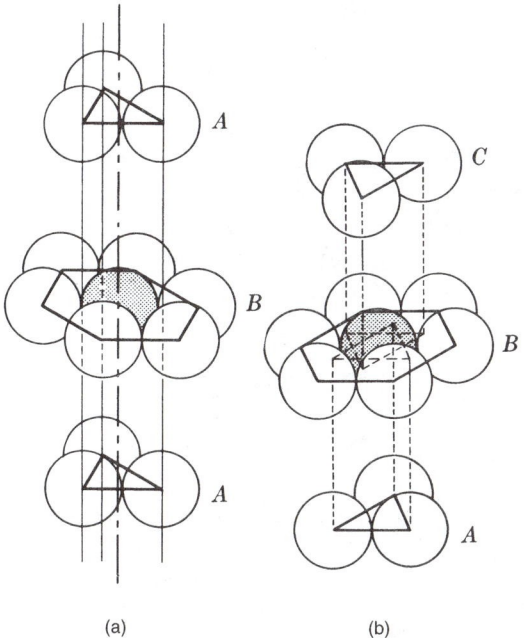

(a)   (b)

Figure 2.3 (a) An exploded view of three layers of close-packed spheres in hexagonal symmetry. Note that the triangles in the top and bottom layers point in the same direction. (Reproduced from Van Vlack, 1964, by permission of Prentice-Hall.) (b) An exploded view of three close-packed layers of spheres in cubic symmetry. Note that the triangles in the top and bottom layers point in opposite directions. (c) Three views of an fcc array of equally-sized spheres. (Reproduced from Smith, 1990, by permission of McGraw-Hill.)

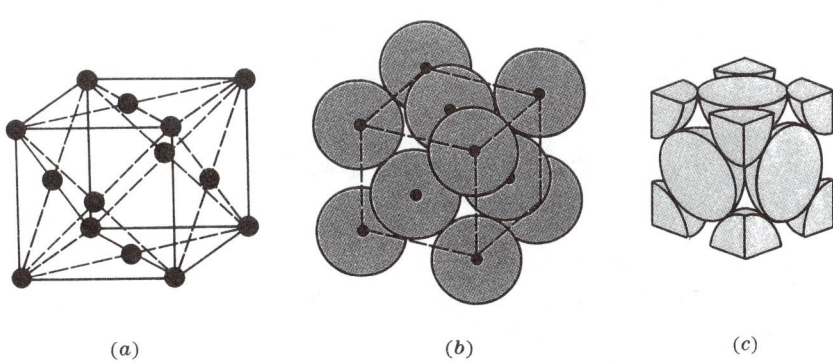

(a)   (b)   (c)

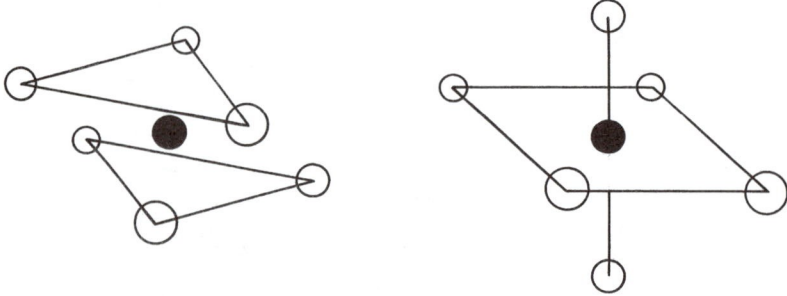

Figure 2.4 Two views of an octahedral coordination sphere.

on the left, the nearest anion neighbors are three anions in the layer above the cation and three in the layer below. They surround the cation at the corners of a regular octahedron (remember that geometric figures are named after the number of faces, not the number of apices), so these are called the octahedral cation sites. As shown in Fig. 2.4, there two different ways to view this unit of a cation surrounded by six anions in octahedral symmetry: either as three anions above and three below with the two anionic triangles oriented in opposite directions, or tipped over to appear as a square of four anions around the cation with two more directly above and below. The two views are identical, but the latter is often the most useful for visualizing crystal structures. A simple calculation shows that a cation that is 0.414 times the size of the close-packed anions exactly fits the octahedral site. The other type of interstice between the close-packed anion layers has four nearest anion neighbors, as shown by the right-hand set of cross-hatched spheres in Fig. 2.2b. The anions are arranged around the cation at the corners of a regular tetrahedron, and these interstices are called the tetrahedral sites. A slightly more complicated calculation shows that the ideal cation size for the tetrahedral sites is 0.225 times the size of the close-packed anions. Thus the tetrahedrally coordinated cation sites are only slightly more than half the size of the octahedrally coordinated cation sites. Both the octahedral and tetrahedral sites surround the cation with a spherically symmetrical array of anions. A number of the most important crystal structures can be derived by various schemes of filling the octahedral and tetrahedral sites in the ccp and hcp anion sublattices with cations.

The Octahedral Sites in Close-Packed Sublattices

One of the most common structures that involves the octahedral sites, and one we will use extensively, is the NaCl structure, sometimes called the rock salt structure. This may be obtained by placing cations in all the octahedral sites in the ccp anion array as in Fig. 2.5. Boundaries of the unit cell are shown in Fig. 2.5a. The anions are labeled as lying in the A, B, and C cubic-close-packed planes in Fig. 2.5b. The cations occupy the midpoints of the cube edges, plus the site at the cube center. The unit cell corresponds to the cube defined by lines drawn between the centers of

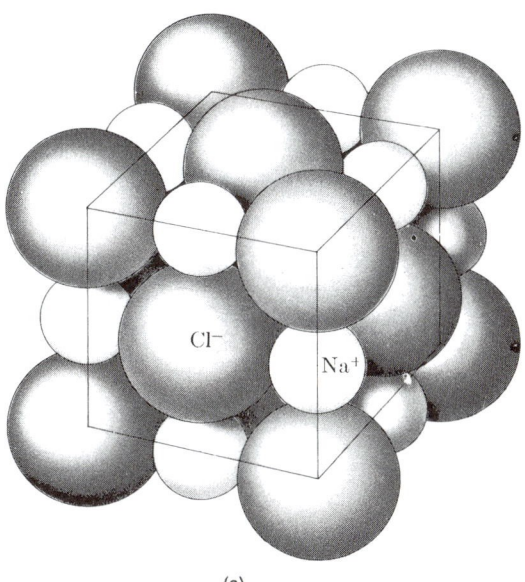

Figure 2.5 The NaCl structure: (a) model showing the boundaries of the unit cell (reproduced from Van Vlack, 1964, by permission of Prentice-Hall.) and (b) model in which letters designate layers of close-packed anions in -A-B-C- sequence (reproduced from Rodgers, 1994, by permission of McGraw-Hill).

(a)

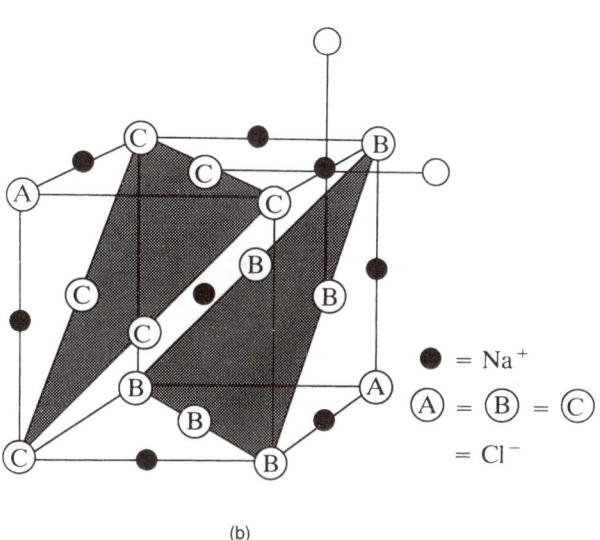

(b)

the corner anions. Thus one-eighth of each of the eight corner anions lie within the unit cell, plus half of each of the six face-centered anions, for a net total of four anions contained within the unit cell. One-quarter of each of the twelve cations on the cube edges lie within the unit cell, plus all the cation at the cube center, for a net total of four cations per unit cell. Thus the NaCl structure accommodates equal numbers of cations and anions and corresponds to an MX type of stoichiometry. This is a typical structure for highly ionic compounds because the cations and anions are optimally dispersed for favorable electrostatic interactions. It is the structure of almost all the alkali halides (LiF, KCl, RbI, CsF, etc.), the alkaline earth oxides (MgO, CaO, SrO, and BaO), and a wide variety of divalent transition metal oxides (MnO, FeO, NiO, etc.). There is considerable tolerance for variations in the cation/anion size ratios, and in many cases oversized cations force the fcc anion sublattice to expand beyond the ideal close-packed configuration. The anions are surrounded by six cations, also in octahedral symmetry; in a compound having equal numbers of cations and anions, these ions must have the same coordination numbers, but not necessarily the same coordination symmetry. The NaCl structure can be characterized as follows:

The NaCl Structure

MX stoichiometry

Cubic-close-packed anions (or at least fcc)

Cations in all octahedral sites

CA_6 octahedral symmetry around the cations

AC_6 octahedral symmetry around the anions

In the last two lines CA_6 denotes a cation C with its nearest coordination shell of six anions A, while AC_6 indicates the cation coordination around the anions.

As the reader may have surmised, there is a hexagonal analog of the NaCl structure: an hcp anion sublattice with all the octahedral sites occupied by cations. This is called the NiAs structure, and it is an alternative structure for compounds having MX stoichiometry and cations that can fit reasonably into octahedrally coordinated sites. However, a structure named after such an obscure compound as NiAs is clearly not one of the big winners in the structural sweepstakes. The problem is in the relative positions of the cations located between the close-packed anion layers. As can be seen from Fig. 2.6, the cations in successive layers are located directly above each other, whereas in the NaCl structure they are in staggered layers. Thus in the NiAs structure, the cations are in parallel strings of octahedral sites, perpendicular to the

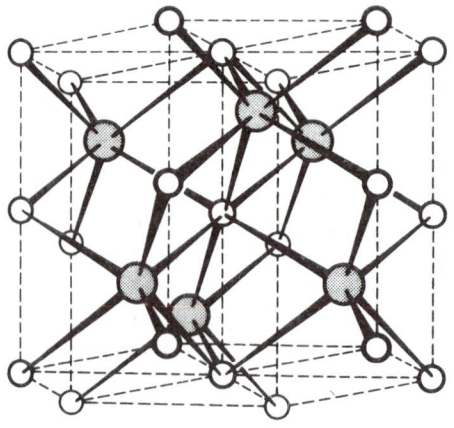

Figure 2.6 The NiAs structure. The larger spheres represent the hcp anion array, and the small, open spheres represent cations in the octahedral sites. Note that the cations are in linear columns and that the cation coordination around the anions is trigonal pyramidal, rather than octahedral as in the NaCl structure. (Reproduced from Wells, 1975, by permission of Oxford University Press.)

close-packed planes. The cation octahedra share common faces, which places the cations in the closest possible situation for adjacent octahedra; the cation–cation separation is greatest for corner-sharing octahedra, intermediate for edge-sharing octahedra, and closest for face-sharing (in the NaCl structure, the octahedra share a combination of corners and edges). While the cations are still in six-coordinate, octahedral symmetry in the NiAs structure, the anions are surrounded by six cations at the corners of a trigonal prism; the three cations above the cation and the three below are in the same triangular orientation, rather than opposed as in the NaCl structure. This is not a spherically symmetrical arrangement. This less effective separation of the cations makes this structure much less favorable for ionic compounds, and it is found only in rather covalent materials such as CrS, MnSb, and FeSn. By itself, it is of no interest to our study of relatively ionic compounds, but it serves as a convenient starting point for the description of some related structures.

If half the cations are removed from each layer between the close-packed anion layers in the NiAs structure, the result is nearly the important rutile structure, as shown in Fig. 2.7. Removing half the cations causes the coordination number of the anions to be reduced from 6 to 3, and the stoichiometry to be changed to MX_2. The anions sit at the apex of a very squat trigonal pyramid with a base of the three closest cations. To obtain a more symmetrical arrangement, the anions move into the plane of the three cations to achieve triangular coordination. The rutile structure is the most stable structure of TiO_2, MnO_2, and MgF_2, among others. The full characterization is as follows:

The Rutile Structure

MX_2 stoichiometry

Approximately hcp anions

Cations in half the octahedral sites (half in each layer)

CA_6 octahedral symmetry (somewhat distorted)

AC_3 triangular (planar) symmetry

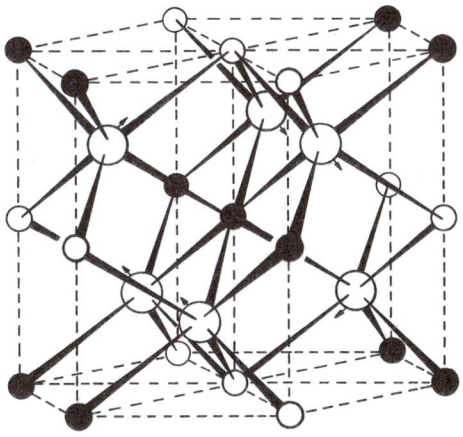

Figure 2.7 The rutile structure shown as a modification of the NiAs structure. The small spheres are located at the cation positions in the NiAs structure, but only those that have been blackened are occupied in the rutile structure. The movement of the anions into planar triangular coordination is indicated by the short arrows.

A closely related structure with the same MX_2 stoichiometry is obtained if all the cations from every other cation layer in the NiAs structure are removed; that is, the same fraction of the cations is eliminated, but in a different way. This is the CdI_2 structure, and it is appropriate only for relatively covalent structures because it leaves adjacent anion layers facing each other without any intervening cations. This arrangement would obviously be untenable for highly ionic compounds and is favorable only when a high degree of covalency reduces the actual magnitude of the anionic charges. There are no compounds with this structure that are of current interest to defect chemistry. There is a cubic analog, the $CdCl_2$ structure, in which alternate layers of cations in the NaCl structure are unoccupied.

Tetrahedral Sites in Close-Packed Sublattices

The occupation of the tetrahedrally coordinated sites in the close-packed anion sublattices is now considered. If all the tetrahedral sites in a ccp anion sublattice are occupied by cations, the resulting structure is known as the antifluorite structure. One can quickly assume that there is a much more important structure called the fluorite structure, but more about that later. The ccp unit cell contains eight tetrahedral sites, all completely within the cell. Thus there are eight cations in the unit cell to combine with the four anions. This means that the compound stoichiometry must be M_2X. Therefore, there can be no halides with this structure, and only those oxides in which the cation is monovalent. This is the structure of the alkali metal oxides (Li_2O, Na_2O, etc.), and since these oxides are extraordinarily sensitive to moisture, and are increasingly unstable with increasing cation size to further uptake of oxygen to peroxides and superoxides, they have no applications of interest. However, the inverse of this structure, the fluorite structure, is of great interest and will be discussed shortly. The hexagonal analog of the antifluorite structure, with all the tetrahedral sites in an hcp anion sublattice filled with cations, is not known to be used by any compound. As in the case of the octahedral sites in the hcp sublattice, the tetrahedral sites line up in columns perpendicular to the close-packed planes. The full occupation of all the tetrahedral sites in close-packed anion sublattices is a real loser among crystal structures.

If only half the tetrahedral sites are occupied by cations in the ccp and hcp anion sublattices, the results are the zincblende and wurtzite structures, respectively (Fig. 2.8 a and b). With only half the tetrahedral sites occupied, that is, a total of four cations per unit cell, the stoichiometry is reduced to MX, and both anions and cations are in four-fold tetrahedral coordination. These structures do not differ very much in their energetic stability, and a number of compounds have both crystal structures in different temperature ranges. For example, CuCl, CuBr, and CuI transform from the zincblende (low temperature) structure to the wurtzite structure at 435, 405, and 390°C, respectively. These structures are generally used not so much because the cation is small enough to prefer a tetrahedral site but to accommodate a tendency toward tetrahedrally directed covalent bonds by sp^3 hybridization. Thus the major users are the smaller d^{10} transition metal cations such as Cu^+ in its halides, as just described, Zn^{2+} in its oxide (wurtzite) and sulfide (both zincblende and wurtzite are

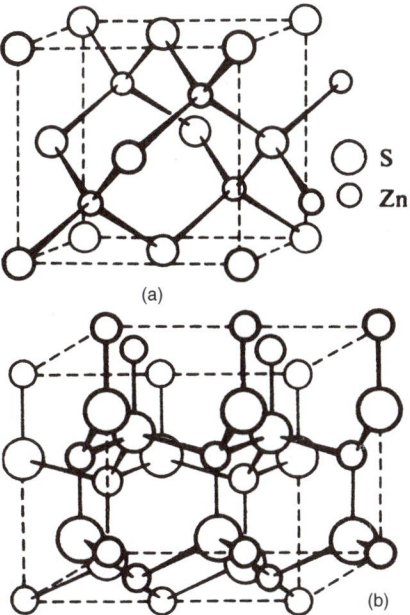

Figure 2.8 (a) The zincblende structure, which is equivalent to the antifluorite structure with half the tetrahedral sites left unoccupied. The anions are in cubic array. (b) The wurtzite structure, the hexagonal analog of the zincblende structure. (Reproduced from Wells, 1975, by permission of Oxford University Press.)

○ S
○ Zn

(a)

(b)

the names of naturally occurring mineral forms of ZnS), and the III-V compounds such as GaAs and InSb (zincblende). The latter are important semiconducting analogs of Si and Ge, which have the diamond structure. The diamond structure is the same as the zincblende structure, except that all the atoms are of the same kind. Thus these structures generally accommodate quite covalent compounds, and the cations are usually big enough to stretch the anion sublattice substantially beyond the ideal close-packed case.

STRUCTURES FOR EIGHT-COORDINATE CATIONS

In the structures described so far, the cations have been in either six-coordinate octahedral sites or four-coordinate tetrahedral sites. If the cation is big enough, it is possible for it to occupy eight-coordinate sites with cubic symmetry.

When the relatively unimportant antifluorite structure was mentioned earlier, the reader was promised that the much more important fluorite structure was yet to come. These two structures have an inverse relationship, with the fluorite structure consisting of anions in all the tetrahedrally coordinated sites of an fcc cation sublattice. Since the anions are larger than the cations, the tetrahedral sites are expanded accordingly, and the cation sublattice is stretched far beyond the close-packed case and can be referred to only as fcc. This structure is shown in Fig. 2.9 (the antifluorite structure is the same except that the positions of the cations and anions are reversed). The large expansion from the close-packed state makes this a very open structure, and it is relatively easy for ions to move around in it. Oxides and halides having the

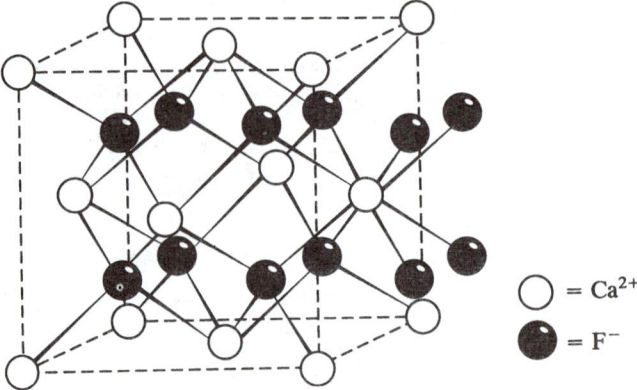

$$\bigcirc = Ca^{2+}$$
$$\bullet = F^-$$

Figure 2.9 The fluorite structure: open circles, cations in fcc array; black circles, anions in all the tetrahedral sites. (Reproduced from Rodgers, 1994, by permission of McGraw-Hill.)

fluorite structure are usually good ionic conductors, and some of them have important applications as a result. Since all the tetrahedral sites lie entirely within the unit cell, the structural stoichiometry is MX_2, and the fluorite structure is in competition with the isocompositional rutile structure. The major distinction is in the size of the cation. If the cation is big enough to be eight-coordinate, then it will prefer the fluorite structure; if not, it will use the six-coordinate rutile structure. Thus while MgF_2 and TiO_2 have the rutile structure, CaF_2 (the mineral fluorite) and the high temperature form of ZrO_2 prefer the fluorite structure. The latter can be characterized as follows:

The Fluorite Structure

MX_2 stoichiometry

fcc cations

anions in all of the tetrahedral sites

CA_8 cubic coordination

AC_4 tetrahedral coordination

More for completeness than for its importance, the CsCl structure is described, as well. This is an alternative to the NaCl structure and, as in the case of the rutile and fluorite structures, the distinction is primarily one of cation size. If the cation is big enough to be eight-coordinate, it will prefer the CsCl structure, as shown in Fig. 2.10; if not, it will use the six-coordinate NaCl structure. Actually, as seen in Table 2.1, there is considerable overlap in the two structures for the alkali halides in terms of the relative sizes of the cations and anions, and only CsCl, CsBr, and CsI assume the CsCl structure. Relative to Table 2.1, when the cation-to-anion size ratio exceeds unity, as in the fluorides of K, Rb, and Cs, one should look at the inverse ratio; that is, it is really the size ratio of the smaller ion to the larger ion that is important, because the NaCl structure actually consists of two identical, interpenetrating fcc sublattices, and

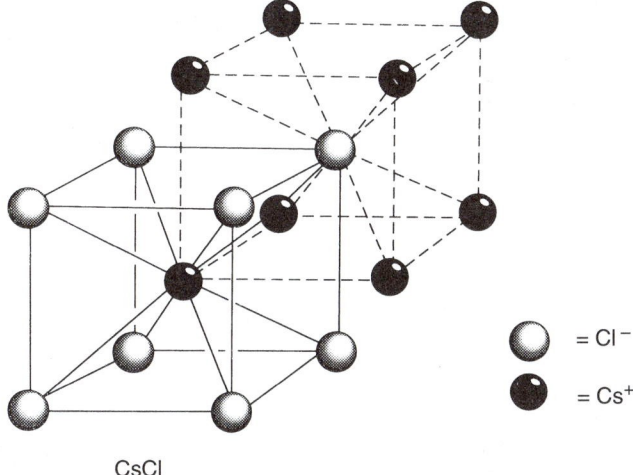

CsCl

Figure 2.10 The CsCl structure shown as two equivalent interpenetrating simple cubic sublattices. (Reproduced from Rodgers, 1994, by permission of McGraw-Hill.)

TABLE 2.1
The ratio of the cation ionic radii to the anion ionic radii for the alkali halides

	Li	Na	K	Rb	Cs
F	0.56	0.77	1.04[1]	1.12[1]	1.28[1]
Cl	0.41	0.57	0.77	0.83	0.94[2]
Br	0.38	0.52	0.71	0.76	0.87[2]
I	0.34	0.47	0.64	0.69	0.79[2]

1. Since the cations are larger than the anions in these compounds, the inverse ratio is the better criterion for the selection of structure.
2. These compounds have the CsCl structure.

Adapted from Wells, 1975.

it is the sublattice of the larger ion, whether it is the cation or the anion, that should be looked at as being derived from a close-packed structure.

In the CsCl structure, each ion sits at the center of a cube of eight ions of the other kind. It thus consists of two interpenetrating simple cubic sublattices. It is not a body-centered structure, since, by convention, in a bcc structure the atoms at the cube corners and the cube center are the same. Both the cations and the anions are eight-coordinated, in cubic symmetry. The ideal size for a cation that sits at the center of a close-packed cube of anions is 0.732 times that of the anion. However, for a compound having this size ratio, the anion will then be much too large to fit in the ideal body-centered site in the cubic array of cations. The best overall compromise for filling space efficiently is for the two ions to be of similar size. There are obviously some considerations other than the size ratio, since RbF and RbCl fall between CsCl and CsI in this regard but have the NaCl structure.

STRUCTURES FOR TERNARY COMPOUNDS

A number of very useful compounds contain three different atomic species and are thus called ternary compounds. Compounds of this class include $BaTiO_3$ and $MnFe_2O_4$. In these examples, the two cations occupy lattice sites with quite different symmetries. Two important structures for ternary compounds are the perovskite and spinel structures.

Perovskite Structure

This structure is named after the mineral perovskite, $CaTiO_3$, which does not have the perovskite structure. That is the crystallographer's idea of a joke. Actually, the structure of $CaTiO_3$ is only slightly distorted from the ideal cubic perovskite structure, and this discrepancy is too small to have been noticed by the early mineralogists. This is the preferred structure for the generic formula ABO_3, when one of the cations, traditionally designated as the A ion, is substantially larger than the other, the B ion in this case. This size difference is typically close to a factor of 2.

There are several alternative ways to view the perovskite structure, as shown in Fig. 2.11. In Fig. 2.11a, it is shown as a cube with the larger cation on the corners, the anions on the face centers, and the smaller cation at the cube center. This view emphasizes the octahedral coordination of the small cation. In Fig. 2.11b, the cube is moved along its diagonal so that the small cations are on the corners, the anions are located at the midpoint of each edge, and the large cation is at the cube center. This shows more clearly the 12-fold coordination around the large cation, and, by extension of the cell, the octahedral coordination around the small cation is also apparent. This very large coordination number for one of the cation sites, the defining feature of this structure, explains why one of the cations must be much larger than the other. It also explains why the mineral perovskite has a slightly distorted perovskite structure; Ca^{2+} is too small to be comfortable in an ideal 12-coordinated site. The

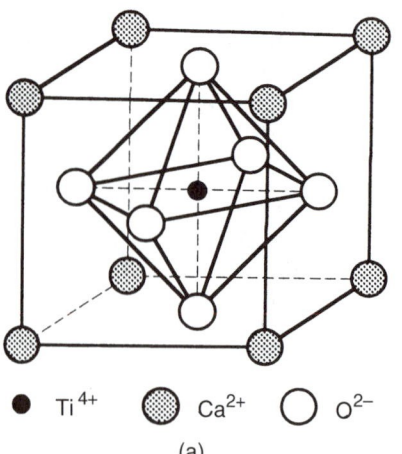

Figure 2.11   The perovskite structure (a) with the small cation at the cube center (reproduced from Barsoum, 1997, by permission of McGraw-Hill), (b) with the large cation at the cube center (reproduced from Evans, 1964, by permission of Cambridge University Press), and (c) showing the close-packed layers that contain the large cations and the anions (reproduced from Wells, 1975, by permission of Oxford University Press).

● Ti^{4+} ◉ Ca^{2+} ○ O^{2-}

(a)

Figure 2.11 (*continued*)

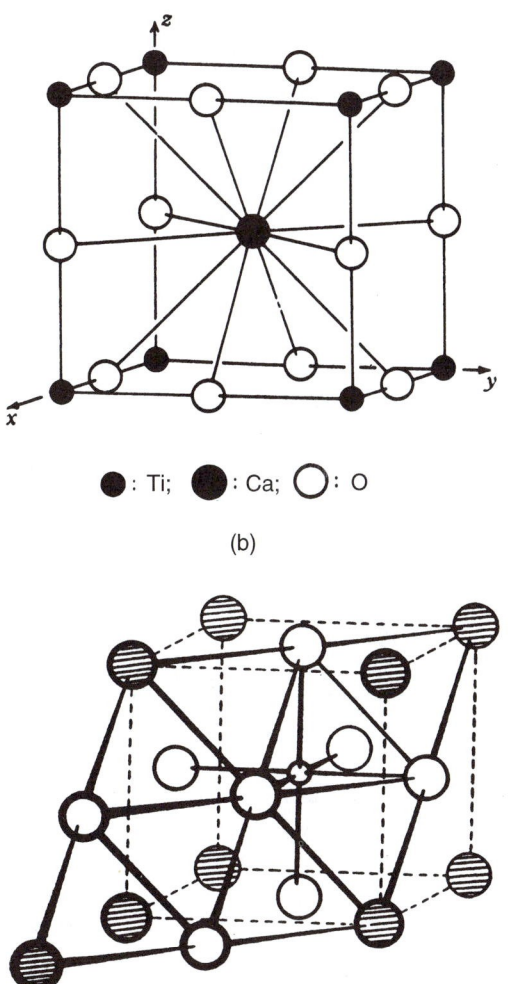

● : Ti; ⬤ : Ca; ◯ : O

(b)

(c)

larger alkaline earth ions, Sr^{2+} and Ba^{2+}, are more suitable. Figure 2.11c shows the perovkite structure as a version of a close-packed structure. The two larger ions, the anions and the larger cations, form a ccp sublattice in which both ions are present in each layer. The smaller cation then occupies all the octahedrally coordinated sites that are surrounded only by anions. Octahedrally coordinated sites that have both large cations and anions as nearest neighbors remain unoccupied. A very large family of materials have the perovskite structure, or its close relatives; apparently it is a very favorable arrangement.

The Spinel Structures

In the spinel structures, the two cations occupy lattice sites of different types. The spinels have the generic formula AB_2O_4. This structure is named after the naturally occurring mineral $MgAl_2O_4$. This compound can be viewed as made up of equal amounts of MgO and Al_2O_3, not in the sense of a solid solution, but in a specific structural arrangement. In addition to this compound, having one divalent and two trivalent cations, there are spinels with one tetravalent and two divalent cations and with one hexavalent and two monovalent cations—almost any combination that adds up to eight positive charges to balance the eight anionic charges. The two types of cation do not usually differ greatly in size. An enormous variety of cations can participate in these compounds, and the anions can be sulfide, selenide, or telluride, in addition to the more common oxide. Some of these materials have important applications, such as the magnetic ferrites that were the key memory elements in early computers.

In a normal spinel of the two-three type, such as $MgAl_2O_4$, the A^{2+} cations, the Mg^{2+}, occupy one-eighth of the tetrahedral sites in a ccp anion array, while the B^{3+} cations, the Al^{3+}, occupy half of the octahedral sites. To have a repeating structural unit, the unit cell contains $A_8B_{16}O_{32}$, and is thus approximately eight times larger than the unit cell of the NaCl structure. This structure is shown in Fig. 2.12.

In some cases, the cations are distributed differently among these same sites because of specific site preferences of certain cations. This is the case for Fe_3O_4 or $FeO–Fe_2O_3$, the mineral magnetite. The electronic structure of the Fe^{2+} ion is $Ne/3d^6$, where Ne represents the underlying electronic arrangement of the rare gas neon. In an octahedral environment, the crystal field splitting gives three equivalent levels of lower energy, the t_{2g} levels whose orbitals are directed in between the six closest anions, and two equivalent levels of higher energy, the e_g levels whose orbitals point directly at nearest neighbors. The energy separation in this case is sufficient to promote a violation of Hund's rule (that "equivalent" electron states are singly filled before any of them are doubly occupied), and the six d electrons fill the three lower levels, as shown in Fig. 2.13. The two empty e_g levels can then hybridize with the empty 4s and 4p orbitals to give six equivalent d^2sp^3 hybrid orbitals that are ideally directed toward the six octahedral anion neighbors. These empty cation orbitals can

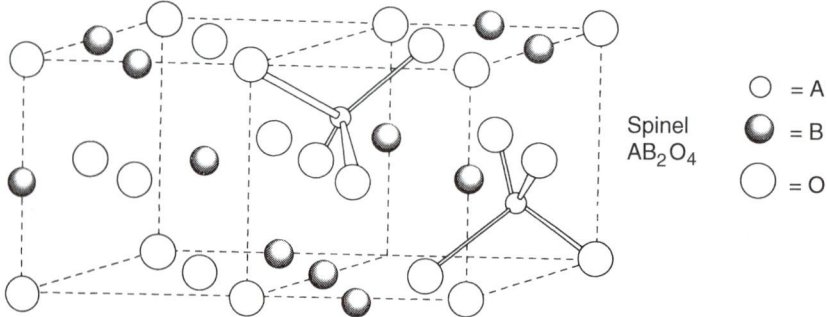

Figure 2.12 The normal spinel structure. The small spheres are the A cations in one-eighth of the tetrahedral sites of the fcc anion sublattice, while the shaded spheres represent the B cations in one half of the octahedral sites. (Reproduced from Rodgers, 1994, by permission of McGraw-Hill.)

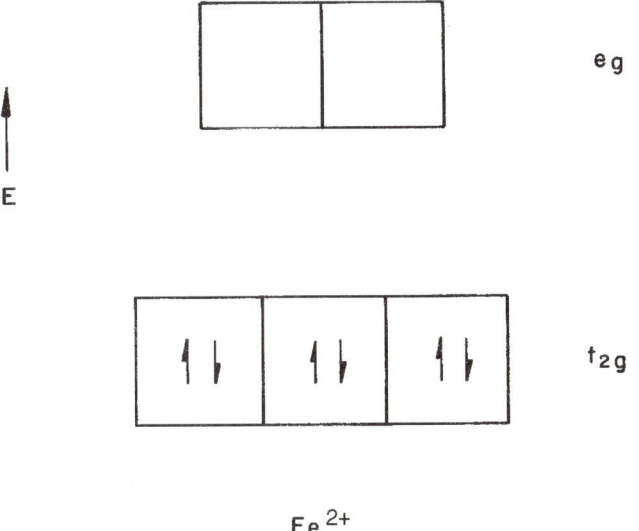

Figure 2.13 The effect of crystal field splitting on the energy levels of the 3d orbitals of Fe^{2+} in octahedral coordination. The three orbitals that are directed between the surrounding anions (t_{2g})are at a lower energy relative to the two orbitals that point directly at the surrounding anions (e_g). In this case the energy splitting is sufficient to cause an apparent violation of Hund's rule that equivalent orbitals are singly occupied before any are doubly occupied.

interact covalently with filled orbitals on the surrounding anions to enhance the total bond strength. This interaction effectively raises the electronic structure of the Fe^{2+} ion to that of the next rare gas, argon, so this is a particularly favorable situation. As a result, the Fe^{2+} prefers to occupy the octahedral sites, rather than the tetrahedral sites occupied by the divalent cations in the normal spinel structure. They displace half of the Fe^{3+} ions, which move over to tetrahedral sites. This is called the inverse spinel structure. The A ions occupy one-quarter of the octahedral sites, while the B ions are split between another quarter of the octahedral sites and one-eighth of the tetrahedral sites. Since Fe in two different oxidation states occupy equivalent octahedral sites, it is very easy for electrons to hop from Fe^{2+} ions to Fe^{3+} ions, and Fe_3O_4 is a black, semiconducting material. In Mn_3O_4, the Mn^{2+} ions have the electronic structure $Ne/3d^5$ (isoelectronic with Fe^{3+}). Thus the same type of bonding just described for Fe^{2+} will still leave the electronic structure around the Mn^{2+} one electron short of the argon structure, and this is not favorable enough to cause a shift in sites. So Mn_3O_4 has the normal spinel structure where the two different oxidation states of the Mn are on sites of different types. This makes electronic transfer much more difficult, and Mn_3O_4 is a light-colored insulator. Co_3O_4, where the Co^{2+} ion has one more electron than Fe^{2+}, behaves similarly to Mn_3O_4. Such subtle changes in the distribution of the ions can lead to enormous differences in physical properties.

Some spinels (e.g., $MgFe_2O_4$ and $MnAl_2O_4$) have cation distributions intermediate between the ideal normal and inverse versions. This is referred to as different degrees of inverseness, and is commonly characterized by a parameter λ, the fraction of B ions in tetrahedral sites. Thus for the ideal normal spinel, $\lambda = 0$, while for the ideal inverse spinel, $\lambda = 0.5$. For the two compounds just listed, $\lambda = 0.45$ and 0.15, respectively.

Why all this detailed discussion about various permutations of the spinel structure? Because it leads us in a logical progression to the concepts of crystalline disorder and lattice defects. The partially inverse spinels can be viewed as partially disordered versions of either end member, the ideal normal spinel or the ideal inverse spinel. The disordered structures contain lattice defects relative to either end member in that cations are located on lattice sites where they do not appear in the ideal reference structure. Thus ionic defects are defined relative to a specific reference structure.

CONCLUSION

With this detailed description of crystal structures in a way that is most useful to our subject, and with this initial introduction to the concepts of crystalline disorder and lattice defects, we are prepared to move on to the principal aspects of defect chemistry.

REFERENCES

Barsoum, M. W. *Fundamentals of Ceramics*. New York: McGraw-Hill, 1997, Fig.3.9.

Evans, R. C. *An Introduction to Crystal Chemistry*. Cambridge: Cambridge University Press, 1964, Fig. 8.20.

Greenwood, N. N., and A. Earnshaw. *Chemistry of the Elements*. Oxford: Pergamon Press, 1984, Fig. 21.4.

Rodgers, G. E. *Introduction to Coordination, Solid State, and Descriptive Inorganic Chemistry*. New York: McGraw-Hill, 1994, Fig. 7.22a.

Smith, W. F. *Principles of Materials Science and Engineering*, 2nd ed. New York: McGraw-Hill, 1990, Fig. 3.6.

Van Vlack, L. H. *Elements of Materials Science*, 2nd ed. Reading, MA: Addison-Wesley, 1964, Figs. 3–10 and 3–34.

Wells, A. F. *Structural Inorganic Chemistry*, 4th ed. Oxford: Clarendon Press, 1975, Figs. 3.35, 4.33, and 17.1. This large volume is a standard reference for inorganic crystal structures.

☑ PROBLEMS

2.1. a. Calculate the ideal cation/anion radius ratio for the octahedral sites in an ideally cubic-close-packed anion sublattice.

b. Calculate the ideal cation/anion radius ratio for the octahedral sites in an ideally hexagonal-close-packed anion sublattice.

c. Calculate the ideal cation/anion radius ratio for the tetrahedral sites in an ideally cubic-close-packed anion sublattice.

Lattice Defects and the Law of Mass Action

3

INTRODUCTION

The basic technique of defect chemistry involves the writing of equilibrium reactions for the formation of defects and the interactions between them, and then forming expressions for the relationships between their activities by use of the law of mass action. Well known to chemists, the law of mass action is routinely used for chemical reactions of all kinds under equilibrium conditions, most frequently in aqueous or gaseous phases. It has a firm thermodynamic base and assumes that dilute solution conditions prevail, that is, there are no interactions between the various species except for those explicitly taken into account. In this situation, concentrations can be taken to be adequate representations of the thermodynamic activities of the various species. It will be seen that this assumption is valid for surprisingly high concentrations of defects in solid compounds.

The use of the law of mass action is so central to the analysis of the defect chemistry of solid compounds that it is essential for the reader to understand its origins, and to be comfortable with its use. This chapter derives mass-action expressions by several alternative ways and relates them to the pertinent thermodynamic parameters, the standard enthalpies and entropies of the equilibrium reactions. To illustrate the first of these examples, we will use the simplest possible case of lattice defects, the case of lattice vacancies in elemental solids (e.g., pure metals and crystalline rare gases). The formation of lattice vacancies in a monatomic solid is shown schematically in Fig. 3.1. Figure 3.1a is a two-dimensional representation of the perfect crystalline form of a monatomic solid with a simple square lattice. A jog has been included on one edge for reasons that will be apparent; the surfaces of real crystals are seldom smooth on an atomic scale and do have steps and jogs on their surfaces. Figure 3.1b depicts the situation after an atom from the interior has been moved to a new surface site. In this example, the atom has been placed on a site that extends the surface jog, although it could have been placed on any surface site. Through this formal process a lattice defect, a vacant lattice site, or vacancy, has been introduced into the crystal. For the time being, the mechanism by which the situation shown in Fig. 3.1a is tranformed

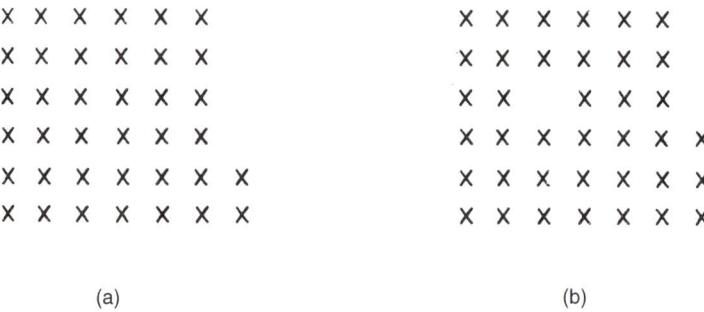

(a)   (b)

Figure 3.1 Two-dimensional schematic representations of a simple elemental crystal: (a) the perfect lattice and (b) lattice with a vacancy.

into that shown in Fig. 3.1b is not important. We are concerned only with these initial and final states. It will be assumed that further disorder of this kind will result in a random distribution of vacancies throughout the bulk.

The creation of vacancies as just described results in the breaking of chemical bonds, and this is an energetically costly process that is not entirely recovered by putting the displaced atom on a new surface site, since such a site does not have the full complement of bonds that are experienced by an atom in the interior. So why should such disorder occur under equilibrium conditions? That is the subject of the next section.

Readers who are familiar with the law of mass action and comfortable with its use for the case of defects in crystalline solids may omit some of the rest of this chapter. However, it is important to understand the role of the configurational entropy in balancing the enthalpic cost of introducing disorder to the lattice.

LATTICE DEFECTS AS PART OF THE EQUILIBRIUM STATE

Why should lattice defects be a part of the equilibrium state of an ideally pure crystal? It is clear that the optimum lattice energy is obtained when all the atoms or ions are present and in their proper places so that the energetic pay back from chemical bonding is maximized. It is now fairly routine to calculate the preferred crystal structure for any given crystalline solid by a process that maximizes the lattice energy for the assemblage of atoms. Any missing or displaced ion can only detract from the lattice energy of the perfect, defect-free crystal. But the equilibrium state is not determined solely by the lattice energy, which is only the enthalpic contribution to the total free energy of the system. Equilibrium at constant pressure is defined as the state in which the system has its minimum value of the Gibbs free energy, so there is an entropy term that must also be taken into account:

$$G = H - TS \tag{3.1}$$

where the enthalpy H is determined primarily by the chemical bonds in a crystal, S is the entropy, and G is the Gibbs free energy. There are two main contributions to the entropy, a vibrational term S_v and a configurational entropy S_c. The latter is a measure of the randomness of the system, sometimes known as the entropy of solution. In a sense, both terms are related to randomness, since a vibrating atom is usually not located precisely at its geometric lattice point at any given instant. The vibrational entropy arises from a change in the vibrational modes in the vicinity of a defect. The atoms around a vacant lattice site can vibrate over a larger amplitude, hence at a lower frequency, while the atoms around an interstitial atom will have their vibrational amplitudes constricted and will move with a higher frequency. The sign of the vibrational entropy term will depend on the direction of change for the vibrational frequencies. In any case, this term is not of major importance for our treatment of intrinsic defects.

The configurational contribution to the entropy can be expressed as follows:

$$S_c = k \ln P \qquad (3.2)$$

where k is Boltzmann's constant and P is the number of possible permutations of the system, the number of equivalent ways that the components of the system can be arranged. There is only one way to arrange a perfect crystal, $P = 1$, and S_c is thus zero for that case; but as defects are introduced, the number of possible arrangements increases. This can be easily demonstrated with the "old shell game." With several shells and one or more peas, what are the odds of guessing which shells are covering a pea? Two situations are shown: four shells and one pea (Fig. 3.2a), and four shells and two peas. These empty shells and filled shells could just as well be empty and filled lattice sites in a crystal. There are four ways to arrange four shells and one pea, and six ways when there are two peas. This can be expressed mathematically as follows:

$$P = \frac{N!}{(N - n)!n!} \qquad (3.3)$$

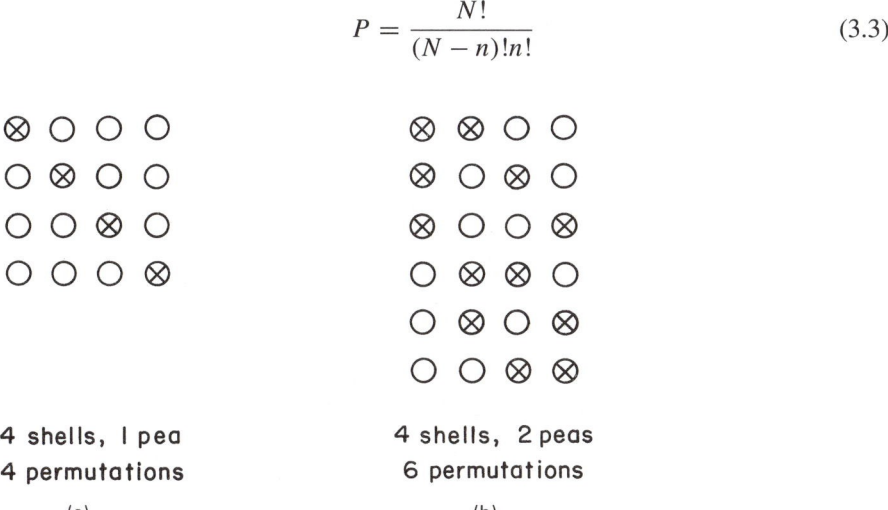

4 shells, I pea
4 permutations

(a)

4 shells, 2 peas
6 permutations

(b)

Figure 3.2 The total possible permutations for shells and peas: (a) four shells with one pea and (b) four shells with two peas.

where N is the total number of sites, or shells, and n is the number of occupied sites, or of peas, so that $N - n$ is the number of vacant sites, or empty shells. $N!$ is, of course, the factorial of N, $1 \times 2 \times 3 \times 4 \times \cdots N$. Substitution of the appropriate numbers into Eq. (3.3) confirms the numbers of permutations shown in Fig. 3.2. It is fortunate that we have a mathematical expression for the total possible permutations, since it would be rather tedious to sketch out all the possibilities for a typical crystal with 10^{16} lattice defects distributed over 10^{20} lattice sites!

The change in free energy that results from the creation of n defects in a crystal is then

$$\Delta G = nh - nT s_v - kT \ln P \tag{3.4}$$

where ΔG is the change in free energy from that of the perfect crystal, h is the enthalpic cost of creating each defect, and s_v is the vibrational entropy that results from each defect. Both h and s_v are treated on a per-defect basis, which is valid as long as the defects are so dilute that they do not interact. Consider a simple two-dimensional lattice of a single species, x, as shown in Fig. 3.1a. The jog that has been included on one of the sides does not change the fact that Fig. 3.1a represents an ideally perfect crystal. In Fig. 3.1b a lattice vacancy has been created by removing an atom from the interior and placing it on a new lattice site at the end of the jog. The net change is that there is now a vacant lattice site, and the total number of lattice sites has been increased by one. This can be confirmed by noting that for each atom in a specific environment in the perfect crystal, there is an exact counterpart in the defective crystal. The only difference is the addition of the vacant site. We will define the following terms:

N_0: the number of lattice sites in the perfect crystal, also the number of atoms in the crystal, and the number of occupied sites in the disordered crystal.

n: the number of vacant lattice sites.

$N_0 + n$: the total number of sites in the disordered crystal.

We can now play the shell game with the disordered crystal. The total number of possible arrangements is

$$P = \frac{(N_0 + n)!}{N_0! n!} \tag{3.5}$$

and this can be substituted into Eq. (3.4). The equilibrium number of defects will be that which causes the largest decrease in free energy from that of the perfect crystal. This we can obtain by differentiating the free energy function with respect to the number of defects and setting it equal to zero (happily, this gives us the minimum free energy state, not the maximum):

$$\frac{d\Delta G}{dn} = h - T s_v - kT \frac{d}{dn} \frac{(N_0 + n)!}{N_0! n!} = 0 \tag{3.6}$$

The first two terms were easy, but the expanded configurational entropy term looks pretty messy. Fortunately, an approximation to Stirling's formula for the case of large numbers is very helpful:

$$\ln x! \approx x \ln x - x \tag{3.7}$$

For $x = 100$, Eq. (3.7) gives a number that is only 0.9% less than that obtained from the full Stirling formula; for $x = 10^{20}$, the number of lattice sites in a typical small crystal, the two forms give the same answer for at least 10 significant figures. The configurational entropy term in Eq. (3.6) thus becomes

$$\frac{ds_c}{dn} \approx \frac{d}{dn} \{ kT [(N_0 + n) \ln(N_0 + n) - (N_0 + n) $$
$$- N_0 \ln N_0 + N_0 - n \ln n + n] \} \tag{3.8}$$

All the terms that derive from the second term of the Stirling approximation cancel out, a frequent indication that a model is on the right track, since nature tends toward simplicity. Completing the differentiation with respect to n, substituting the result back into Eq. (3.6), and setting $d \Delta G / dn = 0$ to find the expression for the minimum value of the free energy (actually the maximum decrease in free energy from that of the perfect crystal), we write

$$\frac{d \Delta G}{dn} \approx h - T s_v - kT \ln \left(\frac{N_0 + n}{n} \right) = 0 \tag{3.9}$$

which can be rearranged to

$$\frac{n}{N_0 + n} \approx e^{s_v/k} e^{-h/kT} = K(T) \tag{3.10}$$

This is an expression of the law of mass action for the creation of a lattice vacancy by removing an atom from the interior of the crystal and placing it on a new lattice site on a surface. The mass-action constant $K(T)$ is a function of temperature only; it is the fraction of lattice sites in the disordered crystal that are vacant at equilibrium. Defects are clearly a part of the equilibrium state, and their concentrations will depend primarily on the enthalpic cost of creating the defects, and on the temperature. Each type of defect in each type of structure will have its own characteristic enthalpic cost, and that will determine the general level of disorder, since the vibrational entropy term is small by comparison.

The right-hand side of Eq. (3.10) can be formally expressed as an exponential in the free energy change associated with the formation of the defects; it does not contain the total free energy change, however, because the configurational contribution to the entropy was explicitly expanded and consumed in the derivation. It does not appear in Eq. (3.10).

The minimization of the free energy by the formation of disorder is depicted graphically in Fig. 3.3. As before, the enthalpic cost and the effect on the vibrational entropy contribution are treated on a per-defect basis, and it is assumed that the

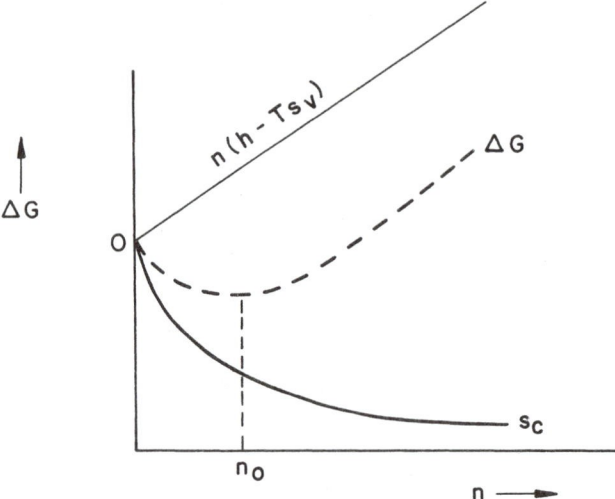

Figure 3.3 Minimization of the free energy G by combination of the enthalpic cost (combined with the vibrational entropy) and the configurational entropy. The minimum value of G determines the equilibrium defect concentration n_0.

enthalpy is the major determining factor. The free energy has been normalized to that of the perfect crystal. It is the initial steep drop in the configurational entropy with the number of defects that causes the minimum in the free energy. A higher enthalpic cost will give a steeper enthalpy line, and will give a minimum in the free energy at a smaller defect concentration, as shown in Fig. 3.4. For $h_2 > h_1$, n_2 will be less than n_1. Temperature has its major effect as a coefficient of the configurational entropy; it will have only a small effect on the enthalpic slope, as a coefficient of the vibrational entropy, assuming that the enthalpy term is dominant.

The enthalpy and vibrational entropy of defect formation are material constants, and they differ with compound and structure. On the other hand, the configurational entropy is just a matter of counting possible arrangements, and while it may depend slightly on the crystal structure and type of defect, it is the same for all compounds with a given structure and defect type. From all these considerations, it is clear that **the concentration of intrinsic defects is determined primarily by their enthalpic cost and the temperature**.

THE LAW OF MASS ACTION

In the preceding section, we used a statistical thermodynamic approach to derive a simple case of the law of mass action for a monatomic crystal. The law of mass action is the heart of equilibrium defect chemistry, and its validity and limitations must be clearly understood if it is to be used with confidence and consistency. In this section we derive a more general expression based on simple thermodynamic arguments.

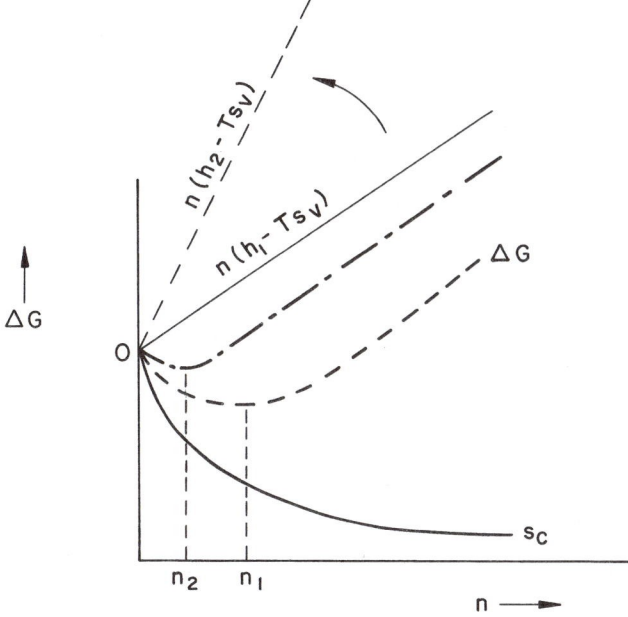

Figure 3.4 The effect of a higher enthalpic cost of defect formation, $h_2 > h_1$. The high cost results in a lower equilibrium defect concentration, $n_2 < n_1$.

Readers who understand the origin of this treatment of equilibrium reactions and are comfortable with its use may skip the rest of this section.

For a general chemical reaction in thermodynamic equilibrium,

$$aA + bB \rightleftharpoons cC + dD \tag{3.11}$$

where the capital letters represent chemical species and the lowercase letters are small integer coefficients, the mass-action expression is

$$\frac{[C]^c[D]^d}{[A]^a[B]^b} = K(T) = e^{-\Delta G^\circ/kT} \tag{3.12}$$

The activities of the components are indicated by enclosing them in brackets, and each activity is raised to the power of its coefficient in the reaction. ΔG° is the standard free energy change for the reaction. An example of such a reaction in aqueous solution might be

$$2AgNO_3 + BaCl_2 \rightleftharpoons 2AgCl(s) + Ba(NO_3)_2 \tag{3.13}$$

where AgCl is an insoluble precipitate. In this case, the identifiable chemical species are the ions dissolved in a matrix, which is the otherwise pure water. This treatment can be transferred directly to defect chemistry, where the identifiable species are lattice defects dissolved in a matrix, which is the otherwise perfect crystal. While most chemists are much more familiar with the use of this relationship in aqueous solutions or in the gas phase, no law of thermodynamics states that it works only in water. In both cases the matrix does not appear explicitly in the reaction equation unless it participates directly in the reaction (e.g., the formation of a hydrated ion in solution).

Where does the mass-action expression come from? It is convenient to introduce the chemical potential, μ_i, also known as the partial molar free energy, of a species i

$$\mu_i = \left(\frac{\partial G}{\partial n_i}\right)_{T,P,n_1,n_2,\dots} \tag{3.14}$$

It is the incremental change in the free energy of the system with a change in the amount of species i in the system, with temperature, pressure, and the amounts of all other species (n_1, n_2, etc.) held constant. The chemical potential of species i is related to its activity a_i

$$\mu_i = \mu_i^\circ + kT \ln a_i \tag{3.15}$$

where μ_i° is obviously the chemical potential at unit activity, and k is the Boltzmann constant. The change in free energy for the reaction described in Eq. (3.11) can then be expressed as follows:

$$\begin{aligned} \Delta G = {} & c\mu_C^\circ + d\mu_D^\circ - a\mu_A^\circ - b\mu_B^\circ + ckT \ln a_C \\ & + dkT \ln a_D - akT \ln a_A - bkT \ln a_B \end{aligned} \tag{3.16}$$

The collection of terms taken at unit activity is, by definition, the standard free energy change for the reaction, ΔG°. When the remaining terms are combined, and the total free energy change is set equal to zero, to correspond to the equilibrium state, the final result can be expressed as follows:

$$\frac{a_C^c a_D^d}{a_A^a a_B^b} = e^{-\Delta G^\circ/kT} = e^{\Delta S^\circ/k} e^{-\Delta H^\circ/kT} = K(T) \tag{3.17}$$

This is a general expression of the mass-action relationship, where $K(T)$ is the temperature-dependent mass-action constant, which is related to the standard free energy change as

$$-\Delta G^\circ = kT \ln K(T) \tag{3.18}$$

Equation (3.17) is formally equivalent Eq. (3.12). This mass-action relationship between the components of an equilibrium reaction will be used repeatedly in our treatment of defect chemistry.

ANOTHER VIEW OF MASS ACTION

At the risk of being tedious, we present a third approach to mass action, which will prove to be very useful for subsequent discussions of the mechanisms of ionic transport. This example involves the movement of an atom from its normal lattice

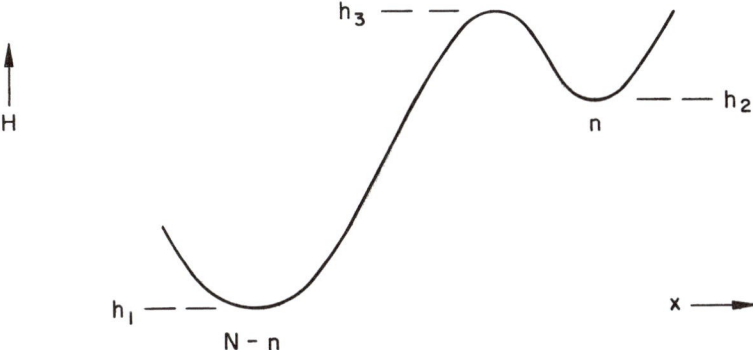

Figure 3.5 Schematic representation of the enthalpy of an atom, or ion, on position x in a lattice: h_1 represents a stable lattice position, h_2 an interstitial position, and h_3 the enthalpic barrier between the two positions; N is the concentration of lattice positions, and n is the number of atoms in interstitial positions.

position to some interstice, or interstitial site, within the crystal. The stable positions for the atoms or ions in a lattice correspond to minima in the periodic lattice potential, as depicted schematically in Fig. 3.5 in the form of the enthalpy, or lattice energy, of the atoms as a function of position. The stable lattice site is at the bottom of the deeper well at the enthalpy level h_1. An interstitial position is represented as a local minimum at a higher enthalpy level designated as h_2. To move from a lattice site to the interstitial site, an atom must pass over an enthalpic barrier of height h_3. At the dynamic equilibrium state, atoms will be moving from lattice sites to interstitial sites at the same rate that they are moving from interstitial sites back to lattice sites. This can be represented by the expression

$$\overrightarrow{J_i} = \overleftarrow{J_i} \qquad (3.19)$$

where J_i is the flux of atom i in atoms per square centimeter per second. The concentration of lattice sites is N, while the concentration of atoms in the interstitial sites is n, so that the concentration of atoms in lattice sites is $N - n$.

The flux of atoms moving toward the right will be proportional to the concentration of atoms in lattice sites that are available for attempts to surmount the barrier, $N - n$, times the frequency of the attempts, that is, the frequency of lattice vibrations v_1, times the probability that any attempt will be successful, $\exp[-(h_3 - h_1)/kT]$. The latter is a Boltzmann factor, an exponential ratio of the enthalpy needed to surmount the barrier $h_3 - h_1$ to the average thermal enthalpy available, kT. This is a measure of the probability that the time-dependent fluctuations of the local enthalpy will result in sufficient concentration of enthalpy on a particular atom as its vibration tries to push it up the enthalpy slope. The flux of atoms to the left will depend on the number of atoms in the interstitial sites available to make the attempt, n, the vibrational frequency of the atom in the interstitial site, v_2, and the appropriate Boltzmann factor to indicate the probability of success, $\exp[-(h_3 - h_2)/kT]$. Substituting these terms into Eq. (3.19) gives

$$(N - n)v_1 e^{-(h_3 - h_1)/kT} = n v_2 e^{-(h_3 - h_2)/kT} \qquad (3.20)$$

This can be rearranged to

$$\frac{n}{N-n} = \frac{\nu_1}{\nu_2} e^{-(h_2-h_1)/kT} = K(T) \tag{3.21}$$

which is in the form of a mass-action expression. Note that the barrier height h_3 disappears from the equilibrium expression because it determines only the rate at which equilibrium is approached. The appropriate enthalpy term, $h_2 - h_1$, which is the difference in enthalpy between the interstitial site and the lattice site, corresponds to an enthalpy of defect formation, ΔH. By comparison with Eq. (3.10), it would appear that the vibrational frequency term is related to the vibrational entropy

$$s_v \sim k \ln \frac{\nu_2}{\nu_1} \tag{3.22}$$

This is, in fact, the usual form of a vibrational entropy term. It is gratifying that this very simplistic model gives the correct form of the mass-action expression, including the entropy term.

This derivation emphasizes the dynamic aspects of mass action and is closely related to the arguments used by the Norwegian chemists Guldberg and Waage as the basis for their introduction of the concept of mass action in 1864. The dynamic background is implied by the incorporation of the word "action" in its name. Similar enthalpy diagrams and derivations will prove to be very useful in Chapter 7 in discussions of ionic transport.

LATTICE DISORDER IN ELEMENTAL SOLIDS

The simplest type of lattice disorder is found in the elemental solids (e.g., the crystalline rare gases and pure metals). These solids contain only a single atomic species and usually form simple structures that are often based on the most efficient packing of close-packed spheres. Disorder almost always is in the form of lattice vacancies, missing atoms, which may be considered as having formed from the perfect crystal by the movement of atoms from normally occupied sites to new sites on the surface, as shown in Fig. 3.1.

The enthalpies of formation of vacancies in a variety of elemental solids are shown as a function of absolute melting point in Fig 3.6, using enthalpies tabulated by Franklin (1972). It is seen that the enthalpies, the thermodynamic cost of disorder, scale with a remarkably linear relationship with the melting points from argon (mp 84 K) to molybdenum (mp 2890 K). The line starts at the origin, since it should require no energy to create defects in a crystal that melts at 0 K! Such a relationship will also be seen to hold for ionic compounds. It is a reasonable correlation, since the formation of vacancies requires the breaking of some of the bonds that hold the atom in the crystalline solid, while melting involves the breaking of all of the bonds in the solid.

The enthalpies shown in Fig. 3.6 are in units of electron volts (eV). This is the traditional and most commonly used energy unit in solid state physics and defect

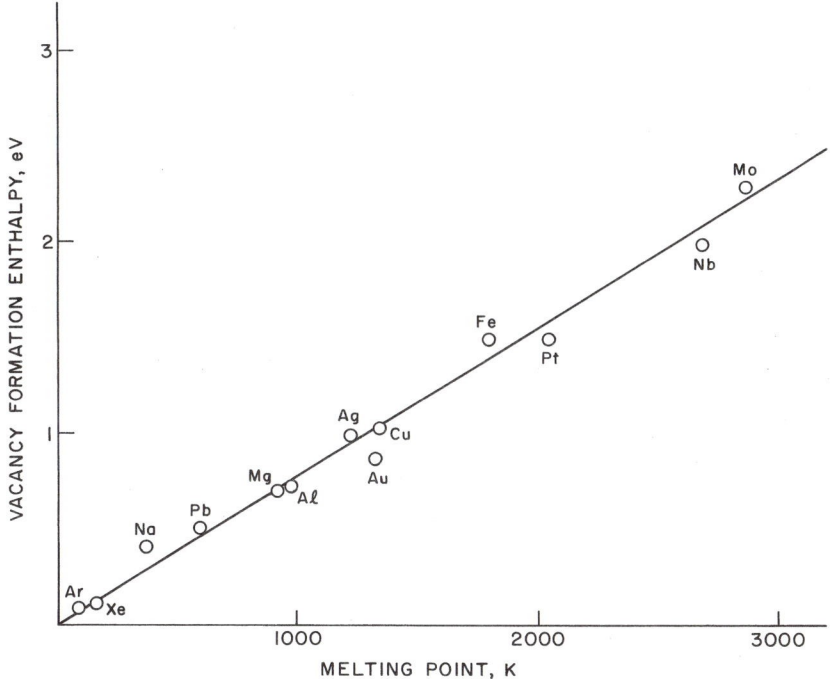

Figure 3.6 The enthalpies of vacancy formation in elemental solids as a function of the absolute melting points. (From values tabulated by Franklin, 1972.)

chemistry. It is convenient because processes in the real world generally range from a few tenths of an electron volt to about 10 eV. Those who are more comfortable in other units will know that there are 96.3 kJ/mol in 1 eV, so a close approximation can be easily obtained by multiplying eV by 100. Also, there is 23.05 kcal/mol in 1 eV. In this text, values given in eV are followed by values in kJ/mol in parentheses.

These thermodynamic parameters have been experimentally determined by a number of techniques. Figure 3.7 shows the data of Losee and Simmons (1968), who took advantage of the increase in the dimensions of a crystal of krypton as the concentration of vacancies increased with increasing temperature. As atoms are moved from the interior to surface sites to leave vacancies in the bulk, the creation of new lattice sites results in an expansion of the crystal. Of course, the dimensions also increase because of thermal expansion, so the total dimensional change was normalized to the change in the lattice parameter. The latter is due only to thermal expansion, so only the additional expansion was attributed to the formation of vacancies. A significant effect due to atomic disorder is seen to begin at about 80 K. These results were then analyzed to obtain an enthalpy of vacancy formation.

Figure 3.8 plots the isobaric specific heat for solid argon against the square of the reciprocal temperature (Beaumont et al., 1961). The three dashed offshoots at the

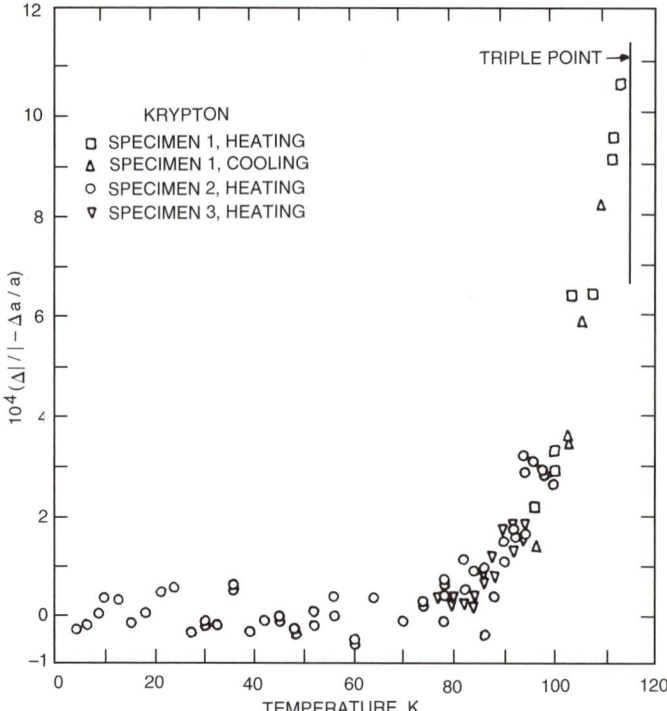

Figure 3.7 The change in length of a single crystal of krypton as a function of temperature, corrected for thermal expansion. The excess dimensional change above about 80 K is attributed to the formation of lattice vacancies. (Reproduced from Losee and Simmons, 1968, by permission of the American Physical Society.)

higher temperatures represent different extrapolations of the lower temperature data according to various theoretical models. The experimental specific heat rises above all these theoretical extrapolations, indicating that the sample is absorbing additional heat. This "excess" specific heat was attributed to the enthalpy of vacancy formation, and a value for the latter could be extracted from the data.

While these two examples do not correspond to our major interest in the defect chemistry of inorganic compounds, they give an opportunity to introduce lattice defects at the simplest possible level. They also correspond closely to the statistical thermodynamic derivation described earlier in this chapter.

SUMMARY

Several approaches have been described for the derivation of the law of mass action. This relationship will be used continuously throughout this book, and it must be understood and accepted. A major conclusion from these derivations is that defects are a part of the equilibrium state of crystalline solids. For pure materials their concentrations are determined primarily by their enthalpic cost and the temperature.

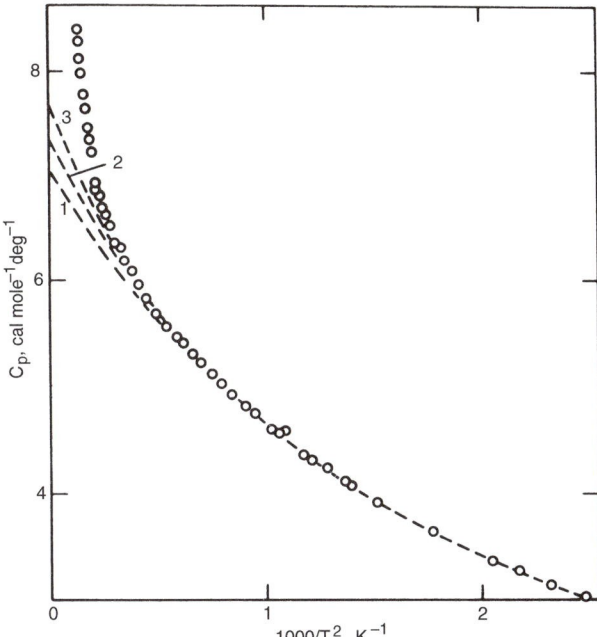

Figure 3.8 The isobaric specific heat of solid argon versus $1000/T^2$. The dashed curves represent the expected results for various ideal models. The actual specific heat rises above all these ideal curves at the higher temperatures because of the absorption of heat due to the creation of lattice vacancies. (Reproduced from Beaumont et al., 1961, by permission of the American Institute of Physics.)

In all cases it has been assumed that temperatures are high enough, and experimental times long enough, to give the diffusive processes by which defects are created time to reach equilibrium. The simple case of the formation of lattice vacancies in monatomic crystalline solids was used to illustrate some of the derivations and to introduce the concept of lattice defects. We are now prepared to consider the main topic of this book: defects in ionic crystals.

REFERENCES

Beaumont, R. H., H. Chihara, and J. A. Morrison. Thermodynamic properties of krypton. Vibrational and other properties of solid argon and solid krypton. *Proc. Phys. Soc. (London)* 78:1462–1481, 1961.

Franklin, A. D. Statistical thermodynamics of point defects in crystals. In *Point Defects in Solids*, Vol. 1, *General and Ionic Solids*, J. H. Crawford Jr. and L. M. Slifkin, Eds. New York: Plenum Press, 1972, pp. 1–101.

Losee, D. L., and R. O. Simmons. Equilibrium vacancy concentration measurements on solid krypton. *Phys. Rev.* 172:934–943, 1968.

Intrinsic Ionic Disorder

4

LATTICE DEFECTS AND REFERENCE STATES

As shown in Chapter 3, intrinsic disorder is part of the equilibrium state of any crystal, including an ideally pure one. This chapter deals with the formation of equilibrium lattice defects in pure, stoichiometric crystals, while in Chapter 5, we will discuss extrinsic ionic defects, those that result from the presence of impurities.

First, we need to define two different types of reference state:

1. The **stoichiometric composition**, which corresponds to integer ratios of the component atoms and ions. For semiconducting or insulating materials, this state corresponds to compositions that have nominally filled valence bands and empty conduction bands.

2. The **reference structure**, from which lattice defects are defined.

It will be very important to distinguish clearly and consistently between these two reference states. For simple compounds such as NaCl, MgO, and MgF_2, these two reference states are obvious and identical. The stoichiometric compositions are the formulas as written, and the reference structures are the NaCl structure for the first two, and the rutile structure for the latter. The definitions are not always so obvious for more complex materials, as in the case of doped compounds such as $Ni_{1-x}Li_xO_{1-x/2}$, which is the stoichiometric composition, while the reference structure is the completely filled NaCl structure. The situation is even less clear when multiple oxidation states are involved, as in the case of $YBa_2Cu_3O_6$, which contains both monovalent and divalent copper. These more complex cases are dealt with later. In this chapter, we concentrate on stoichiometric compositions.

The concept of ionic disorder was introduced in Chapter 2 in the discussion of variations on the spinel theme, specifically in terms of the degree of inverseness. In an ideal normal spinel with the generic formula AB_2O_4, the A ions reside only on tetrahedral sites in a ccp anion sublattice, while the B ions are found only on octahedral sites. But in $MnAl_2O_4$, with an inverseness parameter of 0.15, 15% of the Al ions are located on tetrahedral sites, and the formula can thus be written as follows:

$[Mn_{0.7}Al_{0.3}]_{tet}[Mn_{0.3}Al_{1.7}]_{oct}O_4$, where the subscripts to the brackets indicate the sites occupied by their contents. This can be viewed as a disordered version of the normal spinel structure, where all the Al^{3+} should have been on octahedral sites. The Al^{3+} on tetrahedral sites and the Mn^{2+} on octahedral sites are then lattice defects relative to the reference structure. In this case, the incentive for the system to move toward the inverse spinel structure is to utilize some of the potential for covalent contributions to the bonding. Mn^{2+}, with a d^5 electronic structure, can assume a low-spin configuration that clears two d orbitals that can participate in d^2sp^3 hybridization that is compatible with an octahedral environment. Al^{3+} has only empty s and p orbitals, and they can hybridize to give sp^3 orbitals that fit a tetrahedral environment. Thus some of the Mn^{2+} and Al^{3+} switch sites to maximize the total bonding energy. In $MgAl_2O_4$, Mg^{2+} has no d orbitals available and thus is not as adaptable to covalency in an octahedral environment. Therefore, it prefers to occupy only the tetrahedral sites.

One can easily propose a hierarchy of possible defects in terms of the degree of energetic disruption they cause:

- It should not be very disruptive to move a cation from its normally occupied octahedral site in the normal spinel structure to one of the normally unoccupied octahedral sites. The cation will find itself in a site of the same size, with the same first coordination sphere of six anions. However, it will find itself somewhat closer to other cations than was the case in its proper site.

- Likewise, a cation can be moved from a normally occupied tetrahedral site in the normal spinel structure to one of the normally unoccupied tetrahedral sites.

- A cation can be moved from a tetrahedral site in the normal spinel structure to a normally unoccupied octahedral site. There will certainly be sufficient room, since the octahedral site is larger than the tetrahedral site. The cation will still find itself surrounded by anions, although there will be six anions around it instead of the usual four. When an ion is moved to a site having a different coordination symmetry, there may not be as much opportunity for covalent contributions to the bonding.

- If a cation is moved from an octahedral site in the normal spinel structure to a normally unoccupied tetrahedral site, it will find itself in a smaller site and this may be unfavorable. It will still be surrounded by anions, but by only four rather than the usual six.

- The situation is very similar to that just described when a cation is moved from an octahedral site in the NaCl structure to one of the tetrahedral sites.

- In all these structures that are based on a ccp anion sublattice, it is highly unlikely that anions will move into either of the possible interstitial sites in significant numbers. These sites are much too small to contain an anion, and they are surrounded by other anions, a prohibitive electrostatic penalty in an ionic compound.

- Anions can be displaced more readily in the very open fluorite structure. An anion can be moved from the normally occupied tetrahedral site to an octahedral site in the fcc cation sublattice. It will find itself in an even roomier site, where it is surrounded by six cations.

- It is less likely that a cation will move to the normally empty octahedral sites in the fluorite structure, because of the first coordination sphere of other cations around that site.

Thus from consideration of the space available and the electrostatic environment, one can make quite reasonable intuitive judgments about the relative likelihood of finding defects of different types in various crystal structures. All of these possible processes for creating defects have a price to pay in enthalpy because of the disruption of the ideal bonding, but, as shown in the preceding chapter, this can be more than balanced by the configurational entropy that results from the partial randomization of the lattice.

This chapter deals with intrinsic ionic disorder, of which the types just described are specific examples. These are lattice defects that are not related to an impurity content, or any other extrinsic cause, such as a deviation from stoichiometry. They are a part of the equilibrium state of the ideally pure, stoichiometric crystal. Of course the ideally pure crystal is a fictional concept that does not exist, but if the impurity content is sufficiently low, and the temperature sufficiently high, the resulting extrinsic defect concentrations may be negligible compared with those due to intrinsic disorder. To achieve the equilibrium concentration of defects, the temperature must be high enough to allow the necessary diffusion processes to proceed to equilibrium in reasonable experimental or operational times.

The concentrations of intrinsic ionic defects vary enormously for different materials. In some compounds, such as MgO, the concentrations are so low that impurity effects almost always dominate the properties, while in others, one of the sublattices may become totally disordered (i.e., one sublattice melts within the crystalline matrix of the other sublattice, as is the case of cations in AgI at temperatures above 145°C). In the latter type of situation, the disordered ions become extremely mobile, and the ionic conductivities approach or even exceed those of aqueous solutions.

CONSERVATION RULES

All the examples of lattice disorder described in the preceding section involve the movement of ions within a crystal. These are specific cases that are in accord with a more general set of conservation rules that must be satisfied for a self-consistent treatment of defect chemistry. These rules may be summarized as follows:

- **Conservation of mass**: Atoms are neither created nor destroyed within a system, but must be conserved.

- **Conservation of charge**: The bulk of an ideal crystal is electrically neutral. Charged defects must be created in combinations that are electrically neutral. Matter can be added to or removed from a crystal only in electrically neutral combinations.

- **Conservation of structure** (lattice site ratios): The creation of lattice defects must not violate the inherent ratio of cation sites to anion sites in the structure. Thus cation and anion sites can be created or destroyed only in ratios that correspond to the stoichiometry of the compound (i.e., in electrically neutral combinations).

- **Conservation of electronic states**: The total number of electronic states in a system derives directly from the electronic states of the component atoms, and must be conserved.

The first three rules are essential for the systematic consideration of lattice defects. The last one is not pertinent to ionic disorder and is discussed in Chapter 8.

These conservation rules appear to be so obvious and simplistic that they hardly seem worth enumerating, yet their systematic and rigorous application is the very heart of defect chemistry. In spite of their seeming simplicity, the scientific literature contains many examples of defect models that violate one or more of these conservation rules, particularly the one concerned with the conservation of lattice site ratios in complex structures. Rigorous treatment is not always exciting, but it can be a necessary check on self-consistency.

DEFECT NOTATION

Every scientific field has its own jargon that has been carefully designed to confuse the outsider. Since the development of the initial concepts of defect chemistry in the 1920s, a wide variety of defect notations have been suggested, some of them wonderfully obscure. Over the last 40 years, the logical and self-consistent symbols proposed by Kröger and Vink (1953) have been gradually accepted as the most useful. The use of this notation is now standard, and alternative systems should be avoided. The features of Kröger–Vink notation can be described as follows:

The main symbol: The defect species, which may be an ion, indicated by the atomic symbol for the species, or a vacant lattice site, denoted by V.

The subscript: Indicates the lattice or interstitial site, I, occupied by the defect.

The superscript: Indicates the difference in charge at the defect site relative to the charge at that site in the perfect crystal. A dot is used for an extra positive charge, and a slash denotes an extra negative charge.

Thus in our example of the partially inverse spinel $MnAl_2O_4$, a Mn^{2+} ion sitting on a normally unoccupied tetrahedral site is represented by the symbol $Mn_I^{\cdot\cdot}$; the species is a Mn^{2+} ion; it is sitting on a normally unoccupied interstitial site, subscript I; and it represents an extra double positive charge relative to the absence of charge at that site in the perfect crystal. The latter property is represented by the two superscript dots; the reason for not using + signs here will become apparent. Likewise, a Mn^{2+} ion occupying an octahedral site normally occupied by an Al^{3+} is denoted as Mn'_{Al}.

The species is a Mn^{2+} ion, it is located on an Al site, and it represents one positive charge less than the charge expected at that site in the perfect crystal. That effective defect charge is indicated by a superscript /, one step in the negative direction from the charge in the ideal crystal. In this case the distinction between the absolute electrostatic charge on the ion ($+2$) and the effective defect charge (-1) becomes clear. The absolute charge of the Mn^{2+} ion is clearly $+2$, but when it occupies an Al^{3+} site, its charge difference from that expected at that site (i.e., in the standard state), is one step in the negative direction. This effective defect charge is not arbitrary; the defects will react to an electric field according to their effective defect charge, since the perfect crystal is electrically neutral. Only charges that differ from the standard state will react to an electric field, and only defects that bear an effective charge relative to the perfect lattice will have electrostatic interactions with other defects. Moreover, the condition of charge neutrality requires a balance of these effective defect charges. The use of $+$ and $-$ for these defect charges leads to confusion and inconsistency, and should be avoided.

If the charge at the defect site is the same as that in the perfect crystal (i.e., an isovalent substitution) the superscript is usually omitted. If the absence of effective charge needs to be emphasized, however, a superscript x is sometimes used. For example, a K^+ ion substituted for Na^+ in NaCl may be designated as K_{Na}^x.

Table 4.1 contains a number of important examples that demonstrate the system. It is absolutely essential that the reader become comfortable with Kröger–Vink notation very quickly; otherwise the rest of this book will be incomprehensible.

MAJOR TYPES OF INTRINSIC IONIC DISORDER

We are now prepared to describe specific types of intrinsic lattice disorder that have proved to be important in determining the properties of crystalline compounds. While all conceivable types of disorder that are consistent with the conservation rules are in principle present at equilibrium in amounts that depend on their enthalpic costs and the temperature, in real cases one type of disorder is usually so favored compared with all other possibilities that it is the only one that needs to be considered. For example,

TABLE 4.1
Examples of Kröger–Vink defect notation

	What $^{charge}_{where}$
Cation vacancy in NaCl	$V_{Na}^{/}$
Cation interstitial in AgBr	$Ag_I^{\cdot}$
Anion vacancy in MgO	$V_O^{\cdot\cdot}$
Cation interstitial in Al_2O_3	$Al_I^{\cdot\cdot\cdot}$
Ca^{2+} substituted for Na^+ in NaCl	$Ca_{Na}^{\cdot}$
Mg^{2+} substituted for Ti^{4+} in TiO_2	$Mg_{Ti}^{//}$
O^{2-} substituted for F^- in CaF_2	$O_F^{/}$

the dominant term in the mass-action constant, $\exp(-\Delta H / kT)$, differs by a factor of 50,000 at 800°C for values for ΔH of 1 and 2 eV (100–200 kJ/mol). (We will follow the commom practice in this field of expressing enthalpies in electron volts, eV. This is a convenient unit because most of defect chemistry occurs between 0.1 and 10 eV. In this case k has the value 8.63×10^{-5} eV/K. There are 96.3 kJ/mol in one eV. For those more comfortable with kJ/mol, values in eV will be followed by the corresponding value in the former units in parentheses.) Some types of disorder, such as the exchange of lattice sites by cations and anions, usually called place exchange, never occur to a significant extent in highly ionic crystals because the electrostatic penalty is much too high. However, in compounds with more than one cation species (e.g., the spinels), place exchange between the different cation species can be important. Also, place exchange can be important in very covalent crystals, such as the III–V compounds GaAs or InSb, where the absolute ionic charges are very small. In fairly ionic binary compounds, the major types of intrinsic ionic disorder are cation and anion Frenkel disorder and Schottky disorder.

Cation Frenkel Disorder

The silver halides have been extensively studied because of the importance of their defect chemistry in the formation and development of the latent image in the photographic process, and because they are experimentally quite accessible. Since their melting points are modest (455°C for AgCl and 432°C for AgBr), they can be easily purified by zone refinement; and since they are not hygroscopic, no stringent precautions for handling are necessary. Finally, they are extraordinarily soft and ductile for ionic solids. The author began his professional career studying ionic conduction in AgCl as related to its use as the solid electrolyte in a miniature solid state battery. Convenient samples were easily made by casting molten AgCl into slabs that were reduced by a few passes through a set of laboratory rollers to sheets of the order of 200 μm thick. Samples were then punched from the sheet or cut out with a razor blade or scissors. Such a cooperative material is indeed rare!

Frenkel disorder, so named in recognition of the first realization of the significance of lattice defects, involves the displacement of an ion from its normal lattice site to a location within the crystal that is not normally occupied (i.e., to an interstice or interstitial site). The displaced ion can be either a cation or an anion, and the availability of an interstitial site of adequate size and appropriate surroundings is an important consideration. In both AgCl and AgBr, cation Frenkel disorder is the dominant type of intrinsic ionic disorder. Both compounds have the NaCl structure, so the interstitial site in question is the tetrahedrally coordinated site in the ccp sublattice of anions. It is barely more than half the size of the normally occupied octahedral site, and one might well think that it is too small to be occupied by either the cations or the anions to any significant extent. However, Ag^+ has a remarkable ability to move through the narrow confines of even a close-packed lattice, and this is attributed to an unusual degree of "polarizability." In essence, Ag^+ with an outer filled d shell does not behave as a rigid sphere; rather, it can distort to fit into or move through spaces that would seem to be too small. It is the amoeba of cationic species. As a result,

Ag^+ can be accommodated by the tetrahedral sites in the NaCl structure. These sites are generally considered to be surrounded by four anions at the corners of a regular tetrahedron in their first coordination sphere. Actually, they are surrounded by four anions and four cations at alternate corners of a cube, but it is the much larger anions that dominate the environment and that make the site electrostatically comfortable for an interstitial cation. Since silver ions can squeeze into an interstitial site that has an attractive electrostatic environment, cation Frenkel disorder is the predominant form of intrinsic lattice disorder in both AgCl and AgBr.

The formation of cation Frenkel defects in AgBr can be represented by the following equilibrium reaction:

$$Ag_{Ag} + V_I \rightleftharpoons Ag_I^{\cdot} + V_{Ag}^{/} \qquad (4.1)$$

This equation uses the defect notation developed by Kröger and Vink as described earlier in this chapter. The individual symbols can be described as follows:

V_I A vacant interstitial site, this is a normal component of the perfect crystal, so it bears no charge relative to the standard state.

Ag_{Ag} A silver ion, Ag^+, located on a silver site in the lattice. This is also a normal component of the ideal lattice, so no charge difference is indicated.

$Ag_I^{\cdot}$ A silver ion located in an interstitial site (the tetrahedral site of the NaCl structure in this case). The extra positive charge at the normally uncharged site is indicated by a single dot.

$V_{Ag}^{/}$ A vacant silver site. The species is a vacancy, and its location is at a normally occupied silver site. There is a single positive charge at this site in the perfect lattice, but the vacancy has no absolute charge. This difference of one charge in the negative direction is indicated by the superscript /.

This disorder reaction is represented schematically in Fig. 4.1 in the form of a simple two-dimensional square lattice of positively and negatively charged ions. (The jog that is deliberately included on one edge of the lattice will prove to be useful for the description of different types of defects. Real crystals do indeed have steps and jogs on their surfaces.) Equation (4.1) explicitly demonstrates its compliance with the pertinent conservation rules: there is an Ag^+ ion on both sides of the equation, the number of atoms has remained unchanged; there is no net charge on the left-hand side, and equal numbers of positive and negative defect charges appear on the right-hand side; and an interstitial site and a cation site appear on both sides of the equation. In this particular case, conformity to the conservation rules is assured because ions have only been moved from one site to another within the crystal.

It is common practice to omit normal components of the perfect crystal from the equilibrium reaction and to show only the defect species. The starting point for the reaction is then represented by the symbol "nil," meaning no defects, and the designated defects are defined relative to the perfect crystal. Equation (4.1) would then be written as follows:

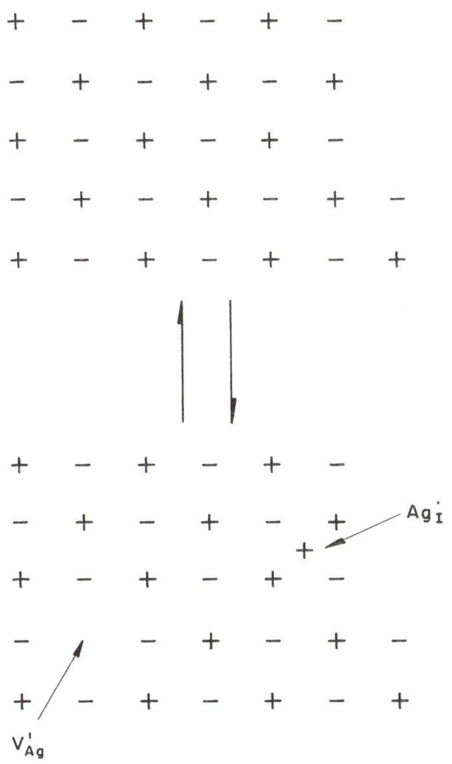

$$+ \quad - \quad + \quad - \quad + \quad -$$

$$- \quad + \quad - \quad + \quad - \quad +$$

$$+ \quad - \quad + \quad - \quad + \quad -$$

$$- \quad + \quad - \quad + \quad - \quad + \quad -$$

$$+ \quad - \quad + \quad - \quad + \quad - \quad +$$

Figure 4.1 Two-dimensional representation of the formation of cation Frenkel disorder in AgBr.

$$+ \quad - \quad + \quad - \quad + \quad -$$

$$- \quad + \quad - \quad + \quad - \quad + \quad \text{Ag}_I^{\cdot}$$
$$+$$

$$+ \quad - \quad + \quad - \quad + \quad -$$

$$- \quad \quad - \quad + \quad - \quad + \quad -$$

$$+ \quad - \quad + \quad - \quad + \quad - \quad +$$

$$\text{V}_{Ag}^{'}$$

$$\text{nil} \rightleftharpoons \text{Ag}_I^{\cdot} + \text{V}_{Ag}^{'} \tag{4.2}$$

Some care is necessary when this shorthand form is used because it does not explicitly demonstrate the conservation of mass and lattice site ratios. If the amount of disorder is high, it is necessary to acknowledge that the concentrations of normally vacant interstitial sites, and of normally occupied cation sites, will be affected. Their concentrations must then be included in the equilibrium formation reaction as shown in Eq. (4.1) and in the resulting mass-action expression.

The application of the law of mass action to Eq. (4.1) yields the following expression:

$$\frac{[\text{Ag}_I^{\cdot}][\text{V}_{Ag}^{'}]}{[\text{V}_I][\text{Ag}_{Ag}]} = K_{CF}(T) \tag{4.3}$$

$$= e^{\Delta S_{CF}/k} e^{-\Delta H_{CF}/kT}$$

where it has been assumed that dilute solution thermodynamics can be used so that concentrations and activities are equivalent. $K_{CF}(T)$ is the temperature-dependent mass-action constant that is made up of the exponential entropy and enthalpy terms as indicated, with the subscripts denoting that these are the thermodynamic parameters for cation Frenkel disorder. Capital letters are used in the subscripts when the full mass-action expression is used—that is, when normal lattice components are not neglected and the concentrations are in fractional units, hence dimensionless. Since the right-hand side is by definition unitless, the left-hand side must also be unitless. That is trivial in this case because the concentration terms appear to the same total power in both numerator and denominator. Typical units are site fraction or defects per cubic centimeter. The left-hand side can be viewed as the ratio of filled to empty interstitial sites times the ratio of vacant to occupied cation sites. For small defect concentrations these are approximately the fraction of occupied interstitial sites and the fraction of vacant cation sites, respectively.

If Eq. (4.1) is the only significant source of ionic defects in the crystal, then it is clear that cation interstitials and cation vacancies are created in equal numbers. This can be expressed by what is in effect an expression of bulk charge neutrality:

$$[Ag_I^\bullet] \approx [V_{Ag}'] \tag{4.4}$$

We note parenthetically that this type of charge neutrality expression must equate the number of the different types of defect. For example, the concentration units could be defects per cubic centimeter, or, if in fractional units, they would have to be normalized to the same base (e.g., to total cation sites). This restriction is demonstrated in this example because there are twice as many interstitial sites as cation sites in the NaCl structure, $[V_I] = 2[Ag_{Ag}]$. Thus if $[Ag_I^\bullet]$ represented the fraction of occupied interstitial sites while $[V_{Ag}']$ represented the fraction of vacant cation sites, the expression for charge neutrality would have to be $[Ag_I^\bullet] \approx 2[V_{Ag}']$. Such an expression would be hopelessly confusing.

Equation (4.4) is written as an approximation because a real crystal will contain other types of intrinsic disorder and also impurity-related defects that, however small, keep this relationship from being a strict equality. However, it can be a perfectly adequate approximation. This relationship can be combined with the mass-action expression to give an expression for the concentrations of the two defect species

$$[Ag_I^\bullet] \approx [V_{Ag}'] \approx \sqrt{2}[Ag_{Ag}]e^{\Delta S_{CF}/2k}e^{-\Delta H_{CF}/2kT} \tag{4.5}$$

(since $[V_I] = 2[Ag_{Ag}]$). Each defect will have a temperature dependence that is characterized by half the enthalpy of the disorder reaction, $\Delta H_{CF}/2$. This does not mean that the enthalpic cost is the same for the two different defects; theoretical calculations clearly indicate that it is not. However, an experiment that measures only some property that is proportional to the defect concentrations is not capable of partitioning the total enthalpy between the individual contributions. Since the defects must always be created in electrically neutral combinations, their temperature dependences derive from an equal division of the total enthalpy between them. The factor 2 appears in the exponentials in Eq. (4.5) because the mass-action expression in this case contains concentration terms with a total of two powers and requires that a square root be taken to obtain expressions for the individual defects. In general, the integer that appears in exponential terms of this type will correspond to the total number of defects created by the disorder reaction. For that reason, the enthalpy of formation per defect (the total enthalpy divided by the number of defects) determines the general level of defect concentrations. By taking the logarithm of both sides of Eq. (4.5), one gets

$$\ln[Ag_I^\cdot] \approx \ln[V'_{Ag}] \approx \ln \sqrt{2}[Ag_{Ag}] + \Delta S_{CF}/2 - \Delta H_{CF}/2kT \qquad (4.6)$$

Then a plot of ln(defect concentration) against $1/T$, the familiar form of an Arrhenius plot, gives a straight line with a slope of $-\Delta H_{CF}/2k$. Plots of $\log[\exp(-\Delta H/2kT)]$ versus $1/T$ for various values of ΔH (Fig. 4.2) show the relative defect concentrations and their temperature dependences. The ordinate in Fig. 4.2 is approximately the

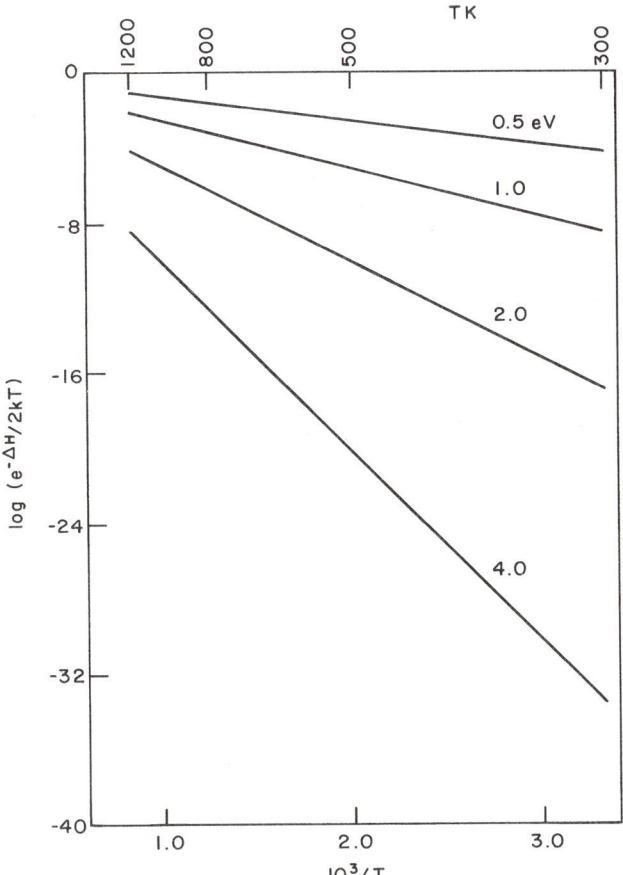

Figure 4.2 Plot of $\exp(-\Delta H/2kT)$ versus $1000/T$ for various values of ΔH. The ordinate represents an approximation to the defect concentration.

"fractional concentration" of defects, modified only by the entropy term, which may change it by as much as two orders of magnitude. It is seen that for an enthalpy of 4 eV (385 kJ/mol), the "fractional concentration" barely rises above 10^{-9}, 1 ppb, even at the highest temperature shown, 1200 K. Even at 2 eV (193 kJ/mol), the temperature must exceed about 600 K to reach 1 ppb, or 800 K to reach 1 ppm.

The enthalpy of formation of singly charged defects should be relatively modest, and because the peculiar deformability of Ag^+ allows it to fit into the small interstitial site with particular ease, the enthalpy of formation for cation Frenkel disorder in the silver halides is unusually small. Various experiments suggest that it is about 1.4 eV (135 kJ/mol) for AgBr. The entropy of formation is much less well known, but it appears that $\Delta S_{CF}/k$ is about 10 (it is usual practice to give entropy values as the dimensionless ratio $\Delta S/k$). Putting these values into Eq. (4.5) gives

$$[Ag_I^{\cdot}] \approx [V_{Ag}'] \approx \sqrt{2}e^{10/2}e^{-1.4/2kT} \tag{4.7}$$

where the defect concentrations are understood to be relative to the concentration of cation sites (e.g., approximately $[V_{Ag}']/[Ag_{Ag}]$ and $[Ag_I^{\cdot}]/[Ag_{Ag}]$). An Arrhenius plot of this relationship is shown in Fig. 4.3. At the melting point of 432°C, the fractional defect concentrations are 2.1×10^{-3} (0.21%), while they drop to 1.5×10^{-4} (0.015%) at 300°C and 3×10^{-10} (0.3 part per billion) at 25°C. The latter value would not be expected to lead to any observable consequences for transport properties and is below any reasonable expectation for the impurity level. Because of the relatively low enthalpy of formation, these are unusually large values for intrinsic defect concentrations at these modest temperatures. It is apparent that while the entropy term can typically contribute an order of magnitude or two to the defect concentrations, the scale of disorder is primarily set by the exponential enthalpy term. In addition to the energetic ease of forming cation Frenkel defects in AgCl and AgBr, these defects are unusually mobile in the crystal lattice. As a result, even the undoped silver halides are quite good ionic conductors.

Schottky Disorder

Walter Schottky and Carl Wagner coauthored papers published in German journals in the late 1920s that first organized the field of defect chemistry and developed the use of dilute solution thermodynamics for defect formation and interactions. Professor Wagner, who followed these pioneering publications over the next five years with additional papers of major importance, can be considered to be the father of defect chemistry. He remained the leading figure in the field until his death in the 1970s. As a result of the work of these two pioneers, one of the major types of intrinsic disorder is named after Walter Schottky.

While NaCl and most of the other alkali halides have the same crystal structure as the silver halides, the NaCl structure, they do not share the same preferred type of intrinsic lattice disorder. The alkali metal ions, with a closed-shell, rare gas electronic structure, behave much more like rigid spheres than do the floppy Ag^+ ions. As a result, there is a greater enthalpic cost for putting an alkali metal ion, such as Na^+,

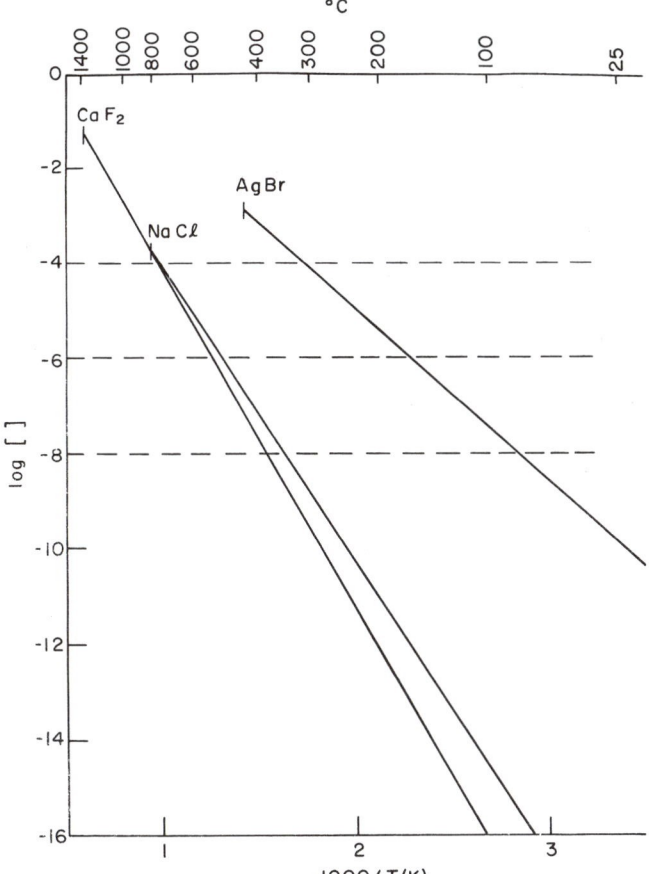

Figure 4.3 Arrhenius plots of the intrinsic defect concentrations in AgBr, NaCl, and CaF₂.

into the small, tetrahedrally coordinated interstitial site in the NaCl structure. The alkali halides prefer to form electrically neutral combinations of vacancies on both the cation and anion sublattices, and this is called Schottky disorder.

The formation of Schottky disorder in a simple two-dimensional lattice is depicted in Fig. 4.4. The components of an electrically neutral combination of ions (equal numbers of cations and anions in this case), are removed from random sites in the interior and placed on new lattice sites on the surface. The surface jog now becomes useful because the new sites merely extend the jog laterally; they do not appear as unusual, as would have been the case if they had been placed somewhere along an atomically flat surface, even though that would have been perfectly valid. Mass is obviously conserved, as is charge, since a neutral combination of ions was moved. New lattice sites have been created, but in the same ratio at which they appear in the ideal structure; thus the lattice site ratio has been conserved. Mass, charge, and structure will automatically be conserved if defects are formed by the movement of

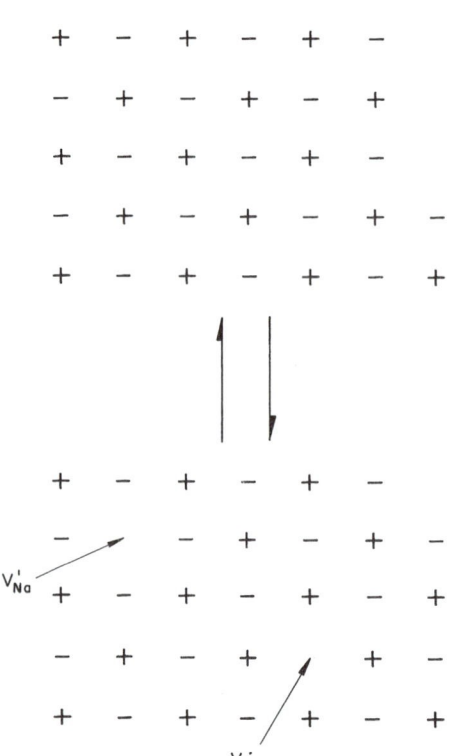

Figure 4.4 Two-dimensional representation of the formation of Schottky disorder in NaCl.

complete formula units of the compound from the interior to a surface. In an ionic compound, the movement to a new surface of a single cation or a single anion site would violate the conservation of both charge and lattice site ratios; these particles must move in neutral, stoichiometric ratios.

It is stated that a stoichiometric combination of cations and anions moves to surface sites, and vacant sites appear distributed in the interior of the crystal. Actually, surface ions move out onto new sites, and the vacated lattice sites, the vacancies, diffuse throughout the crystal. What can serve as a surface for new sites? Obviously, the exterior surface of the crystal will suffice, but other crystal irregularities are also available, such as grain boundaries and dislocations. Since an edge dislocation represents the termination of an extra plane of atoms inserted into a crystal, atoms can be added to the end of an edge dislocation, extending the extra plane. In a similar way, vacancies can be annihilated at the end of an edge dislocation, whereby the extra plane is shortened.

It is not convenient to write a simple equilibrium formation reaction for Schottky disorder that explicitly includes all the starting components of the reaction in defect notation. If one makes a complete count of all the identifiably different types of site in the perfect crystal (e.g., ions on the exterior and interior corners of surface jogs, the ions beside them; those just beneath them, etc., etc.), and compares that with a similar count for the disordered crystal, a one-to-one correspondence for all occupied sites will be observed. The only difference is the appearance of the vacant cation and anion sites distributed throughout the crystal. This can be seen in two dimensions in Fig. 4.4. Therefore, the formation of Schottky disorder is most conveniently shown as starting from "nil," the idealized perfect crystal. For the case of NaCl, this can be written as follows:

$$\text{nil} \rightleftharpoons V'_{Na} + V^{\bullet}_{Cl} \tag{4.8}$$

The mass-action expression for this reaction is

$$[V'_{Na}][V^{\bullet}_{Cl}] = K_S(T) = K^{\circ}_S e^{\Delta S_S/k} e^{-\Delta H_S/kT} \tag{4.9}$$

K°_S will be unity if the defect concentrations are in site fractions, but it must be adjusted for other concentration units. K°_S would then be the product of the concentrations of occupied cation and anion sites, in those units.

If Eq. (4.8) is the only significant source of defects, the two vacancy concentrations must be equal, in accord with the requirement of charge neutrality, and this relationship can be substituted into the mass-action expression to give

$$[V'_{Na}] \approx [V^{\bullet}_{Cl}] = K^{1/2}_S \approx (K^{\circ}_S)^{1/2} e^{\Delta S_S/2k} e^{-\Delta H_S/2kT} \tag{4.10}$$

The factor of 2 appears again in the exponent because two defects are created for each elemental disorder step in this simple 1:1, binary compound.

In a survey of the various experimental determinations of the values of the thermodynamic parameters in Eq. (4.9), it was concluded that the best values for NaCl are $\Delta H_S = 2.45$ eV (236 kJ/mol) and $\Delta S_S/k = 9.3$ (Fuller, 1972). Defect concentrations calculated from Eq. (4.10) with these values are plotted in Fig. 4.3, along with the previously calculated values for AgBr. Whereas 0.019% of the cation and anion sites are vacant at the melting point of 801°C, the concentrations fall to 9 ppm at 600°C and 73 ppb at 400°C. Dashed horizontal lines are included in Fig. 4.3 at concentration levels of 100 ppm (10^{-4} or 0.01%), 1 ppm (10^{-6}), and 10 ppb (10^{-8}). These might be considered to be the level of impurity ions for crystals of modest purity, high purity, and extraordinary purity (usually unachievable), respectively. It will be seen that for intrinsic defect concentrations below the appropriate impurity levels, the impurity content will determine the concentration of lattice defects. This is the realm of extrinsic ionic disorder that is treated in the next chapter. It should be noted that the difference in ΔH for intrinsic ionic disorder in AgBr and NaCl, 1.4 and 2.45 eV, respectively, results in major differences in the level of intrinsic disorder. At 432°C, the melting point of AgBr, the defect concentrations in NaCl are nearly four orders of magnitude lower than those in AgBr; at 200°C the difference increases to nearly a factor of 10^6. Modest differences in the enthalpy of formation per defect have a powerful effect on the defect concentrations. The silver halides have remarkable properties because of their exceptionally low disorder enthalpies. The relative ease of disorder in the silver halides is also reflected in their unusually low melting points.

Anion Frenkel Disorder

Anion Frenkel disorder requires the presence of an interstitial site having an adequate size and a congenial electrostatic environment to accommodate an interstitial anion.

In the case of binary compounds, this type of intrinsic disorder has been firmly established only for compounds having the fluorite structure (e.g., CaF_2 and CeO_2). As described in Chapter 2, this very open structure, having an octahedrally coordinated interstitial site with six cations as nearest neighbors, is larger than the normal anion site. The disorder reaction for anion Frenkel disorder in CaF_2 can be written as follows:

$$F_F + V_I \rightleftharpoons F_I' + V_F^{\cdot}$$ (4.11)

with the corresponding mass-action expression

$$\frac{[F_I'][V_F^{\cdot}]}{[V_I][F_F]} = K_{AF}(T) = e^{\Delta S_{AF}/k} e^{-\Delta H_{AF}/kT}$$ (4.12)

Once again, if Eq. (4.11) is the major source of defects, the expression for charge neutrality is

$$[F_I'] \approx [V_F^{\cdot}]$$ (4.13)

and the defect concentrations are determined by

$$[F_I'] \approx [V_F^{\cdot}] \approx \frac{[F_F]}{\sqrt{2}} e^{\Delta S_{AF}/2k} e^{-\Delta H_{AF}/2kT}$$ (4.14)

since $[F_F] = 2[V_I]$ in the fluorite structure. Experimentally determined values for the thermodynamic parameters for CaF_2 are $\Delta H_{AF} = 2.8$ eV (270 kJ/mol) and $S_{AF}/k = 13.5$ (Ure, 1957). The corresponding defect concentrations are plotted in Fig. 4.3, along with those of NaCl and AgBr. The results turn out to be very similar to those for NaCl, as might be expected from the very similar enthalpies. In fact, the slightly higher enthalpy of formation for anion Frenkel disorder in CaF_2 is nearly balanced by its larger entropy term.

GENERAL COMMENTS ON INTRINSIC IONIC DISORDER

Some aspects of intrinsic ionic disorder are most conveniently discussed by comparing the different types. Thus we will deal with the greater complexity possible for Schottky disorder than for Frenkel disorder, comparing the relative amounts of the various kinds of disorder in a single compound and describing the interactions between them. Relative equilibration rates are considered, and the possible differences in macroscopic properties compared. The most important observable properties that result from ionic defects, diffusion and ionic conductivity, will be discussed in Chapter 7.

Schottky Disorder in More Complex Compounds

Frenkel disorder, whether cation or anion, always involves the displacement of a single ion from its normal lattice site to an interstitial site. Thus it does not matter what type of formula the compound has (MX, MX_2, M_2X_3, etc.). The product of the elemental disorder process is the formation of two defects having charges of equal magnitude but opposite sign. For Schottky disorder, on the other hand, an electrically neutral combination of ions is moved from their normal lattice sites to new sites on a surface, leaving an electrically neutral combination of vacancies in the bulk. If the cation and anion do not have the same charge, there cannot be an equal number of cation and anion vacancies. As stated previously, to conserve mass, charge, and lattice site ratios, it is most convenient to write the equilibrium reaction in terms of the displacement of one formula unit of the component ions. Thus the formation of Schottky disorder in Cr_2O_3 can be expressed as follows:

$$\text{nil} \rightleftharpoons 2V_{Cr}''' + 3V_O^{\cdot\cdot} \tag{4.15}$$

with the corresponding mass-action expression:

$$\left[V_{Cr}'''\right]^2 \left[V_O^{\cdot\cdot}\right]^3 = K_S(T) = K_S^\circ e^{\Delta S_S/k} e^{-\Delta H_S/kT} \tag{4.16}$$

where the entropy and enthalpy of formation refer to the creation of five total defects. If Eq. (4.15) is the major source of defects, then the expression for charge neutrality is

$$3\left[V_{Cr}'''\right] \approx 2\left[V_O^{\cdot\cdot}\right] \tag{4.17}$$

[There seems to be frequent difficulty in understanding charge neutrality expressions such as Eq. (4.17), where the coefficients differ from unity. Why are we multiplying the concentration of the defect with the larger charge by the larger coefficient? That is a misreading of the equation, which is merely a mathematical relationship between the concentrations. It states that the concentration of the more highly charged V_{Cr}''' needs to be only 2/3 that of the lesser charged $V_O^{\cdot\cdot}$, as would be expected. That must be clearly understood, but, as a convenience, the correct result is always obtained if each defect in a charge neutrality expression is multiplied by the number of its defect charges.] An expression for the concentration of each defect is then obtained from the combination of Eqs. (4.16) and (4.17):

$$\left[V_{Cr}'''\right] \approx \left(\frac{2}{3}\right)^{3/5} K_S^{1/5} \approx \left(\frac{2}{3}\right)^{3/5} (K_S^\circ)^{1/5} e^{\Delta S_S/5k} e^{-\Delta H_S/5kT}$$

$$\left[V_O^{\cdot\cdot}\right] \approx \left(\frac{3}{2}\right)^{2/5} K_S^{1/5} \approx \left(\frac{3}{2}\right)^{2/5} (K_S^\circ)^{1/5} e^{\Delta S_S/5k} e^{-\Delta H_S/5kT} \tag{4.18}$$

The factor of 5 appears in the exponential terms in this case because a total of 5 defects are formed in the original disorder reaction. The general level of concentration of Schottky defects in Cr_2O_3 will depend primarily on the enthalpy of formation per defect, $\Delta H_S/5$. The defect reactions have been written in terms of integral numbers of each defect. That is both convenient and conventional, but not necessary. If any other standard is used, the thermodynamic parameters will differ proportionally.

Different Types of Ionic Disorder in the Same Compound

It has been stated that each conceivable type of ionic disorder will be present in amounts that are primarily determined by their enthalpic costs. Since however, the mass-action expression for each type must always be obeyed, there are interactions between the various types that have a common defect. While it is difficult enough to experimentally determine the enthalpy of formation for the major type of disorder, it is next to impossible to obtain the enthalpy of formation of the less favored types. However, it is now possible to calculate relatively accurate enthalpies from theoretical treatments. As an example, we can use the case of CaF_2, for which the enthalpies for the various types of disorder have been estimated by theoretical calculations as follows (Catlow et al., 1977):

Anion Frenkel:	2.65 eV	1.32 eV per defect (127 kJ/mol)
Schottky:	7.80 eV	2.6 eV per defect (250 kJ/mol)
Cation Frenkel:	8.40 eV	4.2 eV per defect (404 kJ/mol)

As described earlier, anion Frenkel disorder is the clear winner, with the lowest enthalpic cost per defect. It involves only singly charged defects, with an anion in an interstitial site that is surrounded by cations, while Schottky disorder involves a doubly charged defect, and cation Frenkel disorder includes a cation in an interstitial site that is surrounded by other cations. Thus the hierarchy of defect enthalpies is in accord with reasonable intuition.

The disorder reactions and their mass-action expressions are as follows:

$$F_F + V_I \rightleftharpoons F_I' + V_F^{\cdot} \tag{4.19}$$

$$\frac{[F_I'][V_F^{\cdot}]}{[V_I][F_F]} = K_{AF}(T) = K_{AF}^{\circ} e^{\Delta S_{AF}/k} e^{-\Delta H_{AF}/kT} \tag{4.20}$$

$$nil \rightleftharpoons V_{Ca}'' + 2V_F^{\cdot} \tag{4.21}$$

$$[V_{Ca}''][V_F^{\cdot}]^2 = K_S(T) = K_S^{\circ} e^{\Delta S_S/k} e^{-\Delta H_S/kT} \tag{4.22}$$

$$Ca_{Ca} + V_I \rightleftharpoons Ca_I^{\cdot\cdot} + V_{Ca}'' \tag{4.23}$$

$$\frac{[Ca_I^{\cdot\cdot}][V_{Ca}'']}{[V_I][Ca_{Ca}]} = K_{CF}(T) = K_{CF}^{\circ} e^{\Delta S_{CF}/k} e^{-\Delta H_{CF}/kT} \tag{4.24}$$

The entropy terms are not known, but the general magnitudes of the mass-action constants [i.e., the right-hand sides, for Eqs. (4.20), (4.22), and (4.24)], will be determined primarily by the exponential enthalpy terms. These give the following approximate values for the mass-action constants at 800°C, where it should be understood that the entropy terms might change these values by an order of magnitude or so if they were known and included:

Anion Frenkel: 3.6×10^{-13}
Schottky: 2.4×10^{-37}
Cation Frenkel: 2.8×10^{-42}

If each of these were the only type of disorder, the defect concentrations would be set primarily by the enthalpy term, $\exp(-h/kT)$, where h is the enthalpy of formation per defect: that is, $\Delta H_{AF}/2$, $\Delta H_S/3$, and $\Delta H_{CF}/2$. For the values just stated, at 800°C, the independent exponential enthalpic terms per defect would give the following approximate concentration levels:

Anion Frenkel: 6.0×10^{-7}
Schottky: 6.2×10^{-13}
Cation Frenkel: 1.7×10^{-21}

But these cannot all be independently true, because their mass-action expressions must all be simultaneously satisfied at equilibrium. When these various processes disorder spontaneously, the predominant type of disorder, anion Frenkel, will give 6×10^{-7} for the approximate fractional concentrations of anion interstitials and anion vacancies at 800°C. Compared with the other sources of charged defects, this means that the condition of charge neutrality can be reasonably approximated as follows:

$$[F_I'] \approx [V_F^{\cdot}] \tag{4.25}$$

As a result, the concentration of V_{Ca}'' will be suppressed by a common-ion effect, or, more accurately, a common-defect effect. If the value for the concentration of $V_F^{\cdot}$ is substituted into the mass-action expression for Schottky disorder, it is found that the concentration of V_{Ca}'' at 800°C is only 6.7×10^{-25}, a factor of 10^{12} less than it would have been in the absence of the anion Frenkel disorder. Thus a defect that results from a less-favored disorder reaction can be suppressed even further by the common-ion (common-defect) effect. When this value for $[V_{Ca}'']$ is substituted into the mass-action expression for cation Frenkel disorder, it is found that the concentration of $Ca_i^{\cdot\cdot}$ must be 4.2×10^{-18}, even higher than it would have been, had it been the major source of defects. These apparently anomalous consequences arise because all mass-action expressions that derive from properly written equilibrium disorder reactions for a system in equilibrium must be simultaneously satisfied. This discussion represents a narrative version of the more rigorous case of the solution of four simultaneous equations in four unknowns: the three mass-action expressions plus the approximate expression of charge neutrality, Eq. (4.25).

It is a simple exercise to show that the individual defect concentrations are linked through the various mass-action constants as follows:

$$[F_I'] \approx [V_F^{\cdot}] \approx K_{AF}^{1/2} \tag{4.26}$$

$$[V_{Ca}''] \approx \frac{K_S}{K_{AF}} \tag{4.27}$$

$$[Ca_I^{..}] \approx \frac{K_{CF} K_{AF}}{K_S} \tag{4.28}$$

A graphical representation of the relative defect concentrations as a function of temperature (Fig. 4.5) rather dramatically demonstrates how one type of intrinsic disorder usually dominates the defect chemistry. Even at the melting point of 1432°C, the concentration of the two major defects exceeds that of the next most abundant species by a factor of 10^7. Thus Eq. (4.25) approximates the full expression of charge neutrality to within 0.1 ppm or less. However, it must be kept in mind that even the minority ionic defects can be of importance for processes that depend on ionic diffusion. In that regard, it is interesting to note that the concentration of $Ca_I^{..}$ at 800°C is higher by a factor of 2500 than it would have been if cation Frenkel disorder had been the only type of intrinsic ionic disorder.

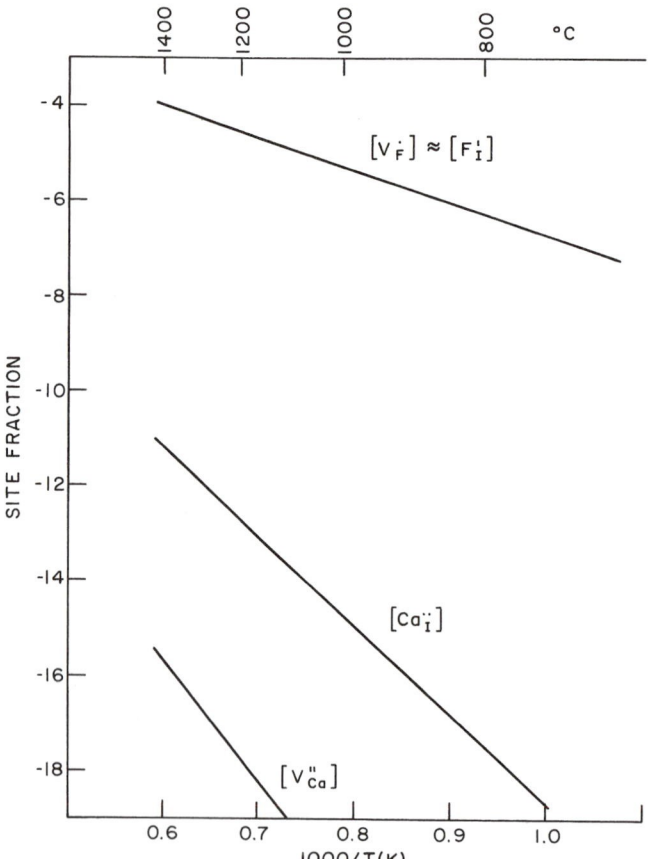

Figure 4.5 Arrhenius plot of the calculated concentrations of various intrinsic ionic defects in CaF_2.

Relative Equilibration Rates

In a discussion of the equilibrium concentration of intrinsic defects, it is implicitly assumed that the experimental conditions involve some combination of high enough temperature and long enough equilibration time to allow the diffusive processes a chance to reach the equilibrium state. In the case of Frenkel disorder, every ion in the crystal has an opportunity to move from its normal lattice site to an adjacent interstitial site. The interstitial ion and the vacancy then need to move only a few lattice spacings apart to be considered randomly distributed in the lattice. The important point is that Frenkel disorder can be uniformly generated (or annihilated) throughout the crystal. However, Schottky disorder is quite different in that vacancies are created at crystal surfaces as ions move out to newly created lattice sites. The vacancies must then diffuse from the surface throughout the crystal. The diffusion distances for random distribution of the defects are clearly much greater for Schottky disorder, and one can anticipate that the equilibration times will be correspondingly greater for the similar temperatures.

With regard to the diffusion distances required to reach equilibrium for Schottky disorder, it is important to consider exactly what is meant by a "surface." For an ideal single crystal, the external surface is the only surface available. However, in a polycrystalline ceramic oxide, the grain boundaries can serve as vacancy sources, and, within the grains, stacking faults, twin planes, and even dislocations can act as effective surfaces for the generation or annihilation of vacancies. The latter three sources can also be present in real single crystals. Thus the external physical dimensions of a sample may not be a valid indication of the diffusion lengths needed for the equilibration of intrinsic ionic disorder.

The Effect on Macroscopic Properties

Ionic defects have their most pronounced effect on transport properties, that is, ionic conductivity and diffusion. These phenomena represent the major applications for lattice disorder, as well as the basis for experimental techniques for their study and characterization. However, ionic transport is best studied in combination with extrinsic disorder, and thus a discussion is postponed to Chapter 7. Lattice defects do have some effects on bulk, static properties, and these are briefly described.

Since Frenkel disorder involves only the movement of ions within a crystal, neither the mass nor the dimensions are changed, and the density remains the same. However, for Schottky disorder, the defect formation reaction results in the creation of new lattice sites and the volume of the crystal increases and the density decreases. Schottky disorder has actually been measured by this technique in alkali halide crystals quenched from various equilibration temperatures. This is obviously an exquisitely delicate experiment. After the surface layer was dissolved from the quenched crystals, they were immersed in a density gradient column. The column contained an inert liquid with a density very close to that of the crystal, and a slight density gradient was established by a small temperature gradient along the column. The density of the crystal was determined by the position at which it became suspended in the liquid. Experimental results for the formation enthalpies for Schottky disorder, ΔH_S, for NaCl (Pelsmaekers et al., 1963) and LiF and KCl (Pelligrini and Pelsmaekers, 1969) gave 1.9, 2.45, and 2.41 eV (180, 236, and 232 kJ/mol), respectively. These values compare reasonably well with those obtained from measurements of the ionic conductivity: 2.45 eV (236 kJ/mol) for NaCl (Fuller, 1972), 2.34 eV (225 kJ/mol) for LiF (Stoebe and Pratt, 1967), and 2.52 eV (243 kJ/mol) for KCl (Fuller, 1972). The values are somewhat less than those obtained from the temperature dependence of ionic conductivity. This finding is attributed to the combination of some of the cation and anion vacancies into electrically neutral pairs. Such neutral pairs will not contribute to the ionic conductivity, but will reduce the density of the crystal. This behavior is discussed further in Chapters 6 and 7.

The refractive index of transparent crystals may also be affected by lattice defects. For example, patterned optical waveguides are prepared by the diffusion of Ti into the surface of $LiNbO_3$ single crystals. Light is confined to the indiffused path because it has a higher refractive index than the bulk of the crystal. In this case, a substitutional impurity center, Ti^{4+} on a lattice site normally occupied by Nb^{5+}, and its associated defects, are the active defects. It will be shown later that a substitutional aliovalent impurity (an impurity ion whose oxidation state differs from that of the one it replaces) requires a compensating defect in order to maintain charge neutrality. This system can be used to construct electrooptical switches, whereby multiple light signals, brought into the waveguide paths by optical fibers, can be switched to adjacent paths by the application of lateral electrical fields. Thus multiple incoming optical signals can be rerouted into various outgoing optical fibers.

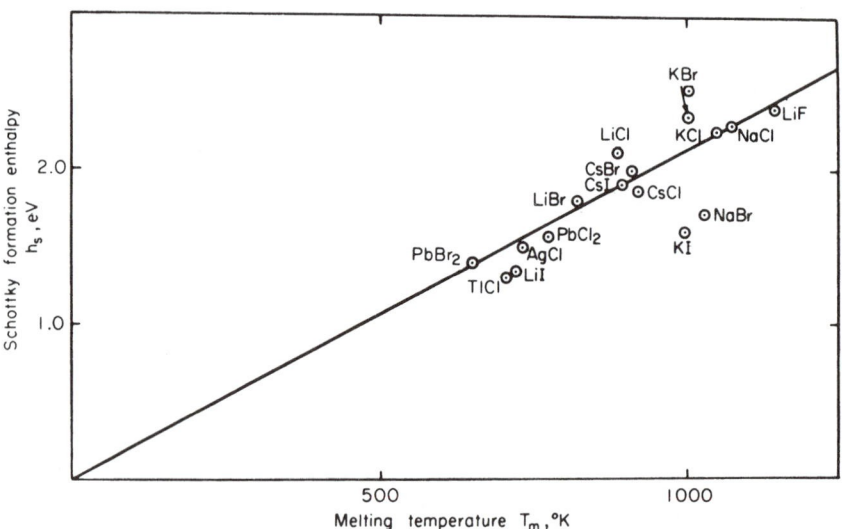

Figure 4.6 The enthalpy of intrinsic ionic disorder in various compounds as a function of the absolute melting point. (Reproduced from Barr and Lidiard, 1970.)

Formation Enthalpies for Lattice Defects

Since the creation of lattice defects involves the breaking of chemical bonds, as does the melting process, it not surprising to find that the enthalpies of formation for intrinsic lattice disorder can be correlated with the melting points of the crystals. This was shown in Fig. 3.6 for monatomic crystals such as metals and the rare gases. A similar plot for various ionic crystals is shown in Fig. 4.6 (Barr and Lidiard, 1970). The correlation is nearly as good. In fact, this correlation has been used to suggest that values that deviate significantly from the communal line are probably in error. Subsequent and more precise experiments have usually confirmed this suspicion. Note that the plot includes compounds with different cation/anion ratios, and thus with different crystal structures. These materials also vary significantly in the ionic–covalent balance in their bonding. These differences in the nature of the compounds have little effect on the correlation demonstrated in Fig. 4.6, where the enthalpies can be closely approximated by the expression $\Delta H = 2.14 \times 10^{-3} T_m$. Efforts to greatly extend this correlation have been only moderately successful, however. For example, attempts to estimate the enthalpy of Schottky disorder in MgO by linear extrapolation of this line to 2800°C, the melting point of MgO, fall substantially short of the values obtained by more recent theoretical calculations.

REFERENCES

Barr, L. W., and A. B. Lidiard. Defects in ionic crystals. In *Physical Chemistry, an Advanced Treatise*, H. Eyring, D. Henderson, and W. Jost, Eds. New York: Academic Press, 1970, p. 177.

Catlow, C. R. A., M. J. Norgett, and T. A. Ross. Ion transport and interatomic potentials in alkaline-earth–fluoride crystals. *J. Phys. C, Solid State Phys.* 10:1627–1640, 1977.

Fuller, R. G. Ionic conductivity (including self-diffusion). In *Point Defects in Solids*, Vol. 1, *General and Ionic Crystals*. J. H. Crawford, Jr. and L. M. Slifkin, Eds. New York: Plenum Press, 1972, Chapter 2. An excellent review of intrinsic ionic disorder in ionic crystals.

Kröger, F. A., and H. J. Vink. In *Solid State Physics*, Vol. 3, F. Seitz and D. Turnbull, Eds. New York: Academic Press, 1956.

Pellegrini, G., and J. Pelsmaekers. Determination of the formation energy of vacancies in lithium fluoride and potassium chloride by quenching. *J. Chem. Phys.* 51:5190–5191, 1969.

Pelsmaekers, J., G. Pellegrini, and S. Amelinckx. A determination of the formation energy of vacancies in sodium chloride by quenching. *Solid State Commun.* 1:92–95, 1963.

Stoebe, T. G., and P. L. Pratt. *Proc. Brt. Ceram. Soc.* 9:171, 1967.

Ure, R. W., Jr. Ionic conductivity of calcium fluoride crystals. *J. Chem. Phys.* 26:1363–1373, 1957.

☑ PROBLEMS

4.1. Write balanced equilibrium reactions and mass-action expressions (include the entropy and enthalpy terms explicitly) for cation Frenkel, anion Frenkel, and Schottky disorder in compounds having the following types of formula:

    a. M_2O

    b. M_2O_5

4.2. Assuming for each case in Problem 4.1 that the reaction as written is the only significant source of charged defects, write the approximate condition of charge neutrality for each defect type, and give an expression for the concentration of each defect species in terms of the appropriate entropy and enthalpy terms.

Extrinsic Ionic Disorder

5

INTRODUCTION

Extrinsic ionic disorder involves the ionic defects that result from the presence of impurity ions dissolved in the lattice. It refers to the single-phase, solid solution region up to the solubility limit of the impurity. Impurity ions usually replace one of the normal ions in the structure of the host compound in substitutional solid solution, as in the case of $CaCl_2$ dissolved in NaCl or Al_2O_3 in TiO_2. The impurity sometimes appears to occupy interstitial sites, but such cases are uncommon. Situations in which the impurity is a cation are much more common than the case of anionic impurities, although the latter are certainly not excluded (e.g., MgF_2 dissolved in MgO, or vice versa).

Isovalent substitutions represent a trivial case of extrinsic ionic disorder because the impurity center is neutral relative to the perfect crystal, and it is not necessary to have any other defect present to maintain charge neutrality. They have no first-order effect on the defect chemistry of the host material. Thus in the case of the NiO–CoO system, the two end members both have the NaCl structure, and the two cations have the same charge and are similar in size. As a result, there is complete miscibility in this pseudobinary system, as shown in Fig. 5.1 (Kingery et al., 1976); there is merely a random substitution of Co^{2+} for Ni^{2+} as the solid solution becomes richer in CoO. Only at temperatures below 800°C does the system separate into two solid phases, but long exposure to temperatures in this region probably is needed for the necessary diffusion processes to occur to significant extent. Whether the impurity center is considered to be Ni_{Co} or Co_{Ni} is arbitrary. Since there is no need for any other defect to be created for charge compensation, the properties of the solid solutions differ very little from those of the end members, and the system is of little interest in terms of transport properties such as diffusion and ionic conductivity.

The main effect of impurity ions on the defect chemistry of a compound results from their difference in charge state relative to the ion they replace. Such impurities are frequently referred to as **aliovalent** ions, to distinguish them from the isovalent case. In the case of aliovalent substitutions, the impurity carries a charge relative to

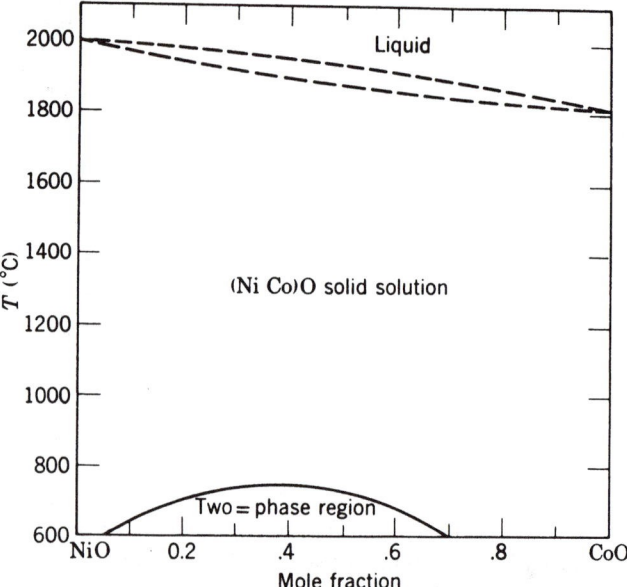

Figure 5.1 The NiO–CoO phase diagram. With the same cation/anion ratio, the same structure, and similar cation radii, there is complete miscibility across the entire compositional range for temperatures above 800°C. (Reproduced from Kingery et al., 1976, by permission of John Wiley & Sons.)

the perfect crystal, and some other defect having the opposite charge must be present to maintain charge neutrality. The presence of these charge-compensating defects can have a profound effect on transport properties. This makes possible the development of desired properties by the deliberate doping of a host crystal with the appropriate amount of a suitable impurity. Diffusion constants and electrical conductivities can often be varied by several orders of magnitude by such techniques, which can be used either to promote desirable properties or to suppress undesirable ones.

This chapter is restricted to solid solutions of stoichiometric binary compounds. In these cases, the aliovalent impurity is charge-compensated by some type of ionic defect, either a vacancy or an interstitial. It will be seen later that electrons and holes can also serve as charge-compensating species, in which case the electrical conductivity may be strongly affected. However, electronic compensation is achieved only by a deviation from stoichiometry, and discussion of this topic is deferred to a later chapter.

THE AgCl–CdCl$_2$ SYSTEM

Since the cations in the AgCl–CdCl$_2$ system have different charges, the compounds have different cation/anion ratios, and they obviously cannot have the same structure. Therefore, by definition, there cannot be complete miscibility across the entire compositional range, and the solid solution regions of CdCl$_2$ dissolved in AgCl, and

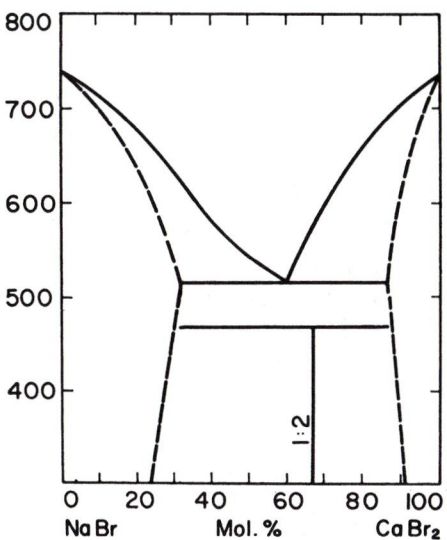

Figure 5.2 The NaBr–CdBr$_2$ phase diagram. With different cation/anion ratios, and thus different crystal structures, solid solubility is limited on both sides. (Reproduced from Levin et al., 1964, by permission of the American Ceramic Society.)

AgCl dissolved in CdCl$_2$, have to be treated separately. Such a situation is depicted in the phase diagram of the closely related NaBr–CaBr$_2$ system shown in Fig. 5.2 (Levin et al., 1964). It is particularly important to note the crystal structures of the two end members, since these will define the structure in the two solid solution regions. AgCl has the NaCl structure, with cations in all the octahedral sites in a ccp anion sublattice, while CdCl$_2$ has the CdCl$_2$ structure (naturally!), which also has a ccp anion sublattice but with only half the octahedral sites filled. This is accomplished by having alternate layers of octahedral sites unoccupied in the ccp anion array. The two structures are thus very closely related; this tends to favor mutual solubility, but at 0.115 nm for Ag$^+$ and 0.095 for Cd^{2+}, the ionic radii of the two cations are sufficiently different to cause the solid solubility ranges to be limited.

The Solid Solution of CdCl$_2$ in AgCl

The solid solutions must form in a way that preserves charge neutrality. There are a number of formal ways to describe the incorporation of an aliovalent ion into a crystalline matrix, but only the end result is important. If one electrically neutral entity is replaced with another neutral entity, charge neutrality is automatically conserved. Therefore, we will replace an integral number of formula units of the host compound with an integral number of formula units of the aliovalent solute compound. If one unit of AgCl is removed from an AgCl crystal and replaced by one unit of CdCl$_2$ (never mind how this is done; thermodynamics does not require us to describe the process by a practical procedure), the Cd^{2+} cation can be placed in the vacated Ag$^+$ site, and one of the Cl$^-$ anions can be accommodated by the vacated Cl$^-$ site. The only place the second Cl$^-$ anion can be located within the crystal is in an interstitial site. This process can be summarized by the following incorporation reaction:

$$CdCl_2 \xrightarrow{(AgCl)} Cd^{\cdot}_{Ag} + Cl_{Cl} + Cl'_I \qquad (5.1)$$

where it is understood that the compound on the left replaces the neutral entity in parentheses above the arrow. The latter notation is convenient because it explicitly shows the lattice sites that are available for the impurity compound. The divalent Cd^{2+} has one more positive charge than the Ag^+ it replaces, and that is balanced by the extra negative charge of the Cl^- in the interstitial site. Conservation of mass and charge is clearly shown, and conservation of structure is indicated by the correspondence between sites cleared by removal of the host compound and the lattice sites occupied by the impurity compound (absolute rigor would require that an empty interstitial site be shown on the left-hand side, but that is not common practice). The reaction arrow has been shown as pointing only to the right, rather than in both directions as was done for the formation of intrinsic ionic defects, because the impurities are almost never introduced under equilibrium conditions. That would require equilibration of the host crystal with a fixed activity of the impurity compound. While these conditions can be met in principle, especially when the impurity has significant volatility, as in the cases of PbO or Bi_2O_3, it is a most unusual approach. For the solution of $CdCl_2$ in AgCl, the usual technique is to melt some AgCl (mp 455°C) and then stir in the appropriate amount of $CdCl_2$. After solidification, the concentration of Cd^{2+} in the crystal is fixed, and is not in equilibrium with anything. If there is no $CdCl_2$ activity in the environment surrounding the solid solution, the $CdCl_2$ should ultimately diffuse out of the crystal; but such diffusion is so kinetically hindered that it can almost always be safely ignored. Again, PbO and Bi_2O_3 in compounds at very high temperatures are notable exceptions.

In the process just described, the cation sublattice remains ideally filled, although some of the occupants are impurity cations; but the anion sublattice is oversubscribed: there are more anions than normal anion sites. The alternative is to adjust the substitutional ratio so that the anion sublattice is ideally filled, and a cation defect is created to maintain charge neutrality. This can be summarized by the following incorporation reaction

$$CdCl_2 \xrightarrow{(2AgCl)} Cd^{\cdot}_{Ag} + V'_{Ag} + 2Cl_{Cl} \qquad (5.2)$$

In this case, one unit of $CdCl_2$ replaces two units of AgCl, to make room for both Cl^- ions. At the same time, two cation sites are emptied, but there is only one Cd^{2+} to occupy them, so one of the cation sites remains vacant. This cation vacancy is the alternative to the interstitial anion in Eq. (5.1) as the charge-compensating ionic defect. The extra positive charge of the impurity center is balanced by the missing positive charge of the vacant cation site. Once again, all the pertinent conservation laws are obeyed. One always has the two options just described: keep the cation sublattice perfect and form an anion defect, or keep the anion sublattice perfect and form a cation defect.

According to the two possibilities described above, the positively charged impurity center, $Cd^{\cdot}_{Ag}$, can be compensated by either of two negatively charged lattice defects:

the cation vacancy V'_{Ag} or the interstitial anion Cl'_i. One is generally sufficiently favored that the alternative need not be considered. A significant concentration of more than one compensating defect is extremely unlikely. The choice can often be predicted. In the case of the interstitial anion, the anion must be squeezed into a site that ideally is only about one-quarter its size; moreover, the environment around the site is predominantly anionic. This combination of inadequate space and unfavorable charge environment is prohibitive. On the other hand, there are no such restrictions on the formation of a cation vacancy, and it is clearly the favored choice, as has been abundantly proven by a wide range of experiments. Moreover, we have already seen in Chapter 4 that the favored form of intrinsic ionic disorder in AgCl is of the cation Frenkel type and involves both interstitial cations and cation vacancies. This indicates that the cation vacancy is an energetically favored defect, so it is reasonable that it is also the extrinsic defect of choice. It will be seen that continuity demands that the compensating defect be one of the preferred intrinsic ionic defects, at least for modest impurity contents. In a structural sense, the presence of cation vacancies in the NaCl structure of the solid solution can be viewed as a move toward the structure of the other end member, $CdCl_2$, which contains alternate layers of unoccupied octahedral sites in a ccp anion sublattice.

Thus the stoichiometric composition of the solid solution can be written as $Ag_{1-2x}Cd_xCl$. It consists of a combination of stoichiometric binary constituents. However, the reference structure from which defects are defined is the ideal NaCl structure, and its composition differs from that of the stoichiometric solid solution.

The author's first professional task was to study ionic conduction in the silver halides related to their use as the electrolyte in subminiature, electrochemical cells that can be combined into high voltage, but very, very low current batteries. The current-delivering capabilities could be significantly improved by doping the silver halides with about 0.02 mol % cadmium halide, a measure that increased the concentration of silver vacancies, which are the major ionic charge carriers in this system. Because these are solid state devices, and contain no liquids, they have the potential for extraordinarily long lifetimes. Ironically, the project came to a successful conclusion at about the time that transistorized circuits began to dominate the field of electronics, and interest shifted to low voltage, high current power supplies. As a result, the devices had only limited application, and they soon faded away. After a few more experiences with similar results, the author decided that he was better suited for a career in academia.

The Solid Solution of AgCl in CdCl₂

If one formula unit of AgCl is substituted for one formula unit of $CdCl_2$, the negatively charged impurity center is compensated by the positively charged anion vacancy.

$$AgCl \xrightarrow{(CdCl_2)} Ag'_{Cd} + Cl_{Cl} + V^{\cdot}_{Cl} \qquad (5.3)$$

The cation sublattice is ideally filled, but there are vacancies in the anion sublattice. The smaller positive charge on the cation site occupied by the impurity is balanced by the missing negative charge at the vacant anion site. The anion sublattice can be ideally filled if one formula unit of $CdCl_2$ is replaced by two formula units of AgCl:

$$2AgCl \xrightarrow{(CdCl_2)} Ag'_{Cd} + Ag_I^{\bullet} + 2Cl_{Cl} \tag{5.4}$$

In this case, there is only one cation site available for the two Ag^+ ions, so one of them has to go into an interstitial site. The smaller positive charge on the impurity center is balanced by the extra positive charge of the cation in the interstitial site.

In this case, the choice between the two alternative positively charged defects is not so clear. A singly charged anion vacancy cannot be excluded, yet there is a certain attractiveness to the interstitial cation. The interstitial sites in the $CdCl_2$ structure are the unoccupied octahedral sites that occur in alternate layers in the ccp anion sublattice. Therefore, the site is of adequate size, and is surrounded by anions. By occupation of these interstitial sites, the $CdCl_2$ structure of the solid solution is moving toward the NaCl structure of the other end member. We are free to speculate because the system has not been studied, and the actual choice is not known.

There is another, minor alternative in this case that involves the interstitial cations. Since the solid solution will contain far more Cd^{2+} than Ag^+, it is possible that there could be some place exchange between the cations on normal and interstitial sites:

$$Cd_{Cd} + Ag_I^{\bullet} \rightleftharpoons Cd_I^{\bullet\bullet} + Ag'_{Cd} \tag{5.5}$$

Thus some of the Cd^{2+} might be located in interstitial sites, while additional Ag^+ occupies normal cation sites. If this exchange went to completion, the corresponding incorporation reaction should be written:

$$2AgCl + Cd_{Cd} \xrightarrow{(CdCl_2)} 2Ag'_{Cd} + Cd_I^{\bullet\bullet} + 2Cl_{Cl} \tag{5.6}$$

Note that the latter creates more defect charges than the alternative, Eq. (5.4). In general, the reactions that create the fewest defect charges are favored because they result in less disruption of the bonding. There is no experimental information available on the choice for this system; the diffusivity of Cd^{2+} might be strongly affected, for example.

THE CaF_2–CaO SYSTEM

Although the impurity ions are anions in the CaF_2–CaO system, the principles remain exactly the same. The end member structures are the fluorite structure for CaF_2, and the NaCl structure for CaO.

The Solid Solution of CaF_2 in CaO

For the solution of CaF_2 in CaO, the two alternative incorporation reactions are

$$CaF_2 \xrightarrow{(CaO)} Ca_{Ca} + F_O^{\bullet} + F_I' \tag{5.7}$$

$$CaF_2 \xrightarrow{(2CaO)} Ca_{Ca} + V''_{Ca} + 2F_O^{\cdot} \qquad (5.8)$$

The impurity center bears a net positive charge and requires a negatively charged compensating defect. The cation sublattice is ideally filled in the first reaction, while the anion sublattice is ideally filled in the second. The first option has the usual problem of finding space for a large interstitial anion in a fully packed structure. The tetrahedrally coordinated interstitial site is too small, and has too much of an anion-dominated first-coordination sphere to be favored for occupation by an anion. With the unique exception of the silver halides, Schottky disorder is thought to be the favored type of intrinsic ionic disorder for compounds having the NaCl structure. Thus the cation vacancy is a logical choice for this solid solution, but experimental confirmation is not available. If the interstitial anion were the preferred choice, we would again face the possibility that the interstitial F^- might displace O^{2-} from normal lattice sites and push these anions into the interstitial sites, although this would result in the formation of more defect charges.

The Solid Solution of CaO in CaF$_2$

The matrix structure in a solid solution of CaO in CaF$_2$ is the fluorite structure, which has large, unoccupied, octahedral sites that are surrounded by six cations. Such a site should be hospitable to interstitial anions, but the electrostatic environment is very unfavorable for cations. Since anion Frenkel disorder is the favored type of intrinsic ionic disorder for this structure, the compensating defect could be of the anionic type. The alternative incorporation reactions are:

$$CaO \xrightarrow{(CaF_2)} Ca_{Ca} + O'_F + V_F^{\cdot} \qquad (5.9)$$

$$2CaO \xrightarrow{(CaF_2)} Ca_{Ca} + Ca_I^{\cdot\cdot} + 2O'_F \qquad (5.10)$$

Based on the discussion above, the fluoride vacancy is expected to be the preferred compensating defect. This is in accord with the much more common case in which a fluorite structure compound is doped with a lesser charged cationic impurity to give a negatively charged impurity center, where it is well known that the compensating defects are anion vacancies. Thus, for ZrO_2 doped with 10–20 mol % CaO or Y_2O_3 to give the impurity centers Ca''_{Zr} or Y'_{Zr}, the compensating defects are oxygen vacancies, and the materials are excellent high temperature oxygen ion conductors. Such materials serve as the electrolyte in sensors that measure the oxygen activity in the exhaust manifolds of modern automobiles. Similar sensors are used in the steel industry to measure the amount of carbon in molten steel. The sensors actually measure the oxygen activity in the steel, but since carbon and oxygen are in equilibrium at those temperatures, their activities are interelated.

THE TiO$_2$–Nb$_2$O$_5$ SYSTEM

It is perhaps useful to analyze a more complex system to demonstrate that a careful application of the same procedures demonstrated for the simpler systems is still applicable. In this case, TiO_2 usually has the rutile structure (it has other structures, but they all revert to rutile at high temperatures). Nb_2O_5 has a complex structure that we need not consider here. Once again there are two alternatives that involve matching up the numbers of cations and anions with the numbers of available sites.

The Solid Solution of TiO_2 in Nb_2O_5

The two alternatives are as follows:

$$2TiO_2 \xrightarrow{(Nb_2O_5)} 2Ti'_{Nb} + 4O_O + V_O^{\cdot\cdot} \tag{5.11}$$

$$5TiO_2 \xrightarrow{(2Nb_2O_5)} 4Ti'_{Nb} + Ti_I^{4\cdot} + 10O_O \tag{5.12}$$

These have been written to maintain integral numbers of formula units. In the first alternative, the cation lattice has been ideally filled, while the anions are matched with available sites in the second. The choice between the two possible compensating defects is not obvious, but the matrix compound, Nb_2O_5, has a relatively close-packed structure, and this is not generally favorable for interstitial defects. Also, the interstitial defect in this case is quite highly charged, another detrimental factor. Finally, oxygen vacancies are known to be the preferred defect in a number of transition metal oxides, and the possibility of their presence can never be rejected as totally unlikely.

The Solid Solution of Nb_2O_5 in TiO_2

The two alternatives in the case of a solid solution of Nb_2O_5 and TiO_2 are as follows:

$$Nb_2O_5 \xrightarrow{(2TiO_2)} 2Nb_{Ti}^{\cdot} + 4O_O + O_I'' \tag{5.13}$$

$$2Nb_2O_5 \xrightarrow{(5TiO_2)} 4Nb_{Ti}^{\cdot} + 10O_O + V_{Ti}^{4/} \tag{5.14}$$

Interstitial anions in a close-packed anion sublattice are extremely unlikely. The cation vacancies are highly charged, but are probably the best choice.

SUMMARY OF IMPORTANT POINTS

- Isovalent substitutional impurities require no extrinsic defects for charge compensation and have no first-order effect on the defect chemistry of the host material.

- Aliovalent impurities require the creation of the charge equivalent of a compensating defect. This may have a major effect on the defect chemistry and the resulting transport properties.

- Usually, only one kind of extrinsic ionic defect results from the impurity content. Two different defects may be energetically very similar, but such exceptions are extremely rare.

- For each case, there are two possible compensating ionic defects. For positively charged impurity centers (donor impurities), the defects are either cation vacancies or anion interstitials, while for negatively charged impurity centers (acceptor impurities), they are either cation interstitials or anion vacancies.

- The compensating ionic defect should correspond to one of the defects that results from the preferred form of intrinsic ionic disorder.

- The choice of compensating defect can often be deduced from the space available and the electrostatic environment in the case of interstitial defects, and from the formal charge on the defect.

- So far we have considered only extrinsic ionic defects as the compensating species. The possible preference for extrinsic electrons or holes, which requires a change in composition from the stoichiometric composition, is dealt with in Chapter 9.

SCHEMATIC REPRESENTATION OF DEFECT CONCENTRATIONS

It is often useful to represent relationships diagrammatically, and defect chemistry is a particularly good example. The information contained in a group of rather sterile equations is much more readily accessible when summarized in a simple diagram. This will be demonstrated by two examples, one simple case and one somewhat more complex example.

Diagrammatic Representation of the Solid Solution of $CaCl_2$ in NaCl

The $CaCl_2$–NaCl system has been thoroughly investigated, and it is known to behave quite ideally. As we saw in Chapter 4, Schottky defects are the preferred type of intrinsic disorder, and positively charged impurity centers such as $Ca_{Na}^{\cdot}$ are thus compensated by cation vacancies. For a complete treatment we need the items represented by Eqs. (5.15)–(5.18).

The Schottky disorder reaction:

$$nil \rightleftharpoons V_{Na}' + V_{Cl}^{\cdot} \tag{5.15}$$

Its mass-action expression:

$$[V_{Na}'][V_{Cl}^{\cdot}] = K_S(T) \tag{5.16}$$

The impurity incorporation reaction:

$$CaCl_2 \xrightarrow{(2NaCl)} Ca_{Na}^{\cdot} + V_{Na}' + 2Cl_{Cl} \tag{5.17}$$

An expression of charge neutrality:

$$[V_{Na}'] \approx [Ca_{Na}^{\cdot}] + [V_{Cl}^{\cdot}] \tag{5.18}$$

The last equation is an equality if there are no other charged defect species with significant concentrations. The defect concentrations as a function of the impurity content at constant temperature can now be obtained by solution of the two simultaneous equations, Eqs. (5.16) and (5.18). However, it is more instructive to take a less rigorous approach. The expression of charge neutrality demonstrates that there are two sources of cation vacancies. For every anion vacancy created by Schottky disorder, a cation vacancy must have been created simultaneously; and for every Ca^{2+} substituted for Na^+, there must also be a cation vacancy. The fractional contributions to the total concentration of cation vacancies will depend on the relative magnitudes of the intrinsic disorder and the impurity content.

The concentrations of the various species as a function of the impurity content at constant temperature are shown schematically in Fig. 5.3 in the form of a log–log plot. For very low impurity concentrations, the fraction of the cation vacancy concentration due to the impurity is negligible. In such cases $[Ca_{Na}^{\cdot}]$ can be neglected relative to $[V_{Cl}^{\cdot}]$, and charge neutrality can be approximated by

$$[V_{Na}'] \approx [V_{Cl}^{\cdot}]; \qquad [V_{Cl}^{\cdot}] \gg [Ca_{Na}^{\cdot}] \tag{5.19}$$

In other words, the behavior is essentially intrinsic. Thus the defect concentrations are not a function of impurity content in this region, and they can be obtained by combination of Eqs. (5.16) and (5.19):

$$[V_{Na}'] \approx [V_{Cl}^{\cdot}] \approx K_S^{1/2} \tag{5.20}$$

On the other hand, at very high impurity contents, virtually all the cation vacancies

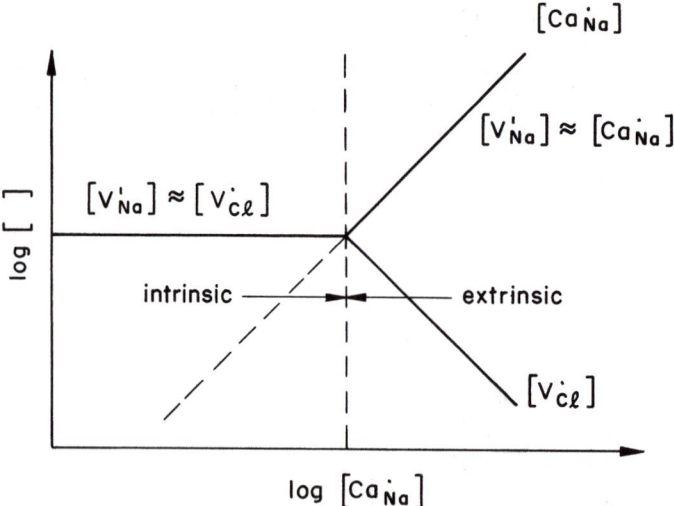

Figure 5.3 Log–log plot of defect concentrations (empty brackets) as a function of the $CaCl_2$ content in NaCl.

will result from compensation of the impurity, and $[V'_{Cl}]$ can be neglected relative to $[Ca^{\cdot}_{Na}]$. Charge neutrality can then be approximated by:

$$[V'_{Na}] \approx [Ca^{\cdot}_{Na}]; \qquad [Ca^{\cdot}_{Na}] \gg [V^{\cdot}_{Cl}] \tag{5.21}$$

This is the region of extrinsic behavior. As the impurity concentration increases, it picks up the intrinsic defect of opposite charge and carries it upward to maintain charge neutrality. The linear relationship results in a slope of $+1$ on the log–log plot. Meanwhile, the mass-action expression for intrinsic disorder must still be obeyed under equilibrium conditions. Thus the behavior of the other intrinsic defect, the anion vacancy, can be obtained for the extrinsic region from Eqs. (5.16) and (5.21):

$$[V^{\cdot}_{Cl}] \approx \frac{K_S}{[Ca^{\cdot}_{Na}]} \tag{5.22}$$

Its concentration is suppressed by the presence of the impurity, and it drops off with a slope of -1.

Nota bene: all valid mass-action expressions must be obeyed simultaneously under equilibrium conditions.

This important point is demonstrated in Fig. 5.4, which expands on Fig. 5.3 by adding lines for some of the minority defects that are involved in the less-favored types of intrinsic disorder (e.g., cation Frenkel and anion Frenkel). Formation reactions and the

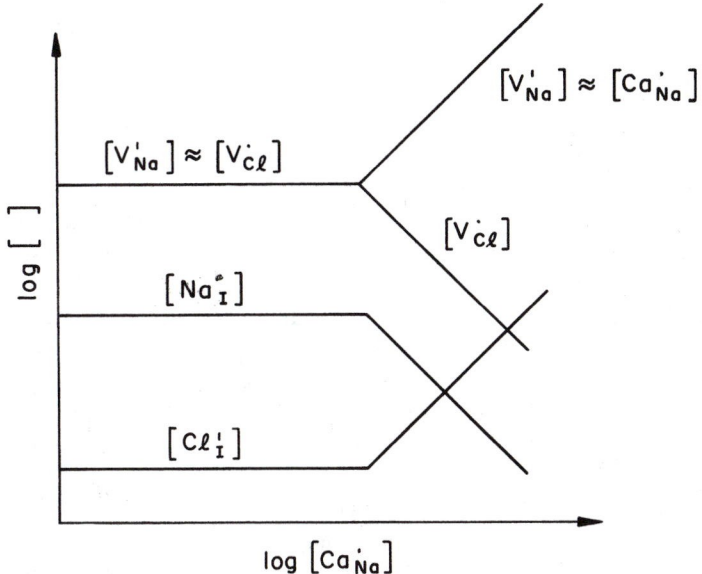

Figure 5.4 Log–log plot of defect concentrations, including minority defects, as a function of the $CaCl_2$ content in NaCl.

corresponding mass-action expressions for these two less-favored types of intrinsic disorder are:

$$\text{nil} \rightleftharpoons \text{Na}_\text{I}^\cdot + \text{V}_\text{Na}' \tag{5.23}$$

$$[\text{Na}_\text{I}^\cdot][\text{V}_\text{Na}'] = K_\text{CF}(T) \tag{5.24}$$

$$\text{nil} \rightleftharpoons \text{Cl}_\text{I}' + \text{V}_\text{Cl}^\cdot \tag{5.25}$$

$$[\text{Cl}_\text{I}'][\text{V}_\text{Cl}^\cdot] = K_\text{AF}(T) \tag{5.26}$$

It has been assumed that $\Delta H_\text{S} < \Delta H_\text{CF} < \Delta H_\text{AF}$, as indicated by theoretical calculations. The concentrations of the cation and anion interstitials in the intrinsic region can be obtained from Eqs. (5.20) and (5.24), and (5.20) and (5.26), respectively:

$$[\text{Na}_\text{I}^\cdot] \approx \frac{K_\text{CF}}{K_\text{S}^{1/2}} \tag{5.27}$$

$$[\text{Cl}_\text{I}'] \approx \frac{K_\text{AF}}{K_\text{S}^{1/2}} \tag{5.28}$$

The concentrations of these defects are suppressed from what they would be if they were a part of the preferred type of disorder, because $\Delta H_\text{S}/2$ is less than either $\Delta H_\text{CF}/2$ or $\Delta H_\text{AF}/2$. In the extrinsic region, their concentrations are:

$$[\text{Na}_\text{I}^\cdot] \approx \frac{K_\text{CF}}{[\text{Ca}_\text{Na}^\cdot]} \tag{5.29}$$

$$[\text{Cl}_\text{I}'] \approx \frac{K_\text{AF}}{K_\text{S}}[\text{Ca}_\text{Na}^\cdot] \tag{5.30}$$

In general, the impurity enhances the concentrations of ionic defects that have the opposite sign of charge and suppresses those having the same sign of charge. Note that the relative amounts of the noncompensating defects can change drastically in this region. In the example shown in Fig. 5.4, the concentration of the least-favored intrinsic defect, Cl_I', eventually exceeds that of one of the most-favored intrinsic defects, $\text{V}_\text{Cl}^\cdot$. However, the solubility limit of the impurity is usually exceeded long before such a change in compensating defect occurs.

Figure 5.4 is a clear demonstration of the need for continuity between the defects formed by the preferred type of intrinsic disorder and the defect that compensates for the charge of the aliovalent impurity center. While $[\text{Cl}_\text{I}']$ has the opposite charge of the impurity, there is no way for it to suddenly leap up from its suppressed level in the intrinsic region to become the initial charge-compensating defect in the extrinsic region. All the defect concentrations vary systematically with the impurity concentration, and their concentrations cannot change abruptly.

In constructing Figs. 5.3 and 5.4, we considered only the extreme cases, where the condition of charge neutrality can be adequately represented by two oppositely charged defects, Eqs. (5.19) and (5.21). These conditions are then projected linearly until they intersect at the boundary between intrinsic and extrinsic behavior. This is obviously inaccurate near the boundary, where all three major defect species must be included in the expression of charge neutrality, Eq. (5.18). This situation can be solved to give the curvature of the defect plots in the transition region, but this is not necessary for the type of general overview needed here, and it is common practice to construct the diagram from straight-line segments. However, if experimental results happen to fall into the transition region, the more rigorous approach is necessary for an accurate analysis of the data.

It is also useful to depict the temperature dependence of the defect concentrations. One way to do this is to add a new set of lines to Fig. 5.3 that show the defect concentrations as a function of impurity content at another temperature. This has been done in Fig. 5.5, where the subscripts on the defect concentrations represent two temperatures, $T_2 > T_1$. In the intrinsic region, the defect concentrations increase with temperature according to Eq. (5.20). The determining factor is $\Delta H_S/2$. At the higher temperature it takes more impurity centers to dominate the charge neutrality equation, so the boundary between intrinsic and extrinsic behavior moves to a higher impurity content. The concentration of the compensating defect in the extrinsic region is fixed by the invariant impurity content, hence has no temperature dependence. However, the concentration of the other intrinsic defect, $V_{Cl}^{\cdot}$, rises with the full enthalpy of the intrinsic disorder reaction, ΔH_S, according to Eq. (5.22). The two intrinsic defects must account for the full enthalpy, so that if one has no temperature dependence, the other gets it all. In many such cases, the relationships are apparent from considerations of plane geometry, independent of the chemistry. Thus, given the slopes of the lines in Fig. 5.5, the vertical distance between the anion vacancy concentrations at the two temperatures in the extrinsic region must be twice the vertical separation of the defect concentrations in the intrinsic region.

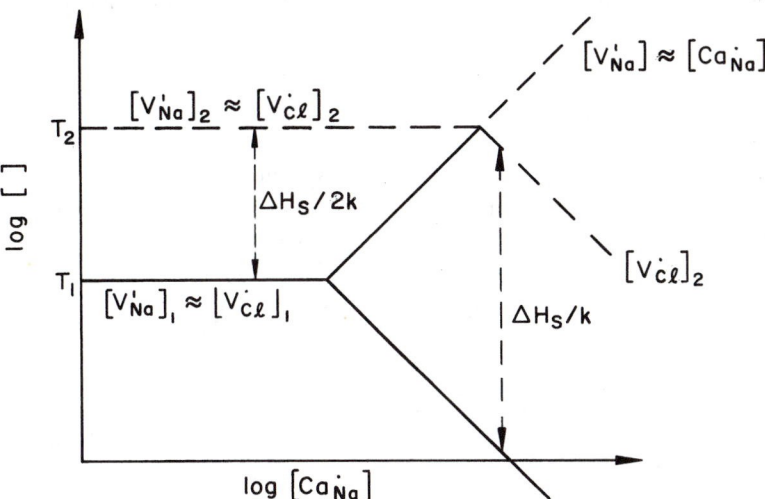

Figure 5.5 Log–log plot of defect concentrations at two temperatures, as a function of the $CaCl_2$ content of NaCl.

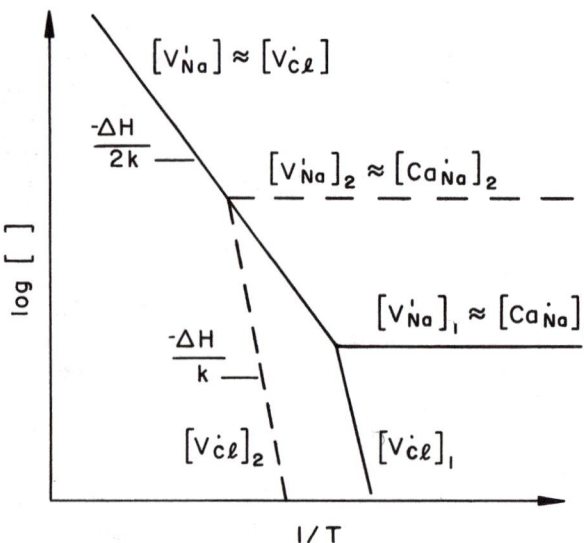

Figure 5.6 Arrhenius plot of defect concentrations as a function of temperature for two concentrations of CaCl₂ in NaCl.

The temperature dependence of the defect concentrations is more usually displayed in the form of an Arrhenius plot, in which one plots the log of the concentrations against the reciprocal absolute temperature at constant impurity content. This is shown in Fig. 5.6 for two impurity concentrations, $[Ca^{\cdot}_{Na}]_2 > [Ca^{\cdot}_{Na}]_1$. In this case, the slopes are related to the enthalpy of Schottky disorder, as seen from Eqs. (5.20) and (5.22). It is from such plots of experimental data that the disorder enthalpies are usually obtained.

The Solid Solution of TiO₂ in Nb₂O₅

In the absence of information on the TiO₂–Nb₂O₅ system, we shall assume that Schottky disorder is the preferred form of intrinsic disorder and that the negatively charged impurity centers, Ti'_{Nb}, are compensated by oxygen vacancies according to Eq. (5.11). The Schottky formation reaction for Nb₂O₅, its mass-action expression, the impurity incorporation reaction, and the expression of charge neutrality are as follows:

$$nil \rightleftharpoons 2V^{5/}_{Nb} + 5V^{\cdot\cdot}_O \tag{5.31}$$

$$[V^{5/}_{Nb}]^2[V^{\cdot\cdot}_O]^5 = K_S(T) \tag{5.32}$$

$$2TiO_2 \xrightarrow{(Nb_2O_5)} 2Ti'_{Nb} + 4O_O + V^{\cdot\cdot}_O \tag{5.33}$$

$$2[V^{\cdot\cdot}_O] \approx 5[V^{5/}_{Nb}] + [Ti'_{Nb}] \tag{5.34}$$

At sufficiently low impurity contents, the behavior will be essentially intrinsic and $[\text{Ti}'_{\text{Nb}}]$ can be neglected relative to $[V^{5/}_{\text{Nb}}]$. The condition of charge neutrality can then be approximated by

$$2[V^{\cdot\cdot}_{\text{O}}] \approx 5[V^{5/}_{\text{Nb}}] \tag{5.35}$$

When combined with the mass-action expression, this gives

$$[V^{5/}_{\text{Nb}}] \approx \left(\frac{2}{5}\right)^{5/7} K^{1/7}_{\text{S}} \tag{5.36}$$

The concentration of the $V^{\cdot\cdot}_{\text{O}}$ in this region is obviously 5/2 times this value. These concentrations are plotted for the intrinsic region in Fig. 5.7. At high impurity concentrations, $[V^{5/}_{\text{Nb}}]$ can be neglected relative to $[\text{Ti}'_{\text{Nb}}]$, and the approximation for charge neutrality becomes

$$2[V^{\cdot\cdot}_{\text{O}}] \approx [\text{Ti}'_{\text{Nb}}] \tag{5.37}$$

This fixes the concentration of oxygen vacancies in the extrinsic region, where $[V^{5/}_{\text{Nb}}]$ is given by Eqs. (5.32) and (5.37) to be

$$[V^{5/}_{\text{Nb}}] \approx \left(\frac{2}{[\text{Ti}'_{\text{Nb}}]}\right)^{5/2} K^{1/2}_{\text{S}} \tag{5.38}$$

The concentration of $[V^{5/}_{\text{Nb}}]$ drops off with a slope of $-5/2$.

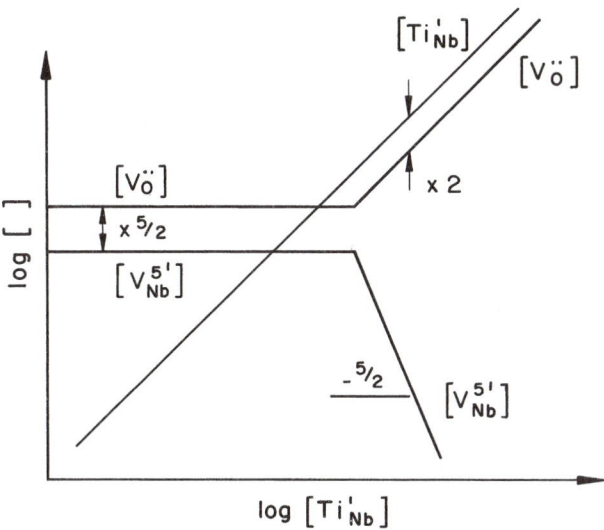

Figure 5.7 Log–log plot of defect concentrations as a function of the TiO_2 content of Nb_2O_5.

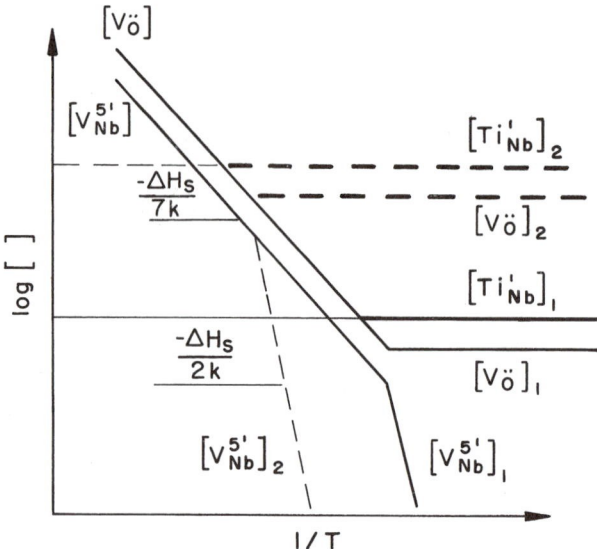

Figure 5.8 Arrhenius plot of defect concentrations as a function of temperature for two concentrations of TiO_2 in Nb_2O_5.

The temperature dependences of the defect concentrations for two different impurity contents are shown in the Arrhenius form in Fig. 5.8. The slopes correspond to the enthalpy terms in Eqs. (5.36) and (5.38).

SUMMARY OF EXTRINSIC IONIC DISORDER

So far, we have considered only the case of solid solutions of stoichiometric binary compounds. By definition, this gives stoichiometric solid solutions, and the compensating defects are limited to ionic species. By loss or gain of one atomic species, usually the anion, nonstoichiometric solid solutions can be obtained in which the compensating defects can be electrons or holes (i.e., extrinsic electronic defects; see Chapter 9).

For the concentrations normally encountered in defect chemistry, the impurity content has no first-order effect on the thermodynamic parameters of the host compound, including the various mass-action constants. In thermodynamic equilibrium, all the valid mass-action expressions must be obeyed simultaneously. The functional dependences of the various defects can then be obtained by a straightforward manipulation of the mass-action expressions and the impurity content such that all the conservation rules are obeyed. The concentration of the impurity can usually be treated as a constant, since it is frozen into the lattice in a nonequilibrium state.

The effect of an impurity on the defect concentrations results from the necessity for including it in the expression for bulk charge neutrality.

If the impurity makes no significant contribution to charge neutrality, it has no significant effect on the defect concentrations.

The initial lattice defect necessary for charge compensation of an impurity center must be one of the lattice defects in the preferred form of intrinsic ionic disorder.

This ensures continuity between the intrinsic and extrinsic regions. Defect concentrations depend on the impurity concentration in a very systematic way, and it is impossible for one of the less-favored defects to suddenly leap up to become the preferred defect for charge compensation.

In many compounds of practical or scientific interest, the energetics for the formation of intrinsic defects are so unfavorable that their concentrations are small compared with the concentrations of naturally occurring impurities. In those cases, intrinsic disorder will never dominate the defect chemistry, which will be controlled by the impurity content and thus may vary somewhat from sample to sample. One way to avoid this sample-to-sample variation is to deliberately add a relatively large amount of a selected aliovalent impurity, so that fluctuations in the naturally occurring impurity content have comparatively little effect.

REFERENCES

Kingery, W. D., H. K. Bowen, and D. R. Uhlmann. *Introduction to Ceramics*, 2nd ed. New York: John Wiley & Sons, 1976, Fig. 7.15.

Levin, E. M., C. R. Robbins, and H. F. McMurdie. *Phase Diagrams for Ceramists*. Columbus, OH: American Ceramic Society, 1964, Diagram 1193.

☑ PROBLEMS

5.1. Assume that Schottky disorder is the preferred type of intrinsic ionic disorder in Al_2O_3. Consider a dilute solid solution of TiO_2 in Al_2O_3 in which the titanium substitutes randomly for aluminum in the lattice.

    a. For the case that the titanium concentration greatly exceeds that of the intrinsic defects, write expressions for each defect species in terms of mass-action constants and the titanium concentration.

    b. Show on a schematic graph how the defect concentrations vary with the titanium concentration. Include both the intrinsic and extrinsic regions.

5.2. Assume that the impurities in the following combinations dissolve in substitutional solid solution. In each case, write two incorporation reactions, one yielding a compensating cation defect and the other yielding a compensating anion defect. Indicate which you think would be the more likely choice in each case.

Impurity		*Material*
$SrCl_2$	in	KCl (NaCl structure)
$PbBr_2$	in	AgBr (NaCl structure)
CaO	in	ThO_2 (fluorite structure)
Y_2O_3	in	ThO_2 (fluorite structure)
Nb_2O_5	in	TiO_2 (rutile structure)
MgF_2	in	MgO (NaCl structure)

Defect Complexes and Associates

6

INTRODUCTION

So far, we have implicitly assumed that ionic defects are totally independent and noninteracting. That is the essence of the requirement for the use of dilute solution thermodynamics and the mass-action approach. Whenever charged defects are present, however, there must be defects with opposite charges to preserve charge neutrality. There is then always the possibility of electrostatic attraction between the oppositely charged defects. Thus in NaCl doped with $CaCl_2$, some of the compensating cation vacancies may be attracted to sites adjacent to the charged impurity center. Also, in AgCl the oppositely charged cation interstitials and cation vacancies formed by intrinsic cation Frenkel disorder may be electrostatically bound together. There is also the possibility for elastic interactions due to the local stresses created by defects. Thus an oversized substitutional impurity cation may attract a vacancy to relieve the local stresses associated with both defects. It is possible to extend the range of dilute solution thermodynamics by explicitly taking into account such bound combinations of defects. This is analogous to describing an aqueous solution of Ni^{2+} and Cl^- ions in terms not only of the isolated ions, but also the complex ions formed between them (e.g., $NiCl_4^{2-}$). In this chapter, we call the bound combination of an impurity center with an oppositely charged ionic defect a **defect complex**, while the combination of two intrinsic ionic defects is referred to as a **defect associate**. The distinction is arbitrary but convenient. The defect complexes and associates are treated as separate and distinct defect species. The discussion is limited to bound pairs of defects, which is by far the most important case. More complicated combinations are possible but are not discussed here.

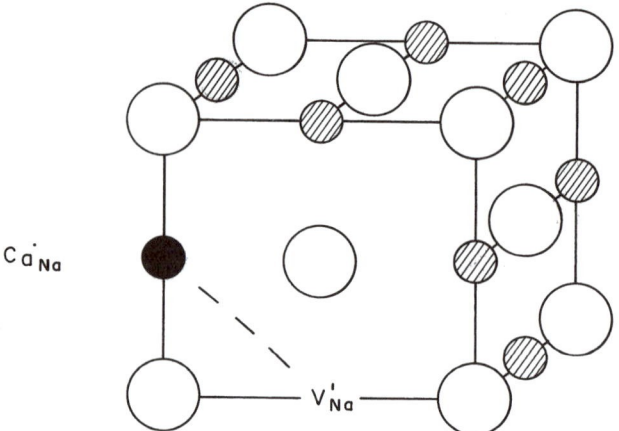

Figure 6.1 Defect complex between a divalent impurity cation and a singly charged cation vacancy in the NaCl structure: open circles, anions; cross-hatched circles, cations; dashed line represents the electrostatic attraction between the oppositely charged defects. The ions are undersized relative to the unit cell.

COMPLEXES CONTAINING AN IMPURITY CENTER AND AN IONIC DEFECT

Stability of the Complexes

Consider the case of NaCl doped with $CaCl_2$, with cation vacancies as the compensating defect. There will be some tendency for the more mobile vacancies to be attracted to the vicinity of the oppositely charged impurity center, as shown in Fig. 6.1. The formation of the defect complexes from the individual defects can be written as an equilibrium reaction

$$Ca_{Na}^{\cdot} + V_{Na}^{\prime} \rightleftharpoons (Ca_{Na}^{\cdot} V_{Na}^{\prime}) \tag{6.1}$$

The complex is indicated by enclosing the component species in parentheses. In this case, the complex is electrically neutral, but it has a dipole moment. The extent to which the defects combine into such complexes depends on the usual balance between enthalpic cost and the configurational entropy. While the formation of complexes reduces the total number of defects, and thus directly impacts the entropy, the pairs introduce an orientational degeneracy that contributes to the configurational entropy; that is, the complex can be oriented in different directions.

A mass-action expression can be written for Eq. (6.1):

$$\frac{[(Ca_{Na}^{\cdot} V_{Na}^{\prime})]}{[Ca_{Na}^{\cdot}][V_{Na}^{\prime}]} = K_C(T) = K_C^{\circ} e^{\Delta S_C/k} e^{-\Delta H_C/kT} \tag{6.2}$$

Since the complexes are expected to break up with increasing temperature, their enthalpy of formation, ΔH_c, should be negative (i.e., exothermic). The reverse of Eq. (6.1) represents the decomposition of the complex, and the enthalpy of decomposition would be positive. The enthalpy of formation is essentially the electrostatic energy that tends to bind the two defects into the complex. As a first approximation, it should be the Coulomb energy between the two charges, but this must be modified to take into account the resulting polarization of the surrounding ions in the lattice. In effect, there is a total amount of electrostatic energy available; some of it causes polarization of the lattice, while the rest binds the defects together. The greater the amount of polarization, the smaller the amount of energy that remains to bind the complex. The dielectric constant of the material is a measure of the polarizability of the lattice. Because the complexes persist for relatively long times, the static dielectric constant is the appropriate value to use. It is not certain, however, whether the bulk dielectric constant is an adequate indication of the polarization on the atomic scale around the complex. If it is, the enthalpy of formation of the complex should be the Coulomb energy modified by the static dielectric constant

$$\Delta H_C \sim \frac{Z_1 Z_2 e^2}{k_s r} \tag{6.3}$$

where e is the unit electronic charge, Z_1 and Z_2 are the number of unit charges on each defect with their sign ($+1$ and -1 in the present example), r is the distance between the defects in the complex, and k_s is the static dielectric constant of the material. For $(Ca_{Na}^{\cdot} V_{Na}')$ in NaCl, ΔH_c is calculated to be about -0.5 eV (50 kJ/mol), which is surprisingly close to the experimentally determined values for such complexes in the alkali halides. Apparently the bulk dielectric constant is an adequate indication of the local polarization. As seen in Fig. 6.1, the two defects in the complex are separated by a fragment of the structure in that the electron clouds of adjacent anions intrude into the space between the defects.

Typical experimentally determined values for ΔH_c in various systems are given below. These values were taken from extensive compilations of thermodynamic parameters by Franklin (1972) and by Fuller (1972).

Example of Sr^{2+} and Ca^{2+} in alkali metal chlorides and bromides: $(Sr_{Na}^{\cdot} V_{Na}')$, -0.5 to -0.6 eV (-50 to -60 kJ/mol)

Example of Cd^{2+} in AgCl: $(Cd_{Ag}^{\cdot} V_{Ag}')$, -0.5 eV (-50 kJ/mol)

Example of O^{2-} or Na^+ in alkaline earth fluorides: $(Na_{Ca}' V_F^{\cdot})$ or $(O_F' V_F^{\cdot})$, -0.3 to -0.5 eV or (-30 to -50 kJ/mol)

Example of Y^{3+} or Gd^{3+} in alkaline earth fluorides: $(Y_{Ca}^{\cdot} F_I')$, -0.45 eV (-43 kJ/mol)

The latter two cases represent negative impurity center-anion vacancy and positive impurity center-anion interstitial complexes, respectively, in these fluorite structure compounds that prefer anion Frenkel disorder.

It is also possible to form complexes that bear a net charge. For example, for Na_2O dissolved in MgO

$$\text{Na}_2\text{O} \xrightarrow{\text{(2MgO)}} 2\text{Na}'_{\text{Mg}} + \text{O}_\text{O} + \text{V}_\text{O}^{\cdot\cdot} \tag{6.4}$$

the following complex formation reaction can occur:

$$\text{Na}'_{\text{Mg}} + \text{V}_\text{O}^{\cdot\cdot} \rightleftharpoons (\text{Na}'_{\text{Mg}}\text{V}_\text{O}^{\cdot\cdot})^\cdot \tag{6.5}$$

If a second cation impurity is bound to the complex, a neutral complex containing three defects results, $(\text{Na}'_{\text{Mg}}\text{V}_\text{O}^{\cdot\cdot}\text{Na}'_{\text{Mg}})$, but this goes beyond the limits of complexity established at the beginning of the chapter. It is apparent from Eq. (6.3) that the binding enthalpy will increase approximately linearly with the product of the charges on the defects.

Accurate use of Eq. (6.3) requires the use of absolute charges on the defects, rather than their nominal charges based on a purely ionic model. This is one example of a situation in which the degree of covalency can have a real impact. However, for a truly quantitative calculation of the formation enthalpies, it is necessary to use a much more sophisticated model than the one just described.

Experimental Evidence for Defect Complexes

Defect complexes make their presence known in a number of experimental situations:

1. Ionic Conductivity. In the region of extrinsic behavior, where the concentration of compensating defects is fixed by the impurity content, an accelerating decrease in ionic conductivity with decreasing temperature is often observed. This phenomenon, which is attributed to a decrease in the concentration of mobile charge carriers as more and more defects are bound into neutral, immobile complexes, is dealt with in some detail in the next chapter.

2. Relaxation Processes. The defect complex has a dipole moment, and its orientation can be influenced by an applied electric field. Thus when the temperature is high enough to make such motion possible, a cation vacancy in NaCl can swing around the $\text{Ca}_\text{Na}^\cdot$ to which it is bound in order to align the dipole with an applied field. This leads to a very useful experiment called thermally stimulated current (TSC), or, more properly, thermally stimulated depolarization (TSD). An electroded crystal is heated to a temperature at which the defects can be readily reoriented but the complexes are not decomposed, and an electric field is applied to align the defect dipoles along the field direction from their original random orientations. The crystal is then cooled, with the field still applied, until the alignment is frozen. The field is removed and the crystal is then attached to a sensitive current meter and gradually warmed. As the defects become mobile, the dipolar orientations gradually randomize, and this movement of charge within the crystal causes a current to flow in the external circuit. Analysis of the current versus time (or current vs. temperature, which is a function of time) gives information on the concentration of complexes, and on the energetics of their reorientation. An example of such a depolarization experiment is shown in Fig. 6.2 for CaF_2 containing about 10 ppm Er^{3+} (Stott and Crawford, 1971). The

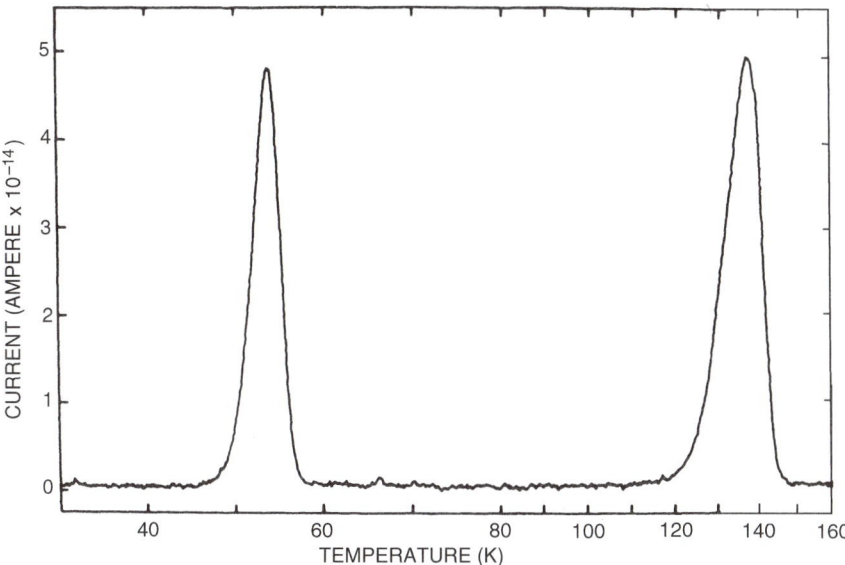

Figure 6.2 Thermally stimulated depolarization currents in CaF_2 doped with 10 ppm Er^{3+}. The peaks are attributed to rotations of the dipolar complex $(Er_{Ca}^{\cdot}F_I')$ with the F_I' in the nearest- and next-nearest-neighbor sites relative to the Er^{3+}. (Reproduced from Stott and Crawford, 1971, by permission of the American Physical Society.)

peaks are attributed to reorientation of the dipolar complex $(Er_{Ca}^{\cdot}F_I')$, with the high temperature peak corresponding to the F_I' being in nearest-neighbor position relative to the Er^{3+}, while for the low temperature peak the F_I' is in the next-nearest-neighbor site. The activation energies of the dipolar motion are 0.380 and 0.167 eV (36.6 and 16.1 kJ/mol), respectively.

A more dynamic manifestation of the same phenomenon can be achieved by the application of an ac field to the crystal at an appropriate temperature. At very low frequencies, the dipoles will be able to keep up with the field, and they will reorient with each half-cycle. The full effect of the polarization reversal will contribute to the dielectric constant, and the energy dissipation (i.e., the dielectric losses), will be low. At very high frequencies, the dipoles will not be able to respond to the field at all; before the dipoles can start to reorient with one field direction, the field will already have reversed. The dipolar contribution to the dielectric constant will be lost, but the dielectric losses will still be low. However, at intermediate frequencies, the dipoles will be able to partially reorient with the changing field direction, but will increasingly lag behind as the frequency increases. The dipolar contribution to the dielectric constant will drop off with increasing frequency, and the resulting "electrical friction" of the frustrated reorientation will result in the dissipation of energy as heat. As a result, the dielectric losses will reach a maximum value at the midpoint of this intermediate frequency range. This represents a nearly ideal case of the classical Debye relaxation phenomenon. Experimental results of this kind are shown in Fig. 6.3 for CaF_2 doped

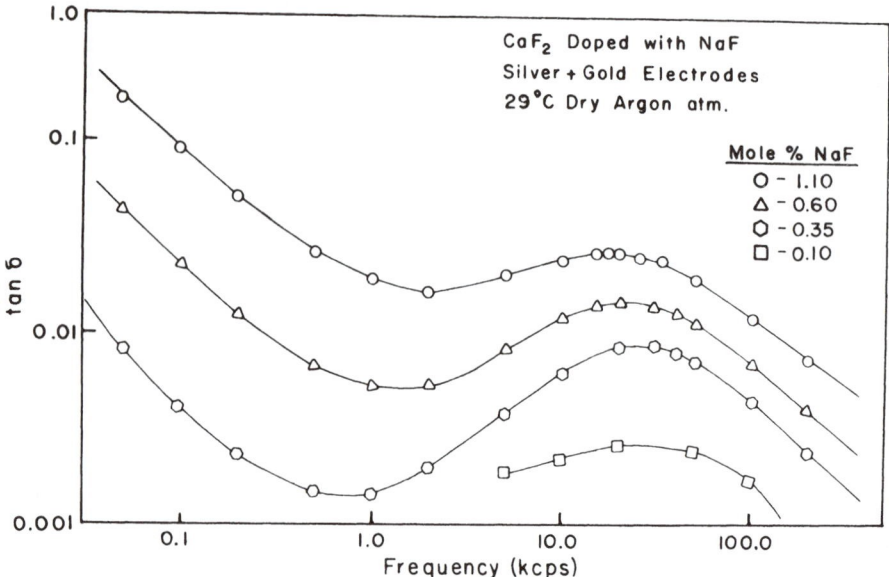

Figure 6.3 Dielectric loss peaks in crystals of CaF_2 doped with various amounts of NaF (Johnson et al., 1969). The peaks are attributed to the reorientation of the defect complex ($Na'_{Ca} V^\bullet_F$). The increase in losses at low frequencies is most likely due to ionic conduction. (Reproduced from Johnson et al., 1969, by permission of Elsevier Science.)

with Na^+ (Johnson et al., 1969). The results are shown as a plot of tan δ (the dielectric loss) versus log frequency at constant temperature for several dopant concentrations. The peaks are attributed to the motion of a fluoride vacancy around the impurity center in the complex ($Na'_{Ca} V^\bullet_F$). For an ideal single relaxation process the maximum in the dielectric loss occurs when the product of the angular frequency, $\omega = 2\pi f$, and the characteristic time constant of the relaxation process, τ, equals unity:

$$2\pi f_{max}\tau = \omega\tau = 1 \tag{6.6}$$

For a thermally activated reorientation process, τ is exponentially related to the temperature:

$$\tau = \tau_0 e^{-\Delta H_0/kT} \tag{6.7}$$

where ΔH_0 is the activation enthalpy for the reorientation process. Combination of Eqs. (6.6) and (6.7) shows that the frequency at which the dielectric losses are at the maximum value is directly related to the temperature

$$\omega_{max} = \frac{1}{\tau_0} e^{\Delta H_0/kT} \tag{6.8}$$

This is an example of "time–temperature superposition," where the behavior of the system as a function of $1/T$ at constant frequency is equivalent to its behavior as a function of $\log f$ at constant temperature. Thus the experiment described above may also be done as a function of temperature at constant frequency, and the results would look very similar. It is apparent that an Arrhenius plot of $\log f_{max}$ versus $1/T$ will give the activation enthalpy of the reorientation process, and this was found to be 0.53 eV (51 kJ/mol) for the CaF_2 doped with Na^+ shown in Fig. 6.3.

There is a direct mechanical analog to the dielectric polarization process just described. A defect complex corresponds to an internal stress field that will have a preferred orientation with an applied mechanical stress. The crystal will respond to an alternating mechanical stress in the same way that it responds to an alternating electrical stress. There is a "storage modulus" that is the analog of the real part of the dielectric constant, and a "loss modulus" that corresponds to the imaginary part of the dielectric constant (i.e., the dielectric losses). The mathematical treatment is exactly the same. Thus a mechanical wave is propagated through the crystal by coupling it to a mechanical transducer such as the moving cone of a small audio speaker connected to a frequency generator, while the transmitted signal is sensed by measuring the voltage generated by a similar speaker whose movable cone is coupled to the opposite side of the crystal. This technique is sometimes referred to as an "internal friction" measurement. Figure 6.4 shows the results of such measurements made on the same Na^+-doped CaF_2 sample used for the dielectric loss measurements shown in Fig. 6.3

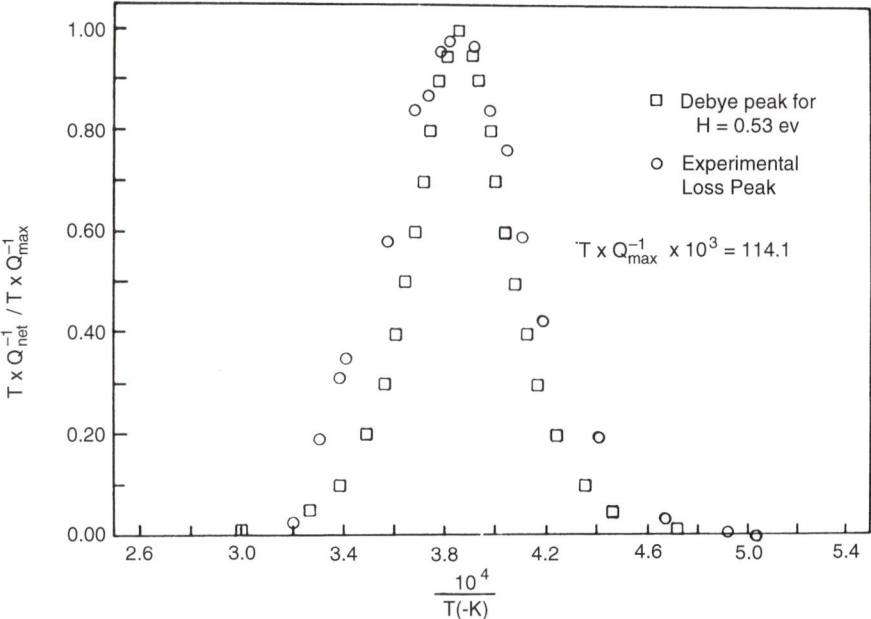

Figure 6.4 Mechanical energy losses (internal friction) for the same CaF_2 crystal doped with 1% NaF shown in Fig. 6.3. The ordinate is a measure of the energy losses. The experimental results are shown as circles, while the squares represent the calculated values for an ideal Debye relaxation with an activation enthalpy of 0.53 eV (51 kJ/mol). (Reproduced from Johnson et al., 1969, by permission of Elsevier Science.)

(Johnson et al., 1969). "Q" is essentially the inverse of the energy loss. The activation enthalpy for the reorientation was found to be the same 0.53 eV (51 kJ/mol) as that from the electrical measurements. Figure 6.4 compares the experimental results with a calculated curve for an ideal Debye relaxation with an activation enthalpy of 0.53 eV (51 kJ/mol).

3. Spectroscopic Techniques. Probably the most direct evidence for defect complexes comes from electron paramagnetic resonance (EPR). In this spectroscopic technique, the resonant condition for energy absorption from a microwave signal by the precessing spin of an unpaired electron is determined as a function of magnetic field and crystal orientation. The resonant condition is very sensitive to the atomic environment around the unpaired electron, and there are also interactions with nuclear magnetic moments. From the results, the detailed atomic environment can be calculated explicitly, and it is possible to determine whether the unpaired spin is near a cation vacancy, and even whether the vacancy is in the nearest-neighbor cation position, or in a next-nearest-neighbor site. Our example of NaCl doped with $CaCl_2$ would not respond to this experiment because there are no unpaired electrons in this system. Thus it is necessary to use a transition metal impurity ion with unpaired spins. Mn^{2+}, with its five unpaired electrons in the d shell, has often been used as the divalent impurity ion for such experiments in the alkali halides.

INTRINSIC IONIC DEFECT ASSOCIATES

Frenkel Pair Associates in Silver Halides

We have treated the formation of cation Frenkel disorder in the Ag halides as the removal of a Ag^+ from its normal lattice site, and its placement in a remote interstitial position. Obviously, the cation does not disappear from its lattice site and then reappear in a distant interstitial site. The first step in this process must be the movement of the lattice cation into the nearest interstitial site, followed by diffusion to a remote location. Movement of the cation into the nearest interstitial site will require some enthalpy, but there will be some residual electrostatic attraction between the vacancy and the interstitial, and additional enthalpy will be required to break the two defects apart and to move them to widely separated sites. Thus we can view this as a two-step process: the formation of the vacancy–interstitial pair on adjacent sites, which we will call an associate, and their subsequent separation to remote sites. Each step can be treated as an equilibrium reaction, with a mass-action expression, and a characteristic enthalpy. For the formation of the intrinsic defect associate, we can write

$$Ag_{Ag} + V_I \rightleftharpoons (Ag_I^{\cdot}V_{Ag}') \tag{6.9}$$

$$\frac{[(Ag_I^{\cdot}V_{Ag}')]}{[Ag_{Ag}][V_I]} = K_A(T) = K_A^{\circ}e^{\Delta S_A/k}e^{-\Delta H_A/kT} \tag{6.10}$$

where ΔH_A is the enthalpy of formation of the associate from the perfect lattice. The decomposition of the associate can then be expressed by

$$(Ag_I^{\cdot}V_{Ag}^{/}) \rightleftharpoons Ag_I^{\cdot} + V_{Ag}^{/} \tag{6.11}$$

$$\frac{[Ag_I^{\cdot}][V_{Ag}^{/}]}{[(Ag_I^{\cdot}V_{Ag}^{/})]} = K_D(T) = K_D^{\circ} e^{\Delta S_D/k} e^{-\Delta H_D/kT} \tag{6.12}$$

where ΔH_D is the enthalpy of dissociation of the associate. The addition of the equilibrium defect reactions, Eqs. (6.9) and (6.11), and the multiplication of the two mass-action expressions, Eqs. (6.10) and (6.11), give the formation reaction and mass-action expression for the direct formation of noninteracting cation Frenkel defects, as described in Chapter 4. This suggests a more graphical representation of the process as shown in Fig. 6.5. The horizontal dimension represents an enthalpy scale, and it is seen that the enthalpy of formation for the unbound defects is the sum of the enthalpy of formation of the defect associate and the enthalpy of its dissociation. The negative of the enthalpy of dissociation, $-\Delta H_D$, can be considered to be the enthalpy of association from the isolated defects. In Chapter 4, we saw that the equilibrium concentration of unassociated cation Frenkel defects was determined largely by the exponential enthalpy term, $\exp(-\Delta H_{CF}/2kT)$, while from Eq. (6.10) it is clear that the concentration of Frenkel associates is determined primarily by $\exp(-\Delta H_A/kT)$. Thus, if ΔH_A is less than $\Delta H_{CF}/2$, there will be more defects in the associates than in unassociated form. As seen from Fig. 6.5, this is equivalent to the case that ΔH_A is less than ΔH_D.

Schottky Defect Associates in NaCl

The same type of treatment is valid for Schottky disorder in NaCl, although the mechanistic approach is somewhat different. It is easy to see that the movement of a Ag^+ into the nearest interstitial site automatically creates a defect associate. In the case of Schottky disorder in NaCl, one can consider either the diffusion of cation vacancy–anion vacancy pairs into the crystal from a surface, along with the individual, unassociated defects, or the diffusion of individual vacancies into the crystal, where

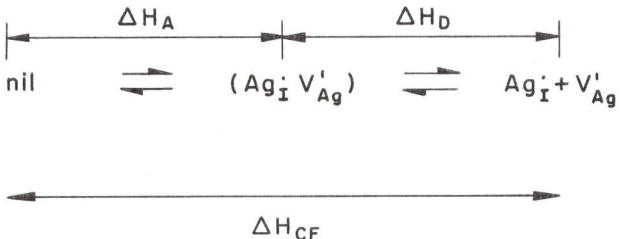

Figure 6.5 Schematic enthalpy scale for the formation of cation Frenkel disorder in both the associated and free states.

some of them combine into associates. This distinction is not important for equilibrium thermodynamics, and the process can again be treated as two steps

$$\text{nil} \rightleftharpoons (V'_{Na} V^{\cdot}_{Cl}) \tag{6.13}$$

$$\left[(V'_{Na} V^{\cdot}_{Cl})\right] = K_A(T) = K^{\circ}_A e^{\Delta S_A/k} e^{-\Delta H_A/kT} \tag{6.14}$$

$$(V'_{Na} V^{\cdot}_{Cl}) \rightleftharpoons V'_{Na} + V^{\cdot}_{Cl} \tag{6.15}$$

$$\frac{[V'_{Na}][V^{\cdot}_{Cl}]}{\left[(V'_{Na} V^{\cdot}_{Cl})\right]} = K_D(T) = K^{\circ}_D e^{\Delta S_D/k} e^{-\Delta H_D/kT} \tag{6.16}$$

The overall process can again be presented along an enthalpy scale as shown in Fig. 6.6, where the relationships are directly analogous to those for the Ag halide system just described.

What can be expected for the binding enthalpy for these intrinsic vacancy associates? A simple Coulombic model of electrostatic attraction, Eq. (6.3), proved to give values close to the experimentally determined, as well as theoretically calculated, enthalpies for complex formation between cation vacancies and impurity centers in the alkali halides. The only change in Eq. (6.3) for the case of intrinsic associates in the same material would be that r, the distance between the defects, is smaller by a factor of $2^{1/2}$. That would suggest an enthalpy of association of about -0.7 eV (-70 kJ/mol). This seems to be too low, and the actual values are closer to -1 eV (-100 kJ/mol). This discrepancy most likely is a result of the inadequacy of the bulk static dielectric constant of the crystal to account for the effect of lattice polarization. Unlike the case of the defect complexes, there is absolutely nothing between the two defects in the intrinsic Schottky associates, because they sit on adjacent lattice sites. In the defect complexes, the two component species sit on nearest-neighbor cation sites, and some of the electron clouds of the adjacent anions intrude into the space between them.

More Complex Systems

The basic principles just described for the Ag and alkali halides can be applied to any system, and there will always be some concentration of intrinsic defect associates,

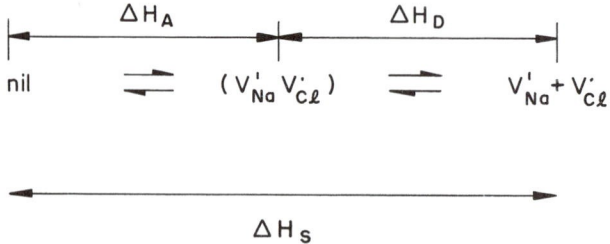

Figure 6.6 Schematic enthalpy scale for the formation of Schottky disorder in both the associated and free states.

although in some cases they will not be at a sufficient level to have any observable effects on measurable properties. In some systems, the associates may bear a net defect charge, and there may be unequal numbers of the oppositely charged intrinsic defects. Thus for an oxide with the generic formula M_2O_3, having Schottky disorder as the preferred intrinsic type, and assuming that the associates are limited to pairs of defects, the formation of defect associates will be represented by

$$\text{nil} \rightleftharpoons 2 \left(V_M''' V_O^{\cdot\cdot} \right)' + V_O^{\cdot\cdot} \tag{6.17}$$

The associates bear a net charge, and there are unassociated defects left over after complete pairing has occurred. This is obviously a much more complicated situation, but the author is not aware of any system for which such a treatment has proved to be necessary.

Evidence for Intrinsic Defect Associates

1. Diffusion versus Ionic Conductivity. In the case of alkali halides with Schottky disorder, ions can diffuse not only by way of individual vacancies, but also by way of the vacancies in bound associates. On the other hand, ionic conductivity is due only to the individual, charged vacancies: the bound vacancy associates have no net charge and do not contribute to the conductivity. Thus if there is a significant concentration of bound associates, there will be more diffusion than would be expected from the observed level of ionic conductivity.

2. Anion Diffusion. In Chapter 5 we learned that when Schottky disorder is found in a compound such as KCl, the anion vacancy concentration decreases steadily with increasing concentration of a substitutional divalent cation impurity, such as Sr^{2+}. For Cl^- diffusion by way of anion vacancies, the diffusion rate should drop off similarly. As shown in Fig. 6.7, however, it is observed that the diffusion rate initially decreases, but then levels off with further impurity additions (Fuller et al., 1968). This behavior has been interpreted to mean that anions also diffuse by way of anion vacancies in the bound intrinsic vacancy associates, because their concentration is not affected by the impurity content. This in turn demonstrates well that the concentrations of neutral defect species cannot be affected by the addition of aliovalent impurities.

THE EFFECT OF IMPURITIES ON THE CONCENTRATIONS OF DEFECT COMPLEXES AND ASSOCIATES

Extrinsic Defect Complexes

Figure 6.8 shows an Arrhenius plot of the calculated concentrations of uncomplexed cation and anion vacancies in KCl doped with various amounts of $SrCl_2$, for assumed values of the thermodynamic parameters (Beaumont and Jacobs, 1966). The enthalpy for Schottky disorder was taken to be 2.26 eV (218 kJ/mol), while the entropy term, $\Delta S_S / k$, was given the value 5.4. The entropy of complex formation was not accurately

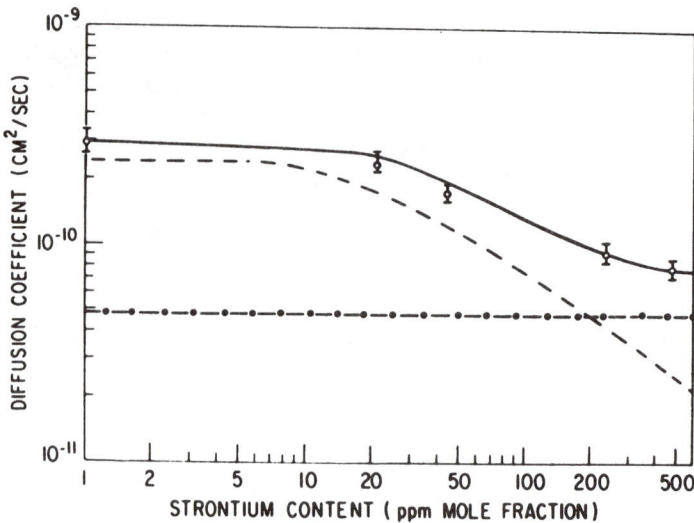

Figure 6.7 The diffusion constant of chlorine in KCl as a function of SrCl$_2$ content: dashed line, diffusion by way of isolated anion vacancies; dot–dash line, diffusion by way of anion vacancies bound in Schottky pairs. (Reproduced from Fuller et al., 1968, by permission of the American Physical Society.)

known, so the free energy was approximated as 0.42 eV (40 kJ/mol); in other words, this is essentially an apparent enthalpy, with the entropy term being neglected. The simple intrinsic and extrinsic regions are clearly seen, and there is a decline in the cation vacancy concentration below the simple extrinsic value at low temperatures and high impurity contents due to complex formation. It is instructive to derive the pertinent dependences.

To analyze the KCl–SrCl$_2$ system we will need the following: the formation reaction for Schottky disorder and its mass-action expression, the impurity incorporation reaction, the reaction for the formation of defect complexes and its mass-action expression, and an expression for charge neutrality. These have all been given for the NaCl-CaCl$_2$ system, but it is convenient to list them again. In the order above, we have:

$$\text{nil} \rightleftharpoons V'_K + V^{\cdot}_{Cl} \tag{6.18}$$

$$[V'_K][V^{\cdot}_{Cl}] = K_S(T) = K_S^{\circ} e^{\Delta S_S/k} e^{-\Delta H_S/kT} \tag{6.19}$$

$$\text{SrCl}_2 \xrightarrow{\text{(2KCl)}} \text{Sr}^{\cdot}_K + V'_K + 2\text{Cl}_{Cl} \tag{6.20}$$

$$\text{Sr}^{\cdot}_K + V'_K \rightleftharpoons (\text{Sr}^{\cdot}_K V'_K) \tag{6.21}$$

$$\frac{[(\text{Sr}^{\cdot}_K V'_K)]}{[\text{Sr}^{\cdot}_K][V'_K]} = K_C(T) = K_C^{\circ} e^{\Delta S_C/k} e^{-\Delta H_C/kT} \tag{6.22}$$

$$[V'_K] \approx [\text{Sr}^{\cdot}_K] + [V^{\cdot}_{Cl}] \tag{6.23}$$

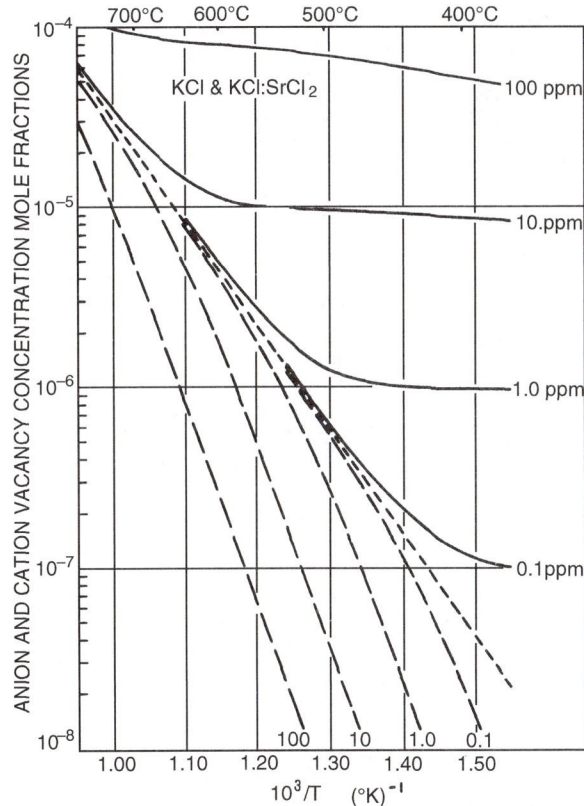

Figure 6.8 Calculated concentrations of cation (solid lines) and anion (dashed lines) vacancies in KCl doped with various amounts of $SrCl_2$. Calculations based on an enthalpy of formation of Schottky disorder of 2.26 eV (218 kJ/mol) and $\Delta S/k$ of 5.4. The free energy of complex formation was taken as 0.42 eV (Reproduced from Beaumont and Jacobs, 1966, by permission of the American Institute of Physics.)

The behavior is discussed for three different regions that are dominated by intrinsic disorder, extrinsic defects, and the defect complexes, respectively.

1. The Intrinsic Region. In this highest temperature region, the impurity content can be neglected relative to the intrinsic Schottky defects, so that charge neutrality can be approximated by

$$[V'_K] \approx [V^{\cdot}_{Cl}]; \qquad [V^{\cdot}_{Cl}] \gg [Sr^{\cdot}_K] \tag{6.24}$$

In combination with Eq. (6.19), this gives the familiar expression for the concentrations of the individual vacancies in the intrinsic region at the highest temperatures

$$[V'_K] \approx [V^{\cdot}_{Cl}] \approx K_S^{1/2} \approx (K_S^{\circ})^{1/2} e^{\Delta S_S/2k} e^{-\Delta H_S/2kT} \tag{6.25}$$

The concentration of defect complexes in this region can be obtained by substituting the cation vacancy concentration from Eq. (6.25) into the mass-action expression, Eq. (6.22), to give

$$
\begin{aligned}
\left[(Sr_K^{\cdot} V_K^{/})\right] &\approx K_C K_S^{1/2}[Sr_K^{\cdot}] \\
&\approx K_C^{\circ}(K_S^{\circ})^{1/2}[Sr_K^{\cdot}]e^{(\Delta S_C + \frac{\Delta S_S}{2})/2k}e^{(\Delta H_C + \frac{\Delta H_S}{2})/2kT}
\end{aligned}
\tag{6.26}
$$

The concentration of complexes increases with increasing impurity content, increasing concentration of intrinsic cation vacancies, and increasing binding enthalpy for the complex. Note that ΔH_C is a negative number, so that tighter binding increases the value of the exponential term.

2. The Extrinsic Region. In the intermediate temperature range, the concentration of anion vacancies can be neglected relative to the impurity content, and charge neutrality can be approximated by

$$
[V_K^{/}] \approx [Sr_K^{\cdot}]; \qquad [Sr_K^{\cdot}] \gg [V_{Cl}^{\cdot}]
\tag{6.27}
$$

and the anion vacancy concentration can be expressed as follows:

$$
[V_{Cl}^{\cdot}] \approx \frac{K_S}{[Sr_K^{\cdot}]} \approx \frac{K_S^{\circ}}{[Sr_K^{\cdot}]}e^{\Delta S_S/k}e^{-\Delta H_S/kT}
\tag{6.28}
$$

So far, the defect complexes have had no effect whatsoever on the concentrations of the individual charged vacancies. The concentration of complexes in the extrinsic region can be obtained by combination of Eqs. (6.22) and (6.27) to give:

$$
\left[(Sr_K^{\cdot} V_K^{/})\right] \approx K_C[Sr_K^{\cdot}]^2 \approx K_C^{\circ}[Sr_K^{\cdot}]^2 e^{\Delta S_C/k}e^{-\Delta H_C/kT}
\tag{6.29}
$$

The complex concentration is increasing more rapidly with impurity content in this region than it did in the intrinsic region.

3. The Complex Region. At very low temperatures, nearly all the cation vacancies and impurity centers will be bound into complexes whose small amount of dissociation corresponds to the reverse of Eq. (6.21). As long as the extent of dissociation is small, the complex concentration can be approximated by the total impurity content, $[Sr]_{total}$:

$$
\left[(Sr_K^{\cdot} V_K^{/})\right] \approx [Sr_K^{\cdot}]_{total}
\tag{6.30}
$$

Charge neutrality will be maintained primarily by the individual charged defects formed by the dissociation of the complex and can still be represented by Eq. (6.27), although $[Sr_K^{\cdot}]$ is no longer the total impurity content, but only a small fraction of it. By combination of Eqs. (6.22), (6.27), and (6.30), the following cation vacancy concentration is obtained:

$$
[V_K^{/}] \approx \left\{\frac{[Sr_K^{\cdot}]_{total}}{K_C}\right\}^{1/2} \approx \left\{\frac{[Sr_K^{\cdot}]_{total}}{K_C^{\circ}}\right\}^{1/2} e^{-\Delta S_C/2k}e^{\Delta H_C/2kT}
\tag{6.31}
$$

and combination of this expression for the cation vacancies with Eq. (6.19) gives the following anion vacancy concentration:

$$[V_{Cl}^{\cdot}] \approx \frac{K_S K_C^{1/2}}{[Sr_K^{\cdot}]_{total}^{1/2}} \approx \frac{K_S^{\circ}(K_C^{\circ})^{1/2}}{[Sr_K^{\cdot}]_{total}^{1/2}} e^{(\Delta S_S + \frac{\Delta S_C}{2})/k} e^{-(\Delta H_S + \frac{\Delta H_C}{2})/kT} \qquad (6.32)$$

The anion vacancy concentration in this region is dropping off even more rapidly than it did in the extrinsic region. According to our assumption of negligible dissociation of the complexes, their concentration is invariant with temperature in this region and is represented by Eq. (6.30).

Comparison of Eqs. (6.27) and (6.31) indicates that while the cation vacancy concentration is increasing linearly with the total impurity content in the extrinsic region, it is increasing only with the square root of the total impurity content in the complex region. As a result, the loss of free cation vacancies becomes increasingly important with increasing total impurity concentration. As shown in Fig. 6.8, with a fractional impurity content of 10^{-7} (0.1 ppm), the extrinsic region is barely reached at the lowest temperature, and there is no evidence of complex formation. With 1 ppm of impurities, there is a substantial extrinsic region, but still hardly any complex formation. With 10 ppm of impurities, the loss of free cation vacancies due to complex formation becomes apparent, and the impurity content affects the defect concentrations up to the highest temperature; that is, there is no purely intrinsic region. With 100 ppm of impurities, the intrinsic region has disappeared, and complex formation is extensive over almost the entire temperature range. However, even for this highest impurity content, only about half the extrinsic cation vacancies are bound into complexes, even at the lowest temperature.

Intrinsic Defect Associates

The case of neutral, intrinsic defect associates such as $(V_{Na}^{\prime} V_{Cl}^{\cdot})$ is much simpler than the case of extrinsic defect complexes just described. Referring to Eqs. (6.13) and (6.14), it is seen that the concentration of intrinsic associates is defined by the latter mass-action expression. It is a function only of the enthalpy and entropy of association and the temperature. The concentration of these associates is unaffected by any impurity content, nor do they in turn affect the concentration of the isolated intrinsic defects. The uncharged associates in this case cannot possibly interact with any other defects, charged or not. They just do their own thing. If the associates bear a net charge, as in the hypothetical case represented by Eq. (6.17), there would be interactions, but there are no such cases of practical interest.

REFERENCES

Beaumont, J. H., and P. W. M. Jacobs. Energy and entropy parameters for vacancy formation and mobility in ionic crystals from conductance measurements. *J. Chem. Phys.* 45:1496–1502, 1966.

Fuller, R. G. Ionic conductivity (including self-diffusion). In *Point Defects in Solids*, Vol. 1, *General and Ionic Crystals*, J. H. Crawford Jr. and L. M. Slifkin, Eds. New York: Plenum Press, 1972, Chapter 2.

Fuller, R. G., C. L. Marquardt, M. H. Reilly, and J. C. Wells. Ionic transport in potassium chloride. *Phys. Rev.* 176:1036–1045, 1968.

Franklin, A. D. Statistical thermodynamics of point defects in crystals. In *Point Defects in Solids*, Vol. 1, *General and Ionic Crystals*, J. H. Crawford Jr. and L. M. Slifkin, Eds. New York: Plenum Press, 1972, Chapter 1.

Hayes, W., and A. M. Stoneham, *Defects and Defect Processes in Nonmetallic Solids*. New York: John Wiley & Sons, 1985.

Johnson, H. B., N. J. Tolar, G. R. Miller, and I. B. Cutler. Electrical and mechanical relaxation in CaF_2 doped with NaF. *J. Phys. Chem. Solids.* 30:31–42, 1969.

Stott, J. P., and J. H. Crawford Jr. Dipolar complexes in calcium fluoride doped with erbium. *Phys. Rev. Lett.* 26:384–386, 1971.

Ionic Transport

7

INTRODUCTION

The preceding chapters have presented the background necessary for a detailed discussion of ionic transport in the form of diffusion and ionic conductivity. Both these important processes depend on ionic defects, and in the absence of defects there would be no significant diffusion or ionic conductivity in ionic crystals. Diffusion is the basic process involved in solid state reactions that produce a wide variety of important products, from refractory bricks for the lining of blast furnaces to ferroelectric ceramics used as the dielectric layers in multilayer ceramic capacitors. Ionically conducting compounds are used as solid electrolytes in electrochemical oxygen activity sensors in the exhaust manifolds of modern automobiles, and as the electrolytes in high temperature fuel cells. A combination of diffusion and conductivity is involved in the formation of oxide scales on metals exposed to air at high temperatures. The migration of ionic defects may also be a mechanism for degradation and failure in devices subjected to electrical or mechanical stress. Thus many important technological processes and applications depend on the type and concentration of ionic defects. The ability to control the defect structure of solid compounds offers the opportunity to design for desired properties and to suppress degradation mechanisms.

BASIC CONCEPTS OF DIFFUSION

Fick's First Law

Diffusion is the response of a system to a concentration gradient; in fact, it is the attempt to eliminate the gradient and to return the system to a homogeneous equilibrium state. In the simplest case, the particle flux is proportional to its concentration gradient, and the proportionality constant is called the diffusion constant. This relationship, known as Fick's first law, is represented for unidirectional diffusion by the equation

$$J_i = -D_i \frac{dc_i}{dx} \tag{7.1}$$

Where J_i is the flux of species i, dc_i/dx is its concentration gradient, and D_i is the diffusion constant for i in a particular material at a given temperature. If the units of J_i are particles/cm² · s, and for the concentration gradient, particles/cm⁴ (particles/cm³/cm), then the diffusion constant carries the units cm²/s. It is the particle flux per unit concentration gradient. This is one of a number of linear response functions, where for relatively small impulses, the response of a system is linear with the impulse. Examples include Ohm's law in the form $I = \sigma E$, where the current density I is proportional to the applied electric field E and the proportionality constant σ, the conductivity, is the current density per unit field; and Hooke's law in the form $\epsilon = G\sigma$, where the strain (fractional elongation) ϵ is proportional to the applied mechanical stress σ, and the proportionality constant G, the elastic compliance, is the strain per unit stress. (The classical form of Hooke's law is $\sigma = E\epsilon$, where E is Young's modulus, or the elastic constant; it is the stress per unit strain). Our phenomenological treatment of diffusion will be in terms of Fick's first law, which is adequate for steady state diffusion as in the case of diffusion across a membrane. A detailed mathematical treatment of transient diffusion and time-dependent concentration profiles requires the solution of Fick's second law for specific boundary conditions and is beyond our requirements. Detailed analyses can be found in standard reference works on diffusion.

A simple diffusion process with a constant concentration gradient is depicted in Fig. 7.1. This figure clearly explains the negative sign in Fick's first law. The diffusing particle moves down the concentration gradient, in the negative direction in a vectorial sense, for a positive concentration gradient.

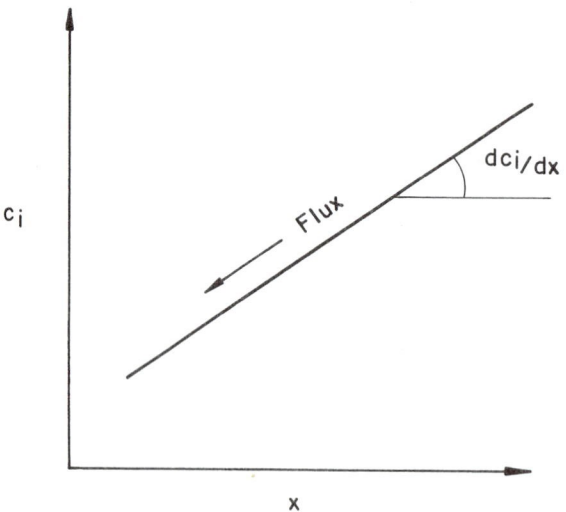

Figure 7.1 The concentration of species i as a function of position for the case of a constant gradient of the species. Diffusion occurs in the negative direction in a positive gradient.

Diffusion Mechanisms

To understand the role of ionic defects in diffusion processes, it is necessary to consider the mechanisms on an atomistic scale. The most important types of diffusion mechanism for ions are as follows.

1. The Vacancy Mechanism. An ion can diffuse by moving into an adjacent vacancy of the same kind. According to this mechanism, the ion is immobile until a vacancy moves up to it and allows it to make a single jump in the diffusion direction. The vacancy may swing around and allow the ion to make another jump in the same direction, or it may move away so that the ion must patiently await the arrival of another vacancy. Figure 7.2a depicts an example of the vacancy mechanism for a single line of ions. Time-lapse photography would show the vacancy moving to the left, and it is convenient to describe the diffusion process in terms of the movement of the vacancy. However, in this case the diffusing ion is moving in the opposite direction. If the diffusing species is an impurity ion, or a dilute solution of radioactive tracer ions of a host species, the vacancies will do a lot more moving around than the identifiable diffusing species. This is the mechanism for diffusion of cations and anions, both native and impurity, in the alkali halides and in most oxides having the NaCl structure.

2. The Interstitial Mechanism. In this mechanism (Fig. 7.2b), the defect and the diffusing species are the same, so they obviously move in the same direction. This process temporarily terminates when the interstitial drops into a vacancy and occupies a normal lattice site. As a result of the dynamic steady state interstitial concentration, however, its turn will come again. This mechanism does not seem to be as important in ionic crystals as the following variation.

3. The Interstitialcy Mechanism. In this variation on the interstitial mechanism

(a)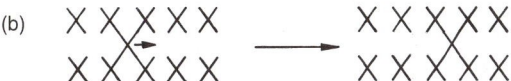

(b)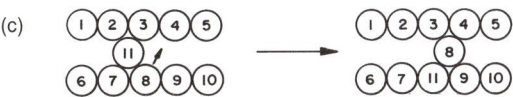

(c) ![diffusion interstitialcy diagram]

Figure 7.2. Stepwise view of diffusion by different mechanisms: (a) the vacancy mechanism, (b) the interstitial mechanism, and (c) the interstitialcy mechanism.

(Fig. 7.2c), each ion is identified by a number, and we can follow the movement of the individual ions. This is a displacement process whereby the interstitial ion displaces a normal lattice ion, takes its site, and pushes the normal lattice ion into an interstitial site. Each interstitial makes only a single move, and then must wait for an approaching interstitial to displace it into another interstitial site. If the ions cannot be distinguished from one another, time-lapse photography would show a sequence similar to that of the interstitial mechanism. However, if the diffusing species can be identified (e.g., by being a radioactive tracer ion), the two situations are quite different. There are two types of interstitialcy movement: the noncollinear type shown in Fig. 7.2c, where the paths of the interstitial and lattice ions are at an angle to each other, and the collinear type, where the interstitial ion pushes the lattice ion straight ahead into an interstitial site in the next row. The interstitialcy mechanism is a prominent mode of cation movement in the silver halides.

The Diffusion Constant

A variation on the type of enthalpy schematic used in Fig. 3.5 to derive the mass-action expression can be used to give insight into the diffusion process. Thus Fig. 7.3 shows the periodic variation of enthalpy along a lattice direction with the distance between stable lattice sites represented as a. The barrier height between sites is h_0. The concentration gradient of the diffusing species i is dc_i/dx, so that if the concentration in site 1 is c_i, the concentration in site 2 is $c_i + a\,dc_i/dx$, and there will be a net diffusion flux of species i toward the left. The net flux will be the total flux toward the right (the positive direction) minus the flux toward the left

$$J_i = \overrightarrow{J_i} - \overleftarrow{J_i} \tag{7.2}$$

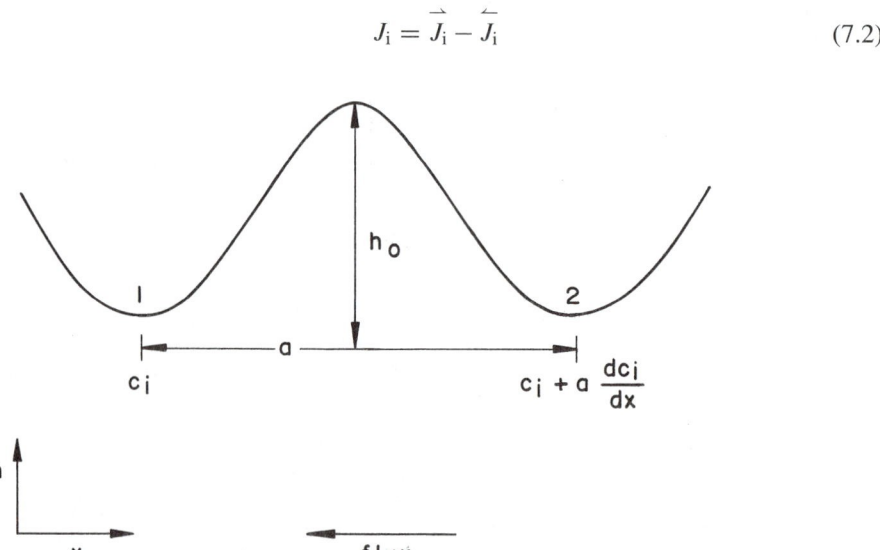

Figure 7.3 Enthalpy as a function of position in a crystalline lattice. The concentration of species i is increasing toward the right.

The flux to the right will be the concentration in site 1 of particles that are trying to surmount the barrier to reach the adjacent site, c_i, times the frequency of their attempts, that is, the lattice vibrational frequency ν_i times the length of a successful jump a times the probability that a jump will be successful. As discussed in Chapter 3, the latter is the probability that a given ion will have enough energy to surmount the barrier and is represented by a Boltzmann term, an exponential of the negative ratio of the enthalpy needed, h_0, to the average thermal energy available, kT. For the flux to the left, the frequency of tries, the jump distance, and the probability of success are all the same as for the flux to the right, but the concentration of ions is different. Thus

$$J_i = c_i a \nu_i e^{-h_0/kT} - \left(c_i + a\frac{dc_i}{dx}\right) a \nu_i e^{-h_0/kT} \tag{7.3}$$

The two terms in c_i cancel, leaving

$$J_i = -a^2 \nu_i e^{-h_0/kT} \frac{dc_i}{dx} \tag{7.4}$$

By comparison with the form of Fick's first law, Eq. (7.1), it is apparent that the diffusion constant can be expressed as follows:

$$D_i = a^2 \nu_i e^{-h_0/kT} \tag{7.5}$$

which clearly carries the correct units (cm²/s). This rather simplistic treatment gives the same result as more sophisticated derivations. Diffusion constants are often expressed in the following general form:

$$D = D_0 e^{-Q/kT} \tag{7.6}$$

and it is seen that D_0 is $a^2\nu$, and the activation energy Q corresponds to the height of the enthalpy barrier between equivalent sites involved in the diffusion, h_0.

For precise analyses of diffusion experiments, some refinement is required of the simple relationship just derived. One of these involves the introduction of a "correlation coefficient" that takes into account the likely sequence of jumps in a more detailed manner. For example, in the vacancy mechanism, the event following the jump of an ion into a vacancy is likely to be a jump back into the same vacancy, so the subsequent jumps are not entirely random. This possibility can be taken into account quantitatively, but it is not necessary for our phenomenological treatment.

IONIC CONDUCTION IN CRYSTALLINE SOLIDS

When a voltage is applied to a solid and current flow is observed, the identity of the charge-carrying species is not apparent. More specific to our present interest, it is not clear whether the charge carrier is electronic or ionic. The first decisive proof of ionic conduction in ionic crystals came from experiments on the silver halides by the German scientist Tubandt (1914, 1920, 1921). He constructed a simple electrolytic

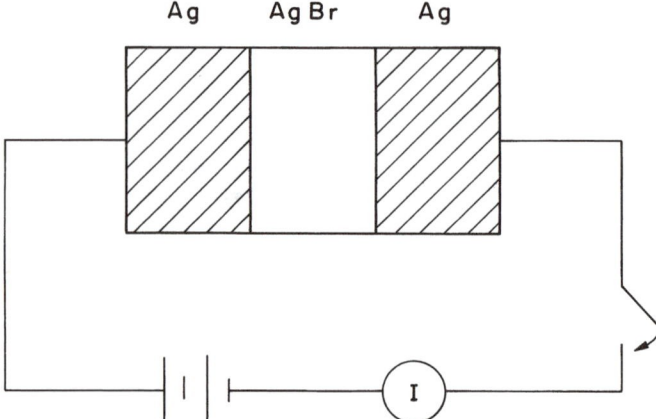

Figure 7.4   The type of cell used by Tubandt (1914, 1920, 1921) to measure the ionic transport number in the silver halides.

cell of the type shown in Fig. 7.4. The cell consisted of an anode and a cathode of metallic silver, and a pressed electrolyte pellet of AgBr or other silver halide. These component parts were weighed separately and then pressed together to complete the cell. A dc current was passed for some period of time at a temperature of about 200°C. The cell parts were then taken apart and weighed again. It was found that the anode had lost weight and that the cathode had gained the same amount of weight, while the AgBr pellet remained unchanged. It was apparent that silver had passed through the AgBr from the anode to the cathode, as would also have been observed with an aqueous electrolyte. Moreover, the weight changes were in good agreement with the expression

$$\Delta W = \frac{Q M_{Ag}}{F} \tag{7.7}$$

where Q is the total charge passed in coulombs, M_{Ag} is the equivalent weight of silver, 107.88 g/equiv, F is the Faraday constant, 96,500 C/equiv, and ΔW is the weight loss of the anode or gain of the cathode in grams. This correspondence indicated that all the current through the AgBr was carried by Ag^+, within the experimental accuracy. The AgBr acts as a nearly ideal ionic conductor, or electrolyte.

This type of experiment was subsequently performed on a variety of crystalline solids for the cases in which the magnitude of the ionic conductivity was high enough to give measurable mass transfers. In some cases, such as Ag_2S, there was a transport of mass, but the amount was less than that expected from the charge transfer. This is expected when the conduction is due to the simultaneous transport of both ions and electrons. These experiments were an important pioneering contribution to the understanding of charge transport in solids.

The total electrical conductivity, σ_T is the sum of the contributions of all charge carriers:

$$\sigma_T = \sigma_{Ag} + \sigma_{Br} + \sigma_e + \sigma_h \tag{7.8}$$

where the right-hand side shows explicitly the partial conductivities due to Ag^+, Br^-, electrons, and holes. It is assumed that the electrodes are nonblocking (i.e., that there is no restriction on the source of ions and electrons or on their discharge). If the anode were replaced by something inert, like platinum, there would be no source of Ag^+ and the current carried by these ions would quickly die out, a process called polarization. Each species i carries a fraction t_i of the total current, called the transport or transference number:

$$\sigma_T = t_{Ag}\sigma_T + t_{Br}\sigma_T + t_e\sigma_T + t_h\sigma_T$$

$$= (t_{Ag} + t_{Br} + t_e + t_h)\sigma_T$$

(7.9)

The sum of the transport numbers for all the mobile species clearly must be unity. The transport number for Ag^+ in the Tubandt experiment on AgBr was unity within the precision of his experiments. The ionic transport number in a copper wire is zero. These transport numbers are a convenient way of expressing the relative contributions of the various species to the total conductivity.

The Conductivity

The conductivity is the proportionality constant in the linear response function that relates the current density to the applied field

$$I_i = \sigma_i E$$

(7.10)

where I_i is the current density for species i (A/cm^2) and E is the applied electric field (V/cm). The conductivity σ_i then carries the units $(\Omega \cdot cm)^{-1}$, its reciprocal, the resistivity, is also frequently used. Equation (7.10) is a normalized form of Ohm's law, in which the conductivity is the current density per unit field.

The conductivity is proportional to the concentration of charge carriers c_i, (carriers/cm^3), the charge they carry $z_i e$ (C/carrier), where e is the unit electronic charge, 1.6×10^{-19} C, and z_i is the number of unit charges on the carrier; and to their mobility μ_i, their ability to move in an electric field

$$\sigma_i = c_i z_i e \mu_i$$

(7.11)

The mobility is seen to have the units of square centimeters per volt-second, which in the form (cm/s)/(V/cm) indicates that it is the velocity per unit field. Equation (7.11) can be compared to the situation of a person standing on an interstate highway overpass measuring the flux of people in cars passing underneath. The people flux will clearly depend on the density of cars on the highway, the number of people per car, and the speed at which the cars are traveling. Since the mobility is the velocity per unit field, the velocity is the mobility times the applied field.

The mobility term implies that the ions move at a constant velocity that is proportional to the field. But an electric field acting on a charge is a force, $F = eE$, and a force acting on a mass causes an acceleration, $F = ma$. So a charged species that can move freely in an electric field should continuously accelerate. This apparent paradox is resolved by realization that the ions do not move continuously, but by a series of individual jumps that are separated in time. The ions are only briefly in motion, and their average velocity is determined by the frequency and distance of the jumps, not by their velocity during the jump.

We can modify Fig. 7.3, the diagram of enthalpy versus position that was used to derive an expression for the diffusion constant, to examine ionic conductivity. Such a modification is shown in Fig. 7.5 for the case of an applied electric field with no concentration gradient. The electric field puts a bias on the periodic enthalpy such that the barrier height to the right is reduced from the zero-field value h_0 and is increased toward the left. The barrier height is changed by the enthalpy that corresponds to a field, E, acting on a charge $z_i e$, over a distance $a/2$, the distance from the equilibrium position to the top of the barrier; $aE/2$ is the potential drop over this distance. The net current to the right I_i will be the difference between the current to the right, over the reduced barrier, and the current to the left, over the raised barrier:

$$I_i = \overrightarrow{I_i} - \overleftarrow{I_i} \tag{7.12}$$

In each direction, the current will be proportional to the concentration of charge carriers c_i, the charge they carry $z_i e$, the frequency of their attempts at the barrier ν_i, the length of a successful jump a, and the probability of success. The latter will be enhanced toward the right by the field and suppressed toward the left:

$$I_i = c_i z_i e a \nu_i e^{-(h_0 - \frac{z_i e a E}{2})/kT}$$
$$- c_i z_i e a \nu_i e^{-(h_0 + \frac{z_i e a E}{2})/kT} \tag{7.13}$$

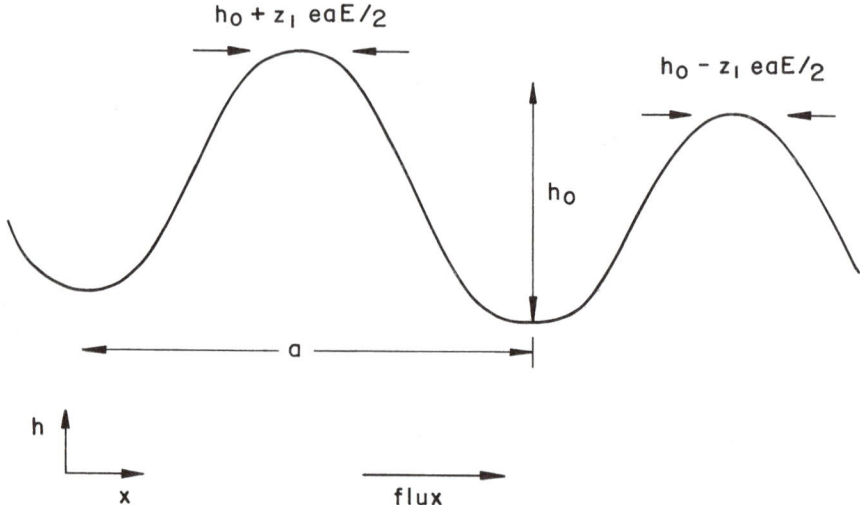

Figure 7.5 Enthalpy as a function of position in a crystalline lattice for the case of an applied field E, but with no concentration gradient.

This can be rearranged to

$$I_i = c_i z_i e a v_i e^{-h_0/kT} \left(e^{\frac{z_i e a E}{2kT}} - e^{\frac{-z_i e a E}{2kT}} \right) \tag{7.14}$$

This can be simplified by use of the identity

$$e^x - e^{-x} = 2 \sinh x \tag{7.15}$$

to give

$$I_i = 2 c_i z_i e a v_i e^{-h_0/kT} \sinh \frac{z_i e a E}{2kT} \tag{7.16}$$

The sinh term can be expanded into a series

$$\sinh x = x + \frac{x^3}{3!} + \frac{x^5}{5!} + \cdots \tag{7.17}$$

where

$$x = \frac{z_i e a E}{2kT} \tag{7.18}$$

If the field is small enough so that

$$\frac{z_i e a E}{2kT} \ll 1 \tag{7.19}$$

then the current density in the low field approximation can be approximated by

$$I_i \approx \frac{z_i^2 e^2 c_i}{kT} a^2 v_i e^{-h_0/kT} E \tag{7.20}$$

By comparison with Eq. (7.10), it is seen that the conductivity is the coefficient of the field

$$\sigma_i \approx \frac{z_i^2 e^2 c_i}{kT} a^2 v_i e^{-h_0/kT} \tag{7.21}$$

while comparison of this with Eq. (7.11) indicates that the mobility is

$$\mu_i = \frac{z_i e a^2}{kT} v_i e^{-h_0/kT} \tag{7.22}$$

When the field is small enough so that the low field approximation given by Eq. (7.19) is valid, the current density is proportional to the field; that is, Ohm's law prevails.

Note that the temperature dependence is primarily determined by the zero-field barrier height, the same barrier that appears in the diffusion constant, Eq. (7.5). The close relationship between the diffusion constant and the mobility can be seen by dividing Eq. (7.22) by Eq. (7.5):

$$\frac{\mu_i}{D_i} = \frac{z_i e}{kT} \tag{7.23}$$

This is known as the Nernst–Einstein relationship, and it can be used to obtain diffusion constants from conductivity measurements, and vice versa. The form of this relationship derives from the fact that while the diffusion constant is a flux (per unit gradient), the mobility is a velocity (per unit field). Once again, a very simple schematic model has given considerable insight into an important physical process.

It is apparent from Eq. (7.21) that the temperature dependence of the conductivity in this low field region is not purely exponential; there is a inverse linear temperature dependence in the preexponential coefficient. Thus an Arrhenius plot of log conductivity against $1/T$ will not give a precise value for the enthalpy of motion. For high values of enthalpy the error is not large, but for more precise evaluations, it is common practice to plot log σT against $1/T$ to obtain a slope that depends on the enthalpy alone. This type of plot would not be appropriate for diffusion experiments, since the temperature does not appear in the preexponential coefficient in the diffusion constant, Eq. (7.5).

At high fields, the approximation indicated in Eq. (7.19) is not valid, and the behavior will not be ohmic. However, the current in the direction opposed to the applied field can then be neglected relative to that in the field direction. That means that the second exponential term in Eq. (7.14) can be neglected to give

$$I_i = z_i e c_i a v_i e^{-\left[\frac{h_0 - z_i (ea/2)E}{kT}\right]} \tag{7.24}$$

Thus, in the high field limit, the current is expected to depend exponentially on the field. This is in fact precisely observed for the electrochemical growth of oxide films on the so-called valve metals, such as aluminum and tantalum. In these cases the oxide films grow by a flow of ions through the films with almost ideal charge efficiency. The ionic current densities are in the milliamperes per square centimeter range for fields of the order of 10^6–10^7 V/cm, which certainly qualify as high fields. This strong exponential dependence of the current on the field results in an approximately limiting film thickness for a given applied voltage. In the case of tantalum, the oxide film thickness is approximately 2 nm per applied volt. These high quality thin, insulating films are the basis of the electrolytic capacitor industry. In the present context we are primarily concerned with the behavior of ionic defects at more modest fields, and our discussion will be limited to the low field, ohmic region.

INTRINSIC AND EXTRINSIC IONIC CONDUCTION

NaCl Doped with CaCl$_2$

The defect chemistry of pure NaCl and of NaCl doped with CaCl$_2$ was treated extensively in Chapters 4 and 5. Since these materials are almost ideal ionic conductors (unless one takes extreme measures such as heating the crystals in metallic Na vapor, which physicists have been known to do), it is convenient to use them as a model system for further discussion of ionic conductivity. To make the process more interesting, we use a real set of experimental data, that of Kirk and Pratt (1967), reproduced in modified form (Hayes and Stoneham, 1985) in Fig. 7.6. The authors

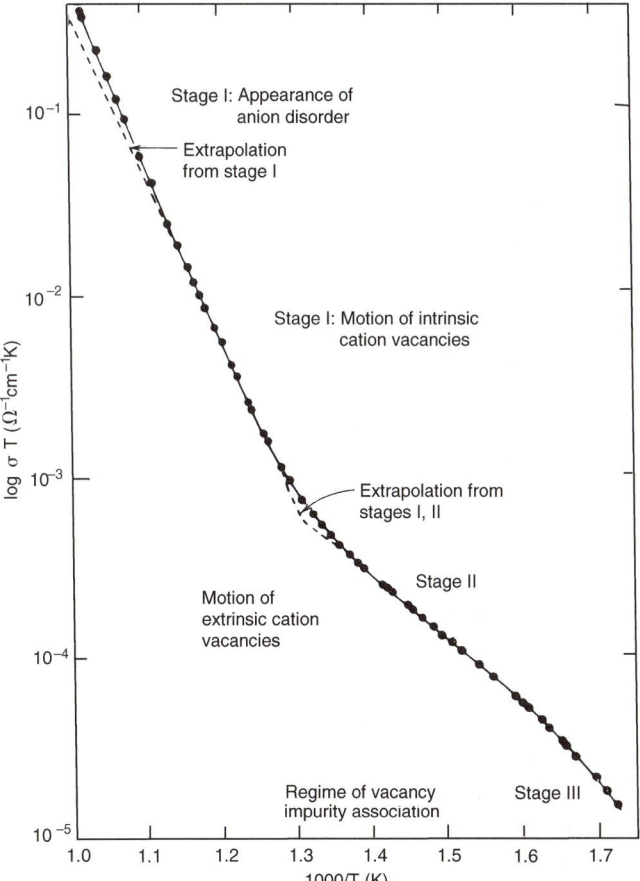

Figure 7.6 Ionic conductivity of NaCl as determined by Kirk and Pratt (1967). (Reproduced from modified version by Hayes and Stoneham, 1985, by permission of John Wiley & Sons.)

refer to their single crystalline sample as "pure" NaCl, whereas there are clearly both intrinsic and extrinsic regions of behavior. This reflects the impossibility of obtaining ideally pure samples and the hazards of describing a material as "pure." The author prefers the adjective "undoped," to indicate that no impurities have been deliberately added. The data in Fig. 7.6 will be treated as those of NaCl with a small residual content of divalent cation impurities, to be collectively referred to as $Ca_{Na}^{\cdot}$.

It will be convenient to repeat some of the pertinent relationships used in the analysis. The mass-action expression for Schottky disorder in NaCl is

$$[V_{Na}^{/}][V_{Cl}^{\cdot}] = e^{\Delta S_S/k}e^{-\Delta H_S/kT} \tag{7.25}$$

In the intrinsic region, the concentrations of the two defects are approximately equal (charge neutrality) and can be represented by

$$[V'_{Na}] \approx [V^{\cdot}_{Cl}] \approx e^{\Delta S_S/2k} e^{-\Delta H_S/2kT} \tag{7.26}$$

In the extrinsic region, the concentration of cation vacancies is fixed by the impurity content (charge neutrality, again):

$$[V'_{Na}] \approx [Ca^{\cdot}_{Na}] \tag{7.27}$$

while the concentration of anion defects is obtained by substituting the latter relationship into the mass-action expression for Schottky disorder:

$$[V^{\cdot}_{Cl}] \approx \frac{1}{[Ca^{\cdot}_{Na}]} e^{\Delta S_S/k} e^{-\Delta H_S/kT} \tag{7.28}$$

According to Eq. (7.22), the mobilities can be expressed as follows:

$$\mu_i = \frac{\overset{\circ}{\mu_i}}{T} e^{-h^{\circ}/kT} \tag{7.29}$$

The following parameters for NaCl were taken from the compilation of Fuller (1972):

$$
\begin{aligned}
\Delta H_S &= 2.45 \text{ eV (236 kJ/mol)} \\
\Delta S_S/k &= 9.3 \\
h_c &= 0.65 \text{ eV (63 kJ/mol)} \\
h_a &= 0.86 \text{ eV (83 kJ/mol)}
\end{aligned}
$$

where the latter two values represent the enthalpies of motion of cation and anion vacancies, respectively. The enthalpies of motion indicate that the cation vacancies are more mobile than the anion vacancies. The defect concentrations in the mass-action expressions are in site fractions (relative to cation sites), and we need volume concentrations to calculate conductivities. These can be obtained with the concentration of lattice sites per cubic centimeter, which can be obtained from the density (2.165 g/cm^3) and molecular weight (58.44 g/mol) of NaCl, which gives 2.23×10^{22} cations (anions)/cm^3.

The only things missing now are the preexponential factors in the expressions for the mobility, Eq. (7.29). For the cation vacancies, the major contributor to the conductivity, the preexponential factor can be obtained from a data point in the intrinsic region of Fig. 7.6. We use the data point where $\sigma T = 0.2$ ($\Omega \cdot$ cm)$^{-1}$ K at $1/T = 1.17 \times 10^{-3}$ K^{-1}; this corresponds to 855 K or 582°C, and a conductivity of 1.17×10^{-5} ($\Omega \cdot$cm)$^{-1}$. The vacancy concentrations at this temperature, calculated from Eq. (7.26), are 6.32×10^{-6} (6.32 ppm) or 1.41×10^{17}cm^{-3}. The cation vacancy mobility can then be calculated from Eq. (7.11) using the value of 1.6×10^{-19} C for the unit electronic charge. The mobility is 5.19×10^{-4}cm^2/V $\cdot$ s, and putting this into Eq. (7.29) gives 3000 (cm^2/V $\cdot$ s)K for the value of $\overset{\circ}{\mu_c}$, the preexponential term for the cation vacancy. There is no comparable way to calculate this term for the anion vacancies from these data, and we will make the rough assumption that it is the same as for the cation vacancies, that is, 3000 (cm^2/V $\cdot$ s)K. From the form of the expression for the mobilities, Eq. (7.22), it is seen that all the terms in the preexponential factor will be the same for the two kinds of vacancies, except possibly the attempt frequencies, ν.

The contributions of the cation and anion vacancies to the ionic conductivity of NaCl in the intrinsic region are given by

$$\sigma_c \approx \frac{e\mu_c^{\circ}}{T} e^{\Delta S_S/2k} e^{-\left(\frac{\Delta H_S}{2} + h_c\right)/kT} \tag{7.30}$$

$$\sigma_a \approx \frac{e\mu_a^{\circ}}{T} e^{\Delta S_S/2k} e^{-\left(\frac{\Delta H_S}{2} + h_a\right)/kT} \tag{7.31}$$

while in the extrinsic region they are

$$\sigma_c \approx \frac{e\mu_c^{\circ}}{T} [Ca_{Na}^{\cdot}] e^{-h_c/kT} \tag{7.32}$$

$$\sigma_a \approx \frac{e\mu_a^{\circ}}{T} \frac{1}{[Ca_{Na}^{\cdot}]} e^{\Delta S_S/k} e^{-(\Delta H_S + h_a)/kT} \tag{7.33}$$

Figures 7.7 and 7.8 show the defect concentrations and the conductivities as a function of the divalent impurity content, symbolized by $[Ca_{Na}^{\cdot}]$, in a log–log plot. Calculated

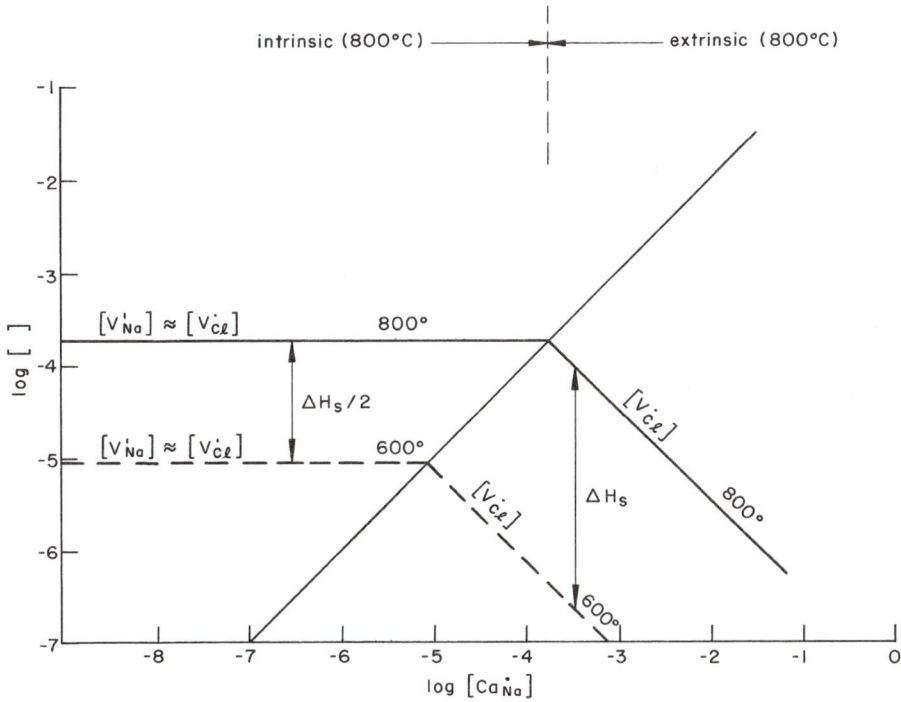

Figure 7.7 Log–log plot of the defect concentrations at two temperatures as a function of the Ca^{2+} concentration in NaCl.

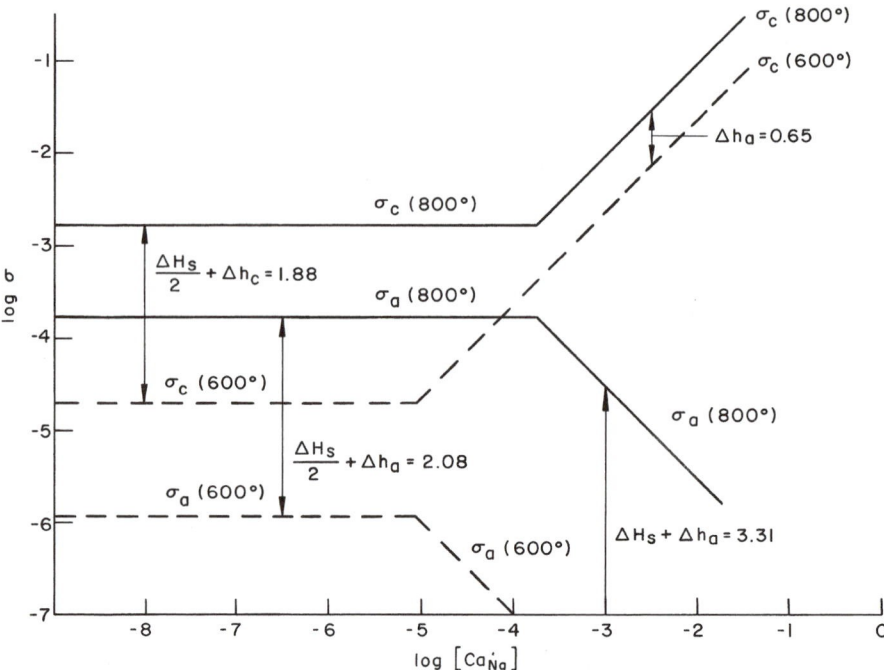

Figure 7.8 Log–log plot of the ionic conductivity due to cation and anion vacancies at two temperatures as a function of the Ca^{2+} concentration in NaCl.

results are shown for two temperatures, 600 and 800°C (since the melting point of NaCl is 801°C, experimental determination of the latter data point would be a very delicate business). The enthalpy terms that determine the temperature dependences are shown for each region. Note that simple plane geometry requires that the line segment labeled ΔH_S in Fig. 7.7 be twice the length of that labeled $\Delta H_S/2$. The impurity scale ranges from 1 ppb to 10%; the former is experimentally unachievable, while the latter greatly exceeds the solubility limits of divalent cations in NaCl. Log–log plots can quickly take you into meaningless regions.

While the cation and anion vacancies share a common line in the intrinsic region of Fig. 7.7, in Fig. 7.8 their contributions to the conductivity split into two parallel lines because of the different mobilities of the two defects. Note that the difference between them is less at the higher temperature, since the anion vacancies, with the higher enthalpy of motion, move up faster with temperature.

The vacancy concentrations and their contributions to the conductivity are plotted in Arrhenius form in Figs. 7.9 and 7.10. In the latter figure the conductivities are plotted as σT to take the temperature term in the preexponential factor out of the slopes. Results are shown for two concentrations of divalent impurity cations, 10^{-5} or 10 ppm, and 10^{-7} or 0.1 ppm. It is actually quite easy to calculate the effective concentration of divalent cations in the sample of Kirk and Pratt: it is about 1 ppm,

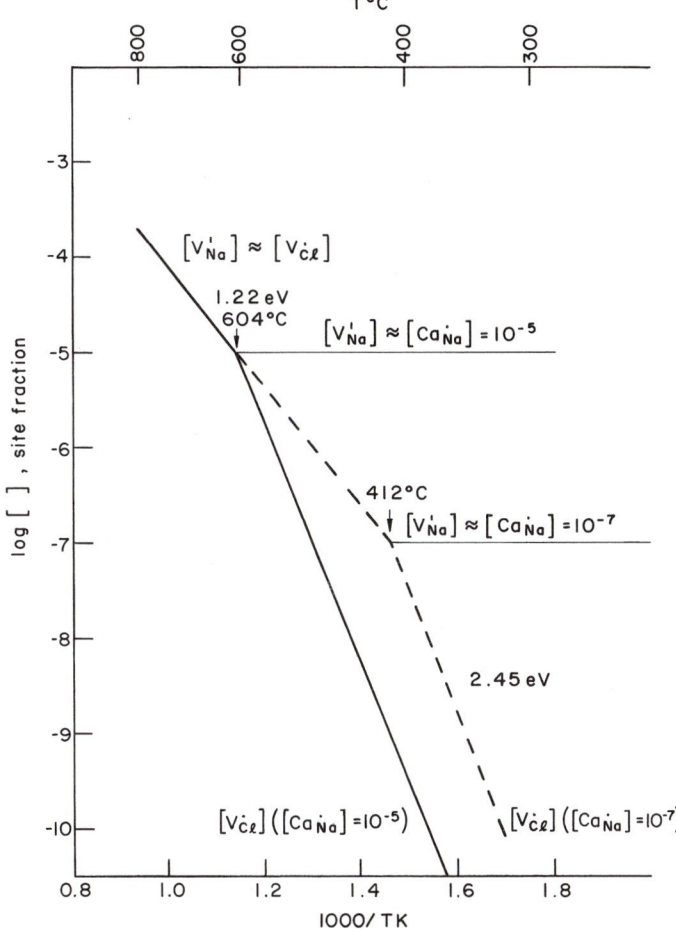

Figure 7.9 Arrhenius plot of the defect concentrations in NaCl at two different concentrations of Ca^{2+}.

which is typical of rather carefully purified alkali halides. It is seen in Fig. 7.10 how fast the contribution of the anion vacancies drops away as the material is cooled into the extrinsic range. While the anion vacancies contribute about 10% of the total conductivity in the intrinsic region, this has dropped to about 1 ppm at 400°C.

Two features of the data of Kirk and Pratt have not been included in this analysis. The first is the slight upturn from linearity at the highest temperatures. This has been attributed to the increasingly significant contribution from the anion vacancies as the temperature is increased. However, Fig. 7.10 indicates that the convergence of the two contributions is not very dramatic. The second feature is the downward curvature at the lowest temperatures. This is due to the formation of complexes between the divalent impurities and the cation vacancies as described in the preceding

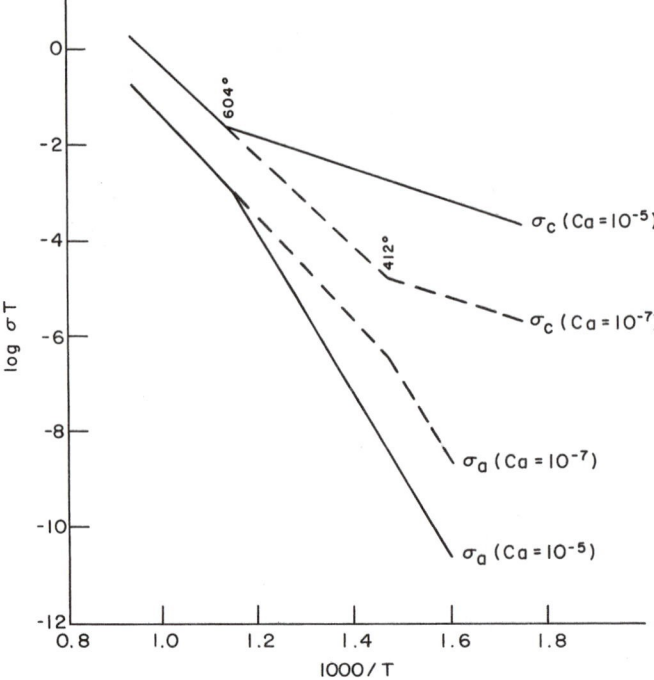

Figure 7.10 Arrhenius plot of the ionic conductivity due to cation and anion vacancies in NaCl at two different concentrations of Ca^{2+}.

chapter. At very low temperatures almost all the impurity centers will be in complexes that will gradually decompose as the temperature is raised:

$$(Ca_{Na}^{\bullet} V_{Na}') \rightleftharpoons Ca_{Na}^{\bullet} + V_{Na}' \tag{7.34}$$

whose mass-action expression is

$$\frac{[Ca_{Na}^{\bullet}][V_{Na}']}{[(Ca_{Na}^{\bullet} V_{Na}')]} = K_D' e^{-\Delta H_D/kT} \tag{7.35}$$

Since charge neutrality will still be expressed by Eq. (7.27), the cation vacancy concentration in this region will be

$$[V_{Na}'] \approx (K_D')^{1/2} [(Ca_{Na}^{\bullet} V_{Na}')]^{1/2} e^{\Delta H_A/2kT} \tag{7.36}$$

where ΔH_D, the enthalpy of dissociation, has been replaced by $-\Delta H_A$, where ΔH_A is the enthalpy of association (i.e., the binding enthalpy of the complex). The conductivity in this region will then be given by

$$\sigma_c \approx \frac{e\mu_c^{\circ}}{T} (K_D')^{1/2} [Ca]_t^{1/2} e^{(\Delta H_A - h_c)/kT} \tag{7.37}$$

where $[Ca]_t$ is the total concentration of divalent cations. The effect of the complexes on the defect concentrations was shown schematically in Fig. 6.8.

One very important lesson to be learned from this analysis of transport data by means of the tools of defect chemistry is that the system does indeed behave ideally in the thermodynamic sense. This justifies the use of the simple mass-action approach that requires the validity of dilute solution thermodynamics. The model quite accurately reproduces the experimental results. In this case the defect concentrations barely exceed 10^{-4}, 0.01%, but it will be seen that the treatment can be successfully extended to much higher concentrations.

AgBr Doped with CdBr$_2$

The silver halides are another very thoroughly studied family of compounds. As mentioned earlier, it is particularly easy to prepare samples from these materials because of their soft, ductile mechanical properties. Thin foil samples can be prepared by rolling and cutting cast slabs. In addition, the unusually low enthalpies for defect formation and for defect motion result in some of the highest known ionic conductivities for solid materials. There has been an additional incentive to study these materials because of the major role of defects in silver halides in the photographic process.

The favored type of ionic disorder in the silver halides is cation Frenkel disorder, and this was discussed in some detail in Chapters 4 and 5. As in the case of NaCl, it will be convenient to repeat some of the pertinent relationships here. The mass-action expression for cation Frenkel disorder in AgBr is

$$[Ag_I^\cdot][V_{Ag}'] = e^{\Delta S_{CF}/k} e^{-\Delta H_{CF}/kT} \tag{7.38}$$

In the intrinsic region, the concentrations of these two defects are approximately equal (charge neutrality) and can be expressed by

$$[Ag_I^\cdot] \approx [V_{Ag}'] \approx e^{\Delta S_{CF}/2k} e^{-\Delta H_{CF}/2kT} \tag{7.39}$$

In the extrinsic region, the approximation to charge neutrality will shift to

$$[V_{Ag}'] \approx [Cd_{Ag}^\cdot] \tag{7.40}$$

and the concentration of interstitial cations will be given by

$$[Ag_I^\cdot] \approx \frac{1}{[Cd_{Ag}^\cdot]} e^{\Delta S_{CF}/k} e^{-\Delta H_{CF}/kT} \tag{7.41}$$

The defect mobilities will again be expressed by Eq. (7.29).

The thermodynamic parameters for defect formation and motion have not been determined as accurately and consistently in the case of AgBr as in the example of NaCl. Nevertheless, the following values appear to be reasonable and are based on a collection of data compiled by Franklin (1972). There was no value given for the entropy of formation, and the value used here is based on two experimental determinations for AgCl. The following parameters were used in the subsequent evaluation:

$$
\begin{aligned}
\Delta H_{CF} &= 1.16 \text{ eV } (112 \text{ kJ/mol}) \\
\Delta S_{CF} &= 10 \\
h_I &= 0.1 \text{ eV } (10 \text{ kJ/mol}) \\
h_V &= 0.3 \text{ eV } (30 \text{ kJ/mol})
\end{aligned}
$$

The defect concentrations as a function of Cd concentration at 200 and 400°C are shown in Fig. 7.11. The general appearance is very similar to the comparable plot for NaCl (Fig. 7.7), with one notable quantitative difference. Even though the temperatures used to illustrate the behavior of AgBr are much lower than those used for NaCl (200 and 400°C vs. 600 and 800°C), the defect concentrations are higher by one to two orders of magnitude for AgBr. This is the result of the much smaller enthalpy of defect formation for the latter.

The contributions of the two intrinsic defects to the conductivity can be calculated from their concentrations and mobilities. For the latter we will use an experimentally determined value for the conductivity of AgBr in the intrinsic region obtained by Teltow (1949): 0.185 $(\Omega \cdot \text{cm})^{-1}$ at 375°C. From this and the calculated concentration of interstitial cations, a mobility of 0.0122 cm²/V · s is obtained for their mobility at

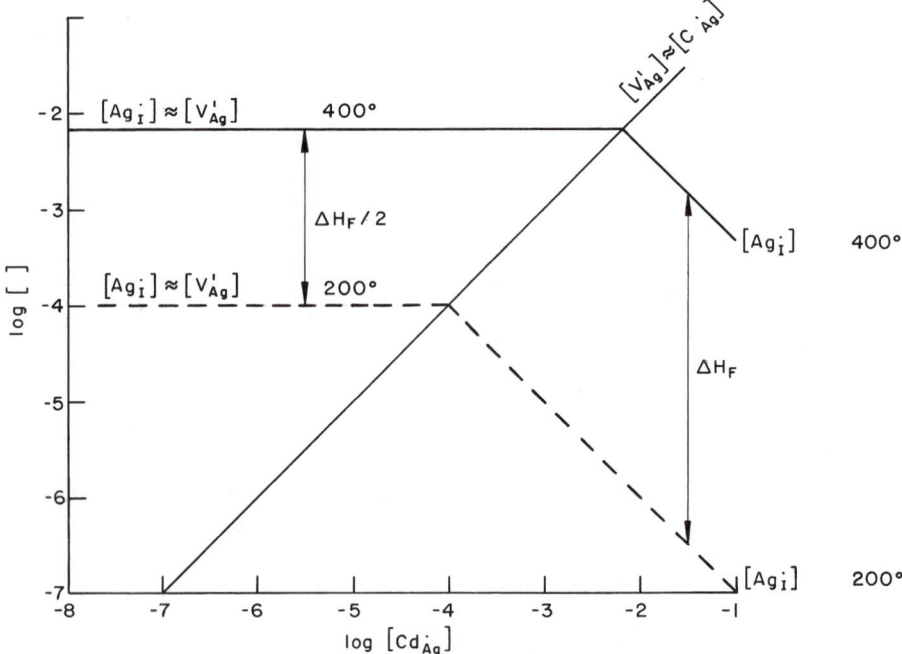

Figure 7.11 Log–log plot of the defect concentrations at two different temperatures as a function of the Cd²⁺ concentration in AgBr.

that temperature, and from Eq. (7.29) and the value above for the enthalpy of motion for the interstitials, the preexponential factor is calculated to be $\mu_I^\circ = 47.4$ cm² K/V · s. In the absence of specific information, it will be assumed that this value is also valid for the cation vacancies.

The calculated contributions to the total conductivity as a function of the dopant concentration are shown in Fig. 7.12. In this case, the behavior is quite different from that of NaCl. In fact the results indicate that there should be a minimum in the conductivity as a function of the Cd concentration. The experimental results of Teltow (1949) are shown in as a linear plot (Fig. 7.13a), and in log–log form (Fig. 7.13b). There is indeed a minimum, whose depth decreases with increasing temperature, as predicted. This correspondence of experimental results with theoretical prediction was taken at that time, 1949, as strong confirmation of the validity of the concepts of defect chemistry. Why is there a minimum in the conductivity in this case but not in the case of NaCl? In the latter case the doping increased the concentration of the more mobile defect, the cation vacancy, and most of the conductivity was contributed by this defect over the entire doping range. In the case of AgBr, the concentration of the more mobile species, the interstitial cation, is decreased by the dopant, while the concentration of the less mobile defect, the cation vacancy, is enhanced, and there is a resulting shift from predominantly interstitial conduction to predominantly

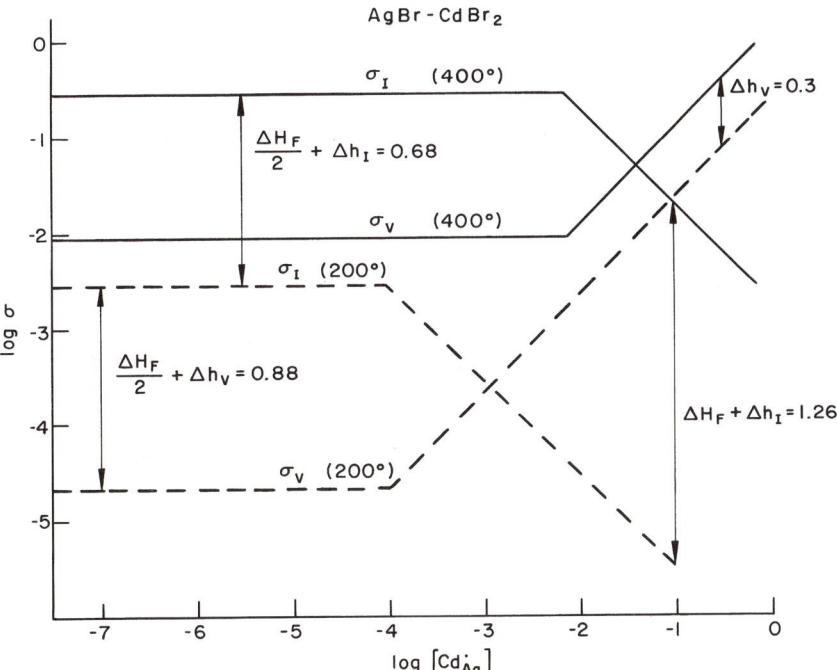

Figure 7.12 Log–log plot of the ionic conductivity due to cation interstitials and vacancies at two temperatures as a function of the Cd²⁺ concentration in AgBr.

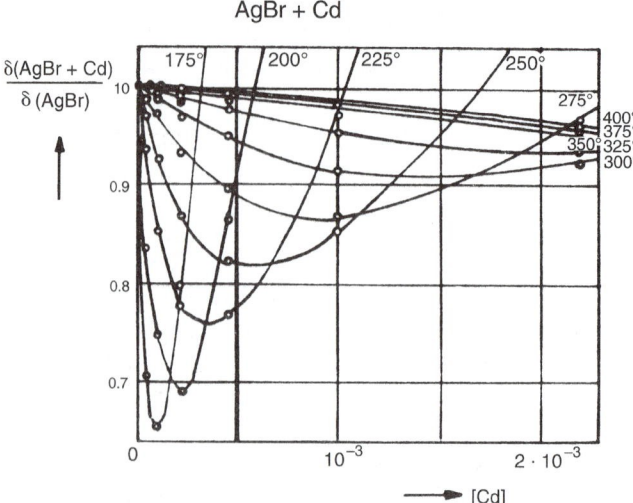

Figure 7.13a The ionic conductivity of AgBr at several temperatures as a function of the Cd^{2+} concentration: linear plot. (Reproduced from Teltow, 1949, by permission of Wiley-VCH, STM.)

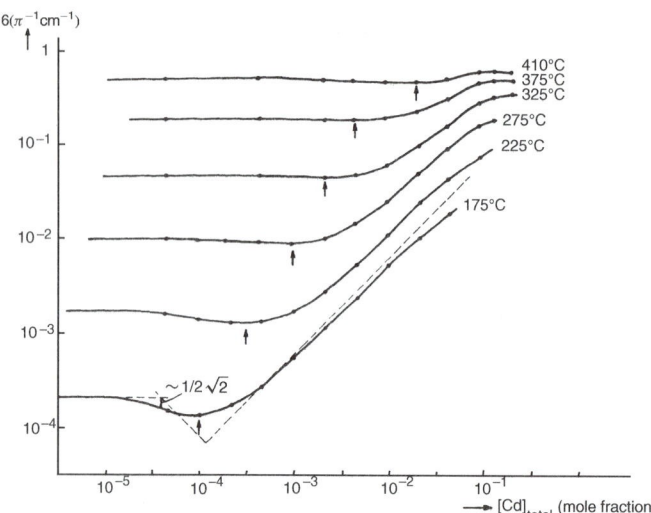

Figure 7.13b The ionic conductivity of AgBr at several temperatures as a function of the Cd^{2+} concentration: log–log plot. (Reproduced from Teltow, 1949, by permission of Wiley-VCH, STM.)

vacancy conduction as the concentration of the dopant increases. Both the intrinsic defect concentrations and their mobilities are much greater in the AgBr system than in NaCl, and the conductivities are much higher in the former, even at the lower temperatures used in these examples.

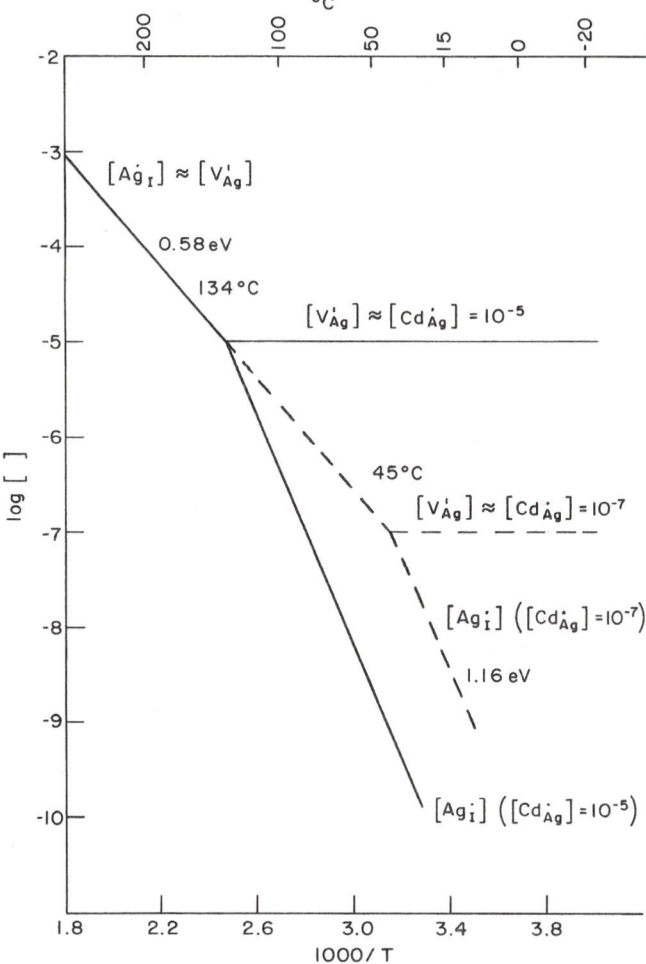

Figure 7.14 Arrhenius plot of the defect concentrations in AgBr at two different Cd^{2+} concentrations, 10^{-5} (10 ppm) and 10^{-7} (0.1 ppm).

The defect concentrations as a function of temperature are displayed in Arrhenius form in Fig. 7.14. The appearance is again very much like that of NaCl. The slopes are deceptively similar because the temperature scale has been compressed by a factor of 2 in the case of AgBr. Note that with a divalent cation content of 10^{-5} (10 ppm), NaCl shows intrinsic behavior down to 604°C, while at the same dopant concentration, AgBr is intrinsic all the way down to 134°C. This reflects the unusually high intrinsic defect concentrations in AgBr.

The temperature dependence of the conductivities, plotted as log σT, is shown in Arrhenius form in Fig. 7.15. Once again, the form is different from that of NaCl because of the shift from predominant interstitial conduction to predominant vacancy

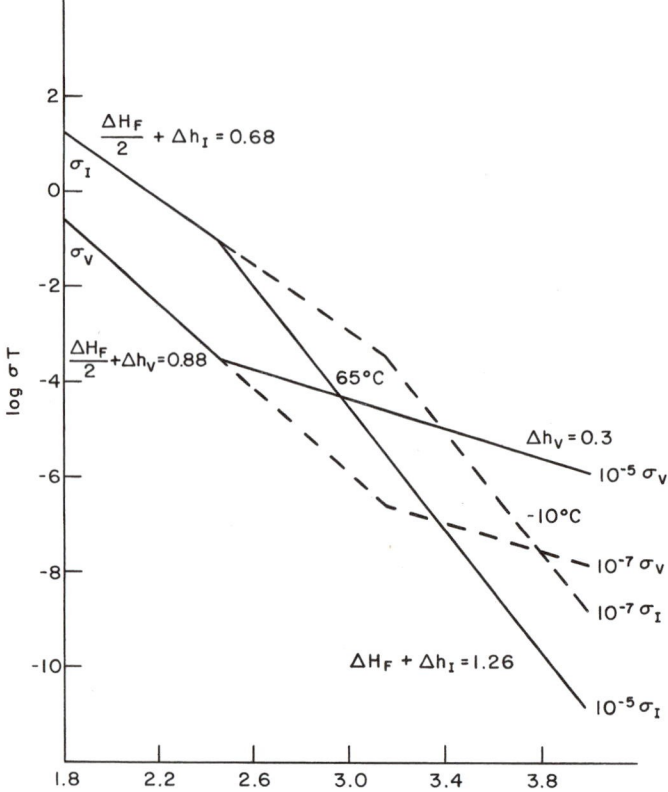

Figure 7.15   Arrhenius plot of the ionic conductivity due to cation interstitials and vacancies in AgBr at two different concentrations of Cd^{2+}.

conduction as the temperature decreases. The interstitial conductivity in the extrinsic region drops away with decreasing temperature with a composite enthalpy of 1.26 eV (121 kJ/mol), while the vacancy conductivity is dropping due only to their enthalpy of motion, 0.3 eV (30 kJ/mol). The shift to vacancy conductivity occurs at 65°C with a Cd concentration of 10^{-5} (10 ppm), and at $-10°C$ with a Cd concentration of 10^{-7} (0.1 ppm).

Finally, Fig. 7.16 compares the NaCl and AgBr systems on the same plot. The conductivity of AgBr is generally four to five orders of magnitude higher than that of NaCl. The closest the latter ever comes is by a factor of a thousand at about 60°C.

FAST ION CONDUCTORS

Some crystalline solids have extraordinarily high ionic conductivities, reflecting very high levels of disorder. The conductivities in some cases rival those of molten salts or aqueous solutions. These materials are collectively referred to as fast ion conductors

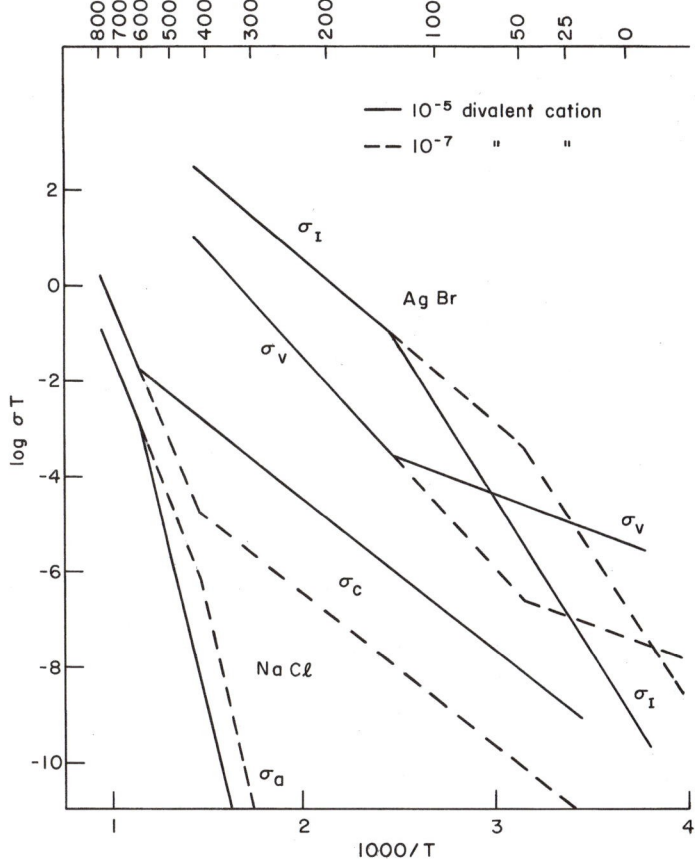

Figure 7.16 The ionic conductivities due to various defects in AgBr and NaCl compared for two different concentrations of divalent impurity cations.

(they have also been called superionic conductors, but that term is misleading because the materials do not share the unique properties of the conventional electronic superconductors, in which the resistivity goes identically to zero below some characteristic transition temperature). A few examples of fast ion conductors are shown in Arrhenius form in Fig. 7.17, which includes values for LiF for comparison with a compound with normal behavior (Hayes and Stoneham, 1985). The large jump in conductivity for the latter compound occurs at its melting point, the transition from the crystalline solid to the molten salt; the values shown for the other materials refer only to the solid compounds. In most cases the high conductivities occur above a transition temperature below which the behavior is normal; unfortunately, these transition temperatures are far above room temperature, which excludes the use of the unusually high conductivities in most practical applications. Only in the case of β-alumina does

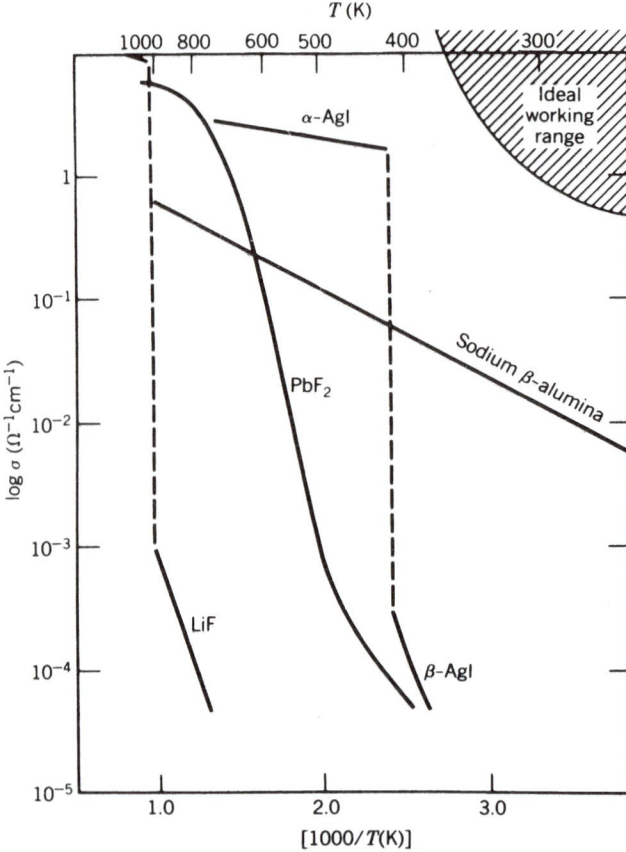

Figure 7.17 Arrhenius plot of the ionic conductivities of several "fast ion conductors" compared with that of LiF. The sharp increase in the conductivity of the latter occurs only at its melting point; in the other cases, the increase occurs within the solid state as the result of a structural transition. (Reproduced from Hayes and Stoneham, 1985, by permission of John Wiley & Sons.)

the high conductivity persist to low temperatures, although the conductivity of this material is not as high as in some of the other compounds.

Silver Iodide

Silver iodide is the original and classic example of a fast ion conductor. Its unusual properties were recognized at a very early date (Tubandt and Lorenz, 1914). AgI has the wurtzite structure at room temperature. The iodide ion is so large that the Ag^+ can fit in the tetrahedral interstices of the hexagonal-close-packed anion sublattice. Between a transition temperature of 147°C and its melting point of 552°C, the structure consists of a body-centered iodide sublattice with the Ag^+ ions distributed among the tetrahedrally coordinated sites. There are 12 of these sites per unit cell, as shown in Fig. 7.18, while there are only two cations. The enthalpic barrier between these sites is

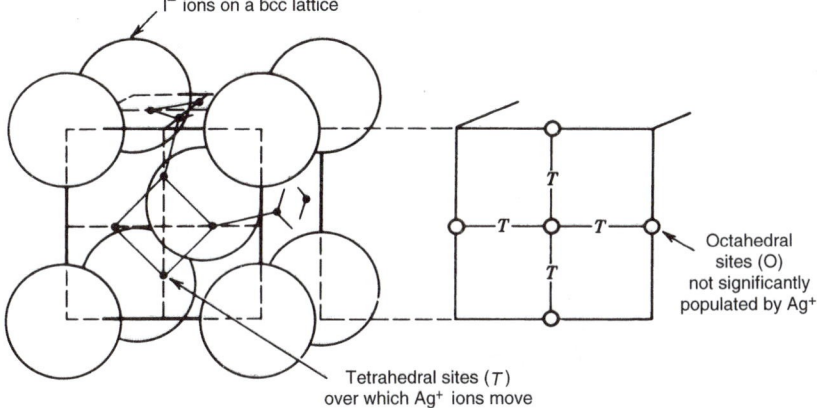

Figure 7.18 The high temperature form of AgI. The 12 equivalent tetrahedrally coordinated cation sites within the cell are randomly occupied by only two cations. (Reproduced from Hayes and Stoneham, 1985, by permission of John Wiley & Sons.)

very small, so that the Ag^+ can roam freely among them. In effect, the cation sublattice has melted within the rigid body-centered anion sublattice. This can be viewed as the ultimate in cation Frenkel disorder. The conductivity actually decreases slightly when the compound melts, because the conducting paths are not so well organized in the melt. There is a large increase in the entropy when a compound melts, because of the transition from an ordered solid to a disordered melt. In AgI, and other compounds that have a transition to fast ion conduction, there is a large increase in entropy at the transition temperature, reflecting the almost complete disordering of one of the sublattices.

Fluorides Having the Fluorite Structure

Some fluorides, and a few chlorides, of large cations (e.g., Ba^{2+}, Sr^{2+}, and Pb^{2+}), have the fluorite structure and undergo a high temperature transition similar to that of AgI. In this case it is the anion sublattice that becomes largely disordered with resulting fast ion conduction. This represents a high degree of anion Frenkel disorder. The conductivity of PbF_2 is shown as an example in Fig. 7.17. The transition temperature is 439°C, which is more than 400° below the melting temperature of 885°C. This is an unusually large range of fast ion conduction for this class of compounds, which generally lies at about 200°C. Again, a large anomaly in the specific heat at the transition temperature results from the large increase in entropy due to the increased disorder.

β-Alumina

β-Alumina is actually a sodium aluminate with the ideal composition $Na_2O–11Al_2O_3$. The structure (Fig. 7.19), consists of layers of the spinel structure separated by

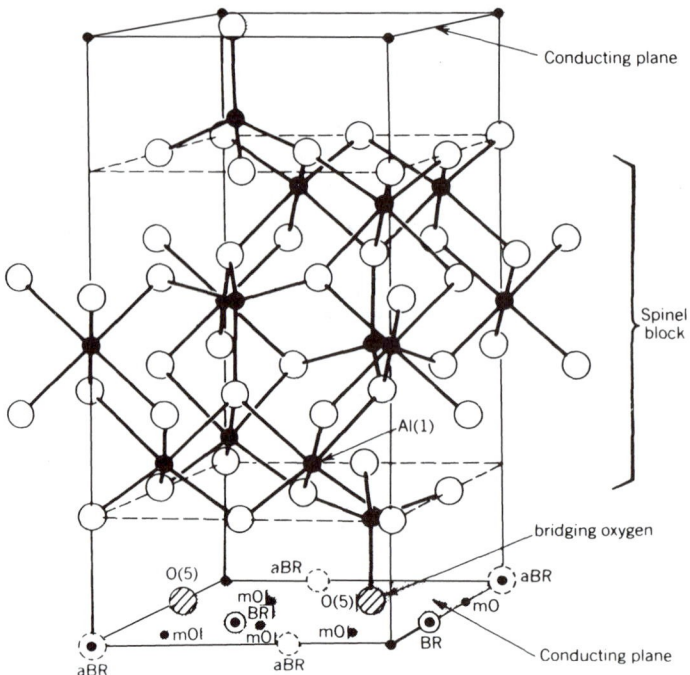

Figure 7.19   The structure of β-alumina showing the spinel-like layers and the conducting planes that are occupied by the highly mobile Na$^+$. (Reproduced from Hayes and Stoneham, 1985, by permission of John Wiley & Sons.)

conducting planes held open by bridging oxygens. The sodium is distributed among sites of several different types in the conducting planes. Again, there are many more sites than cations, and the low energy barriers between cation sites result in very high cation mobilities. The high conductivity is two-dimensional and is negligible perpendicular to the conducting planes. Contrary to the fast ion conductors described earlier, in β-alumina the disorder and high conductivity persist to low temperatures. A closely related compound, β''-alumina, with composition Na_2O–MgO–$5Al_2O_3$, has a slightly higher conductivity, undoubtedly owing to its higher sodium content and to the fact that some of the different sodium sites in β-alumina have become equivalent in β''-alumina. The latter compound has been extensively investigated for use as the electrolyte in sodium–sulfur batteries. In both materials, the sodium content is significantly higher than that of the ideal composition, and this may be necessary for the high conductivity.

It is possible to replace the Na$^+$ content of these compounds with other cations by ion exchange (i.e., by immersion in a molten salt of the other cation). The K$^+$ analog has a somewhat lower conductivity, and this is attributed to the large size of the cation, which cannot move as freely in the conducting slot. The Li$^+$ analog also has a lower conductivity; in this case the smaller cation moves from the center of the conducting plane and becomes more strongly attached to the oxygens of the adjacent spinel blocks.

REFERENCES

Franklin, A. D. Statistical thermodynamics of point defects in crystals. In *Point Defects in Solids*, Vol. 1, *General and Ionic Crystals*, J. H. Crawford Jr. and L. M. Slifkin, Eds. New York: Plenum Press, 1972, Chapter 1.

Fuller, R. G. Ionic conductivity (including self-diffusion). In *Point Defects in Solids*, Vol. 1, *General and Ionic Crystals*, J. H. Crawford Jr. and L. M. Slifkin, Eds. New York: Plenum Press, 1972, Chapter 2.

Hayes, W. and A. M. Stoneham. Lattice defects. In *Defects and Defect Processes in Nonmetallic Solids*. New York: John Wiley & Sons, 1985, Chapter 3.

Kirk, D. L., and P. L. Pratt. *Proc. Br. Ceram. Soc.* 9:215, 1967.

Nowick, A. S. Defect mobilities in ionic crystals containing aliovalent ions. In *Point Defects in Solids*, Vol. 1, *General and Ionic Crystals*, J. H. Crawford Jr. and L. M. Slifkin, Eds. New York: Plenum Press, 1972, Chapter 3.

Teltow, J. *Ann. Phys.* 5:63, 1949

Tubandt, C., and S. Eggert. *Z. Anorg. Chem.* 110:1969, 1920.

Tubandt, C., and S. Eggert. *Z. Anorg. Chem* . 115:105, 1921.

Tubandt, C., and E. Lorenz. *Z. Phys. Chem.* 87:513, 1914.

Intrinsic Electronic Disorder

8

INTRODUCTION

Intrinsic electronic disorder involves the thermal excitation of electrons from the chemical bonds of a solid material into higher energy states. In a pure, stoichiometric semiconducting or insulating compound, it is the only source of electronic charge carriers. It should be recalled that a stoichiometric material is either a simple compound (whether binary, such as MgO, ternary, such as $MgAl_2O_4$, or of higher order, with simple integer ratios of the atomic components) or a solid solution of stoichiometric components, such as $(1-x)TiO_2 + x/2Al_2O_3$ to give $Ti_{1-x}Al_xO_{2-x/2}$. As we shall see, the ground state corresponds to a nominally filled valence band and a nominally empty conduction band for both insulating and semiconducting materials. Electrons in the conduction band and holes in the valence band will be considered to be defects relative to the standard state in which all electrons are in their lowest available energy states.

To discuss electronic disorder, it is necessary to have an understanding of the basic concepts of the band structure of solids. We will not need a highly sophisticated or mathematical level of understanding, and the description will be very much phenomenological. Some of the author's students have referred to this as a thirty-minute version of the band theory of solids. For more advanced treatments, the reader is referred to standard texts on introductory solid state physics for supplemental reading.

THE DEVELOPMENT OF ENERGY BANDS

It is instructive to first consider how the electrons in two simple atoms combine to form a chemical bond. Consider two hydrogen atoms that are initially remote from one another, as shown in Fig. 8.1. Neither atom has any effect on the other, and they both have identical 1s electronic states that contain one electron; that is, each state is half-filled. As the two atoms are brought closer together, their 1s orbitals will eventually begin to overlap. Since the Pauli exclusion principle does not allow two identical electronic states in the same system, the two 1s-type states must become

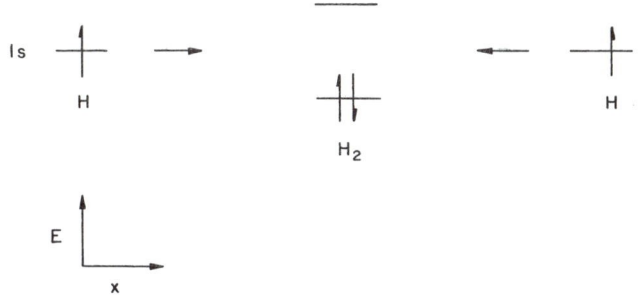

Figure 8.1 Development of the covalent bond between two hydrogen atoms.

different in energy. This is accomplished by having one electronic state move to a higher energy while the other drops to a lower level, as shown in Fig. 8.1 for the H_2 molecule. The two atomic orbitals are transformed into two molecular orbitals that are no longer identified with specific atoms in the molecule. The two electrons, one from each atom, fill the lower level, or bonding orbital, with opposed spins, and form the covalent bond that holds the hydrogen molecule together in its ground state. The upper level, or antibonding orbital, is then empty. The presence of the empty upper level can be detected by excitation of one of the electrons from the lower electronic state by irradiation of the molecule with sufficient energy, and the system will emit characteristic radiation as it returns to its ground state. This simple diatomic molecule demonstrates our fourth important conservation law:

4. The conservation of electronic states: The total number of electronic states in a system derives directly from the electronic states of the component atoms, and must be conserved.

We started with two atoms with a total of two electronic states, and we obtained a molecule with the same total number of electronic states. The number of electronic states in the combined system is the sum of the number of electronic states in the isolated atoms. This conservation of states is valid for any number of atoms in the system.

Let us apply this principle to the formation of metallic sodium from a collection of sodium atoms. Each atom has the electronic structure $1s^2/2s^22p^6/3s^1$, which corresponds to the stable electronic structure of the inert gas neon, plus one valence electron. The 3s orbitals interact in much the same way as the 1s electrons in the hydrogen atoms in the example described above. This interaction is demonstrated in Fig. 8.2. Initially there is a single Na atom, with a single 3s state that contains a single electron. When a second atom is added to the system, the original 3s states must split apart in energy, and the two valence electrons will occupy the lower energy level. As more atoms are added to the system, the number of separate energy levels derived from the 3s states continues to equal the number of atoms, and their separation gradually decreases, so that the number of levels within an increment of energy increases. It requires the combination of 2.5×10^{22} atoms to make 1 cm^3 of metallic Na. In such a

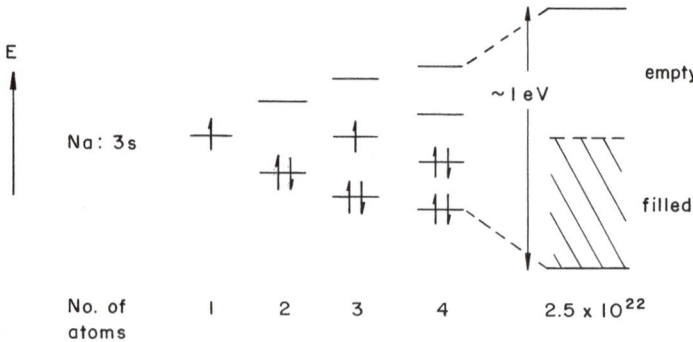

Figure 8.2 Development of an energy band from the 3s electrons of sodium.

piece of metal there will then be 2.5×10^{22} energy levels derived from the individual 3s atomic levels, and they will be spread out over an energy increment of the order of 1 eV (100 kJ/mol). The individual energy levels must be very, very close together, and their separation is meaningless at room temperature, where the average thermal energy is $kT = 0.025$ eV (2.4 kJ/mol). Thus this collection of energy states represents an effective continuum of states that is called a band. In this case, it will be identically half-filled with electrons, as were the 3s levels of the individual component atoms.

For an electron to be able to contribute to electrical conduction, it must be able to increase its energy. A field acting on a charge creates a force, $F = Ee$, and a force acting on a mass causes an acceleration, $F = ma$. Acceleration results in an increasing velocity, which requires an increase in the kinetic energy, $E_k = 1/2mv^2$. Thus if an electron cannot increase its energy, it cannot respond to the applied field. Since the electrons at the top of the filled part of the band shown in Fig. 8.2 have empty energy states directly above them, they can move upward through a continuum of energy states. The electrons that are far down in the filled, lower half of the band cannot respond to the field. Only the electrons within a few kT of the top of the filled states can contribute to the conductivity. This situation of a partially filled band is the defining characteristic of a metal. The concentration of charge carriers is nearly temperature-independent.

We can go through the same process with the next element, Mg:$1s^2/2s^22p^6/3s^2$, as shown in Fig. 8.3. In this case the 3s states in the atoms are filled, and the band of energy states derived from these atomic states will be identically filled in the metal. That would make Mg an insulator or a semiconductor, except that the next higher atomic levels, empty 3p states, combine into a band that overlaps the 3s band. Thus there are still contiguous empty energy states that the uppermost electrons can accelerate into in response to an applied field, and Mg responds like a typical metal, as we know it to be. In the case of Al:$1s^2/2s^22p^6/3s^23p^1$, the 3s band is filled and the next 2p band is half-filled, and the material is again a metal.

The situation changes with Si:$1s^2/2s^22p^6/3s^23p^2$. To optimize the bonding in the elemental form, the 3s and the three 3p orbitals hybridize into four equivalent sp^3 orbitals that are directed toward the corners of a regular tetrahedron. Each of these

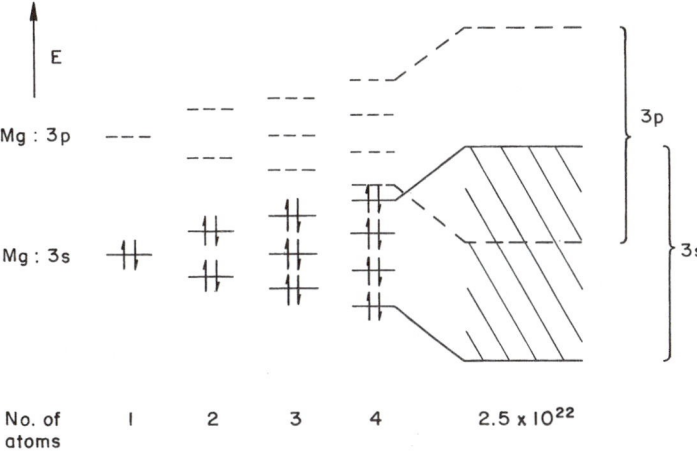

Figure 8.3 Development of energy bands from the 3s electrons of magnesium. The nominally filled band derived from the 3s states overlaps with the empty band derived from 3p states.

hybrid orbitals contains one of the valence electrons from its own atom and shares another electron from an equivalent orbital on an adjacent atom, to form four covalent bonds around each Si atom. These filled states combine into a band of states called the valence band, and the next higher band does not overlap it; there is a gap between the uppermost filled band, derived from the filled sp³ states, and the lowermost empty band, known as the conduction band, as shown in Fig. 8.4. This is called the band gap and is designated as E_g. Mobile charge carriers can be created in this ideally pure system only by thermal or optical excitation of an electron across the band gap into the nominally empty upper band. This is the defining characteristic of a semiconductor or an insulator, the difference between the latter two being only an arbitrary difference in the magnitude of the band gap. The excitation process not only gives an electron in the conduction band, where it can move freely through the empty states, it also leaves an empty electronic state at the top of the valence band. This also allows charge transport, since adjacent electrons can move into this empty state, called a hole. With successive electron movements in an electric field, the hole moves laterally in the opposite direction to the electrons in the conduction band, and it acts as a particle with a positive (missing negative) charge and a positive mass. Both the electrons in the conduction band and the holes in the valence band will be treated as distinct species (i.e., defects) that contribute to charge transport with characteristic mobilities.

THE MASS-ACTION APPROACH

The analogy between intrinsic electronic disorder and Frenkel disorder will be apparent. In both cases, a species moves from its normal condition in the thermodynamic standard state to a higher energy state and leaves behind an empty state, either a

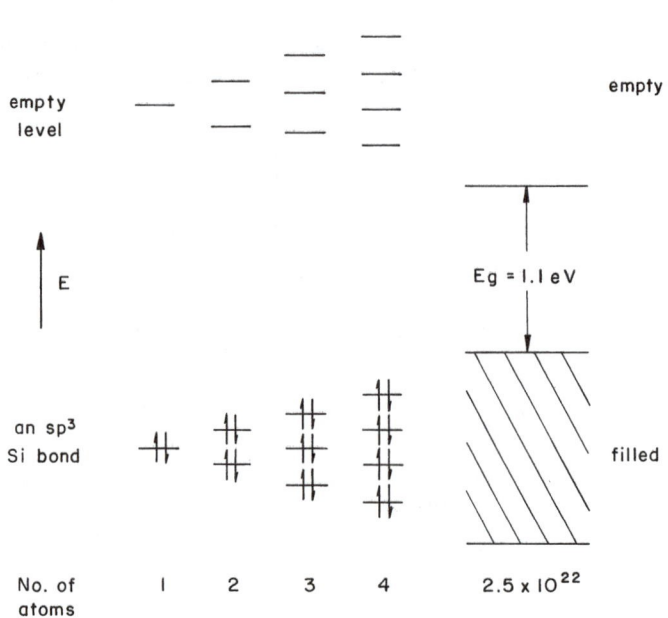

Figure 8.4 Development of the valence and conduction bands from the 3s and 3p states of silicon.

vacancy or a hole. We can write an equilibrium reaction for the electronic excitation process

$$nil \rightleftharpoons e' + h^{\cdot} \tag{8.1}$$

where e' represents an electron in the conduction band, and $h^{\cdot}$ is a hole in the valence band. We have maintained the defect notation for effective charge because we are treating both species as defects. A mass-action expression can be written for any valid equilibrium reaction, which in this case is

$$np = K_I(T) = K_I^{\circ} e^{-E_g/kT} \tag{8.2}$$

where, according to convention, $[e'] = n$ and $[h^{\cdot}] = p$ are used as a shorthand notation, with n and p representing the concentrations of negative and positive charge carriers, respectively. This is a typically thermally activated process, with the band gap, E_g, being the energy that has to be surmounted to create the disorder. If Eq. (8.1) is the only significant source of electrons and holes, then their concentrations must be equal, so that

$$n \approx p \approx K_I^{1/2} \approx (K_I^{\circ})^{1/2} e^{-E_g/2kT} \tag{8.3}$$

The analogy with Frenkel disorder is again obvious.

The concept of a band gap came from the physics community. Before long, it was noted that the band gap is temperature dependent and can be expressed as follows:

$$E_g = E_g^\circ - \alpha T \tag{8.4}$$

where E_g° is called the band gap at absolute zero, and α is its linear temperature coefficient. It took a surprisingly long time to note the similarity between Eq. (8.4) and the familiar expression for the Gibbs free energy, $\Delta G = \Delta H - T\Delta S$. Thus E_g is the free energy, E_g° is the enthalpy, and α is the entropy of the reaction shown in Eq. (8.1). Then E_g° is the enthalpy of intrinsic electron ionization, sometimes referred to as the band gap at 0 K. In the same sense, ΔH is the Gibbs free energy at absolute zero. Equation (8.3) can be written in the more familiar form:

$$\mathrm{n} \approx \mathrm{p} \approx (K_I^\circ)^{1/2} e^{\Delta S_I/2k} e^{-E_g^\circ/2kT} \tag{8.5}$$

This is only one of many examples of two different disciplines describing the same phenomenon with different notation.

Whereas there are many different kinds of intrinsic ionic disorder (cation Frenkel, anion Frenkel, Schottky, etc.), there is only the one type of intrinsic electronic disorder for all nonmetallic solids, and it is expressed by Eq. (8.1).

THE FERMI FUNCTION

The rigorous description of electronic states in solids requires the use of Fermi–Dirac statistics. While this is not necessary for most of the situations dealt with in defect chemistry, its application, and the assumptions that allow its simplification, should be understood. The key relationship is the Fermi function $f(E)$, which is the probability that an available electronic state at energy E will be occupied. Thus, if there are $N(E)$ electronic states at energy E, the number of occupied states at that energy, $n(E)$, is

$$n(E) = f(E)N(E) \tag{8.6}$$

The Fermi function is usually written in the form

$$f(E) = \frac{1}{e^{E-E_F/kT} + 1} \tag{8.7}$$

where E_F is a reference energy, called the Fermi energy or Fermi level. The probability that a state is occupied is a function of its distance in energy above or below E_F. When $E = E_F$, $f(E) = 1/2$, therefore, E_F is the energy level at which the probability of occupation is 50%. The combination of Eqs. (8.6) and (8.7) can be cross-multiplied and rearranged to give

$$\frac{n(E)}{N(E) - n(E)} = e^{-(E-E_F)/kT} \tag{8.8}$$

which is in the form of a mass-action expression giving the ratio of occupied and unoccupied states. For the case $n \ll N$, which is equivalent to $(E - E_F) \gg kT$, the Fermi function can be simplified to a Boltzmann expression for the fraction of occupied states

$$\frac{n(E)}{N(E)} = e^{-(E-E_F)/kT} \tag{8.9}$$

This implies that we are operating out on the high energy tail of the Boltzmann distribution of thermal energy, as shown in Fig. 8.5. The main point to this discussion is that the formulation in terms of the Fermi function is very closely related to the mass-action approach. They are just usually written in different configurations.

For intrinsic electronic disorder, the electrons in the conduction band will be concentrated at the bottom of the band. Thus the concentration of available states N_C near the band edge E_C is of particular interest, since the occupation of these states determines the concentration of charge carriers. Using the Boltzmann approximation, this is given by

$$n = N_C e^{-(E_C-E_F)/kT} \tag{8.10}$$

An analogous expression for holes is given by

$$p = N_V e^{-(E_F-E_V)/kT} \tag{8.11}$$

where N_V is the effective density of hole states (i.e., normally occupied electron states) at the top of the valence band E_V. The probability of a hole state being occupied is a function of its distance in energy below E_F. When Eqs. (8.10) and (8.11) are multiplied together, the Fermi energy drops out, and an expression that is equivalent to the mass-action expression for intrinsic electronic disorder, Eq. (8.2), is obtained:

$$np = N_C N_V e^{-(E_C-E_V)/kT} \tag{8.12}$$

where $E_C - E_V = E_g$. The relationship between Fermi–Dirac statistics and the

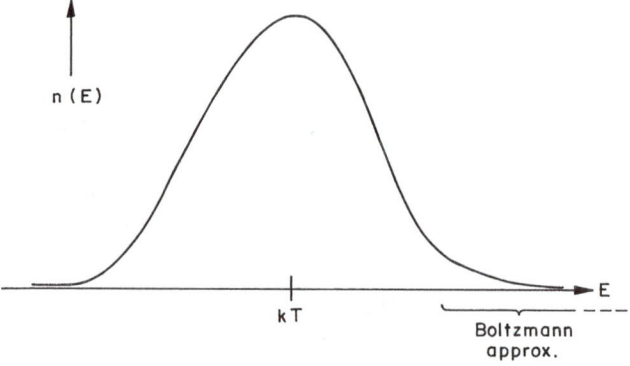

Figure 8.5 A Boltzmann distribution indicating the high energy region, where the Boltzmann approximation is valid.

mass-action approach is clearly demonstrated. When the electronic states of interest are close to the Fermi energy, the Boltzmann approximation is no longer valid, and for accurate results it is necessary to use the complete Fermi function. Note that it would be equally possible to treat intrinsic ionic disorder by Fermi-type formalism. Once again, it is necessary to understand the relationship between the approach that developed within the physics community and that more commonly used by chemists.

Comparison of Eqs. (8.10) and (8.11) with (8.3) suggests that $(E_C - E_F) = (E_F - E_V) = E_g/2$, and that E_F is therefore located at the midpoint of the energy gap for this case of intrinsic electronic disorder.

The Fermi function is symmetrical about E_F and varies from 0 (totally empty states) to 1 (completely filled states). Near absolute zero it is a nearly square function that develops increasing curvature with increasing temperature. It is shown superimposed on a simple band diagram in Fig. 8.6. It is now obvious why E_F must lie near the midpoint of the band gap for intrinsic disorder: the concentrations of electrons in the conduction band and holes in the valence band must be equal for this situation. Although the Fermi function has finite values within the band gap, there are no available states there in a pure material, so the product $f(E)N(E)$, which determines the electron population, is zero in the gap. It is clear from Fig. 8.6 that the concentration of electrons in the conduction band, and the concentration of holes in the valence band, will increase with decreasing band gap. E_F will vary from midgap only insofar as the density of states functions in the conduction and valence bands, $N_C(E)$ and $N_V(E)$, differ in magnitude and shape. Strictly speaking, the concentration of electrons in the conduction band is the integral of the Fermi function and the density of states function

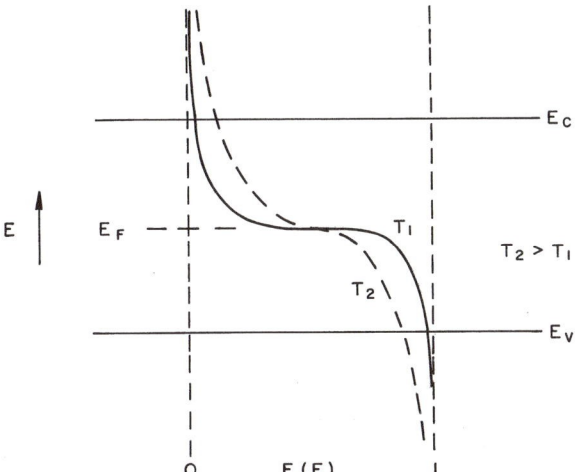

Figure 8.6 The Fermi function for two temperatures, $T_2 > T_1$, superimposed on a simple band diagram. The function varies from 0 at high energies, corresponding to empty states, to 1 at low energies, corresponding to filled states.

$$n = \int_{E_C}^{\infty} f(E)N(E)\, dE \tag{8.13}$$

where the upper limit of integration can be taken as infinity because the Fermi function is asymptotically approaching zero with increasing energy. Using the expression for the density of states obtained by treatment of the electrons as a free electron Fermi gas, and the Boltzmann approximation, the integration of Eq. (8.13) gives

$$n = \left(\frac{4\pi m_e kT}{h^2}\right)^{3/2} e^{-(E_C - E_F)/kT} \tag{8.14}$$

where m_e is the effective mass of the electron, a concept discussed later, and h is Planck's constant. The deviation of E_F from the midgap position will depend on the ratio of the effective masses of the electrons and holes.

HOLES, WAVES, AND EFFECTIVE MASSES

The concepts of holes, waves, and effective masses are sufficiently bizarre to require some further explanation. Why does a hole, a missing electron, act as if it has a positive mass, and why do the effective masses of the electrons and holes differ from their true masses? The answer lies in the effort to describe wave mechanical phenomena in terms of classical mechanics. The following is a simplified version of some of the relationships. This treatment was inspired by the approach used in Kittel's well-known text, *Introduction to Solid State Physics* (1966).

The momentum of a particle, $p = mv$, is related to the kinetic energy $E_k = 1/2mv^2$, and, because of the wave nature of matter, it is also related to a wavelength:

$$p = \sqrt{2m E_k} = \frac{h}{\lambda} = \frac{hk}{2\pi} \tag{8.15}$$

where k is called the wave number, defined as $k = 2\pi/\lambda$. Thus a classical plot of kinetic energy versus momentum takes the form of a smooth parabola, as shown for negative momentum values in Fig. 8.7. However, there will be an interaction between the wave nature of the particle and the periodicity of the lattice. Thus when the Bragg diffraction condition is satisfied

$$n\lambda = 2d \sin \theta \tag{8.16}$$

where d is the interatomic spacing in a particular crystallographic direction and θ is the angle between the propagation direction and a set of planes in the lattice, the wave will not be able to propagate through the crystal in that direction but will be diffracted, as shown in Fig. 8.8b. As the momentum increases from zero, this situation will first occur for $n = 1$ and $\theta = 90°$. However, since the electrons and ion cores are oppositely charged, there will also be an electrostatic potential energy contribution to total energy. As shown in Fig. 8.9, at the wavelength (momentum) that satisfies the Bragg condition, there will be a range of total energies, depending

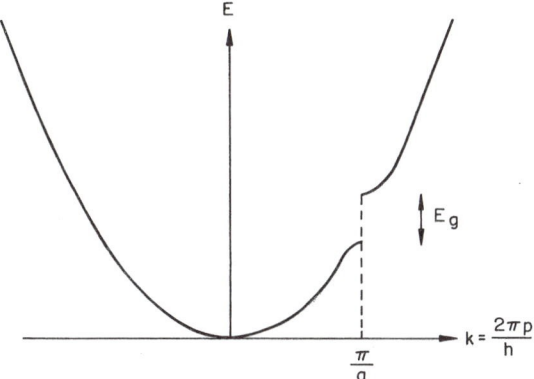

Figure 8.7 Energy as a function of momentum. The left-hand side shows the classical parabolic dependence where only kinetic energy is important. The right-hand side shows the break in the classical curve caused by the potential energy interactions for the conditions of Bragg diffraction. The resulting range of energies at a constant momentum corresponds to the band gap.

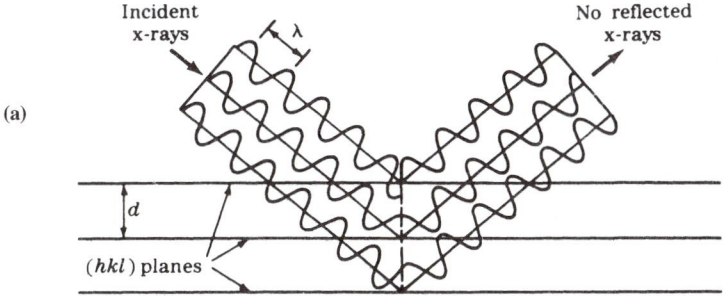

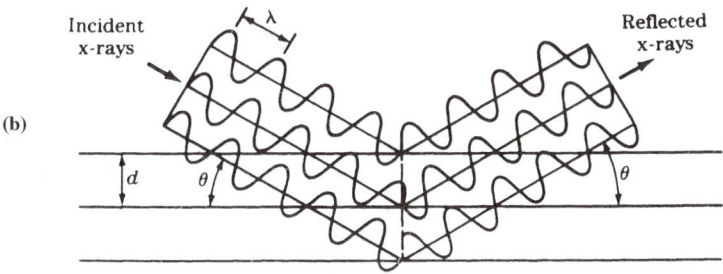

Figure 8.8 Illustration of Bragg diffraction by a crystalline lattice: (a) cancellation of reflected rays by destructive interference, and (b) propagation of in-phase reflected rays. (Reproduced from Smith, 1986, by permission of McGraw-Hill.)

on the phase of the electron wave with respect to the lattice spacing. Since the square of the wave amplitude at a given position indicates the probability of finding the electron at that position, for the phase relationship shown in Fig. 8.9a the electron finds itself mostly between the atom cores, and the electrostatic energy will be at a minimum. In Fig. 8.9b, the electrons are located mostly at the ion cores, and the potential energy contribution will be at a maximum. Thus there is a range of total

(a)

(b)

Figure 8.9 Interaction of the electron wave with the periodic lattice for the condition of Bragg diffraction, with the electrons located primarily (a) between the ion cores and (b) at the ion cores.

energies, all corresponding to a single value of the momentum, for which diffraction prevents continued propagation. Thus the smooth, classical parabola must be broken, as shown on the right-hand side of Fig. 8.7. The range of energies that correspond to a momentum that satisfies the Bragg condition, for which propagation through the crystal is prevented, corresponds to the band gap. Note that the magnitude of the band gap will depend on the direction through the crystal, since the interatomic spacing is a function of direction. For considerations of electrical conductivity, it is the minimum band gap that is of importance

In classical mechanics, the first derivative of the kinetic energy with respect to the momentum gives the particle velocity

$$\frac{d E_k}{dp} = \frac{2p}{2m} = v \tag{8.17}$$

while the second derivative gives the reciprocal mass

$$\frac{d^2 E_k}{dp^2} = \frac{1}{m} \tag{8.18}$$

Thus for the classical case, shown on the left-hand side of Fig. 8.7, as the momentum increases, the velocity (i.e., the slope of the curve), increases steadily with constant mass. When the diffraction condition is taken into account, as on the right-hand side of Fig. 8.7, as the electron momentum increases and approaches the band gap, the velocity begins to steadily decrease and goes to zero at the band edge. That is because the electron wavelength is approaching the diffraction condition; and the harder we push on it, the harder it is to propagate, and it slows down, eventually, to zero velocity. For the classical relationship, $F = ma$, if a positive force results in a negative acceleration, the mass must also appear to be negative. This is reinforced by the observation that the second derivative, Eq. (8.18), is negative in this region (the curve is concave downward), also indicating a negative mass. Insofar as the curvature near the band edge can be approximated by a parabola, concave downward, and with a sharper curvature than the classical parabola, the electron in that region behaves as if it has a negative mass whose absolute value is some fraction of its true mass. Thus a missing electron near the band edge, with a negative mass and a negative charge, becomes our hole, with a positive mass and a positive charge. Above the band gap, the electron velocity picks up again with increasing momentum, the second derivative is positive, and the electron mass again appears to be positive, but is not equal to the true rest mass.

How did we get into this mess? It arises because the treatment of the electrons as a free electron Fermi gas considers only the wavelike nature of the electrons, and that includes only the kinetic energy. When it is realized that there are potential energy contributions due to the electrostatic interactions between the electrons and the positive ion cores, there are two choices. One can either start over and explicitly include the potential energy contribution in the derivations, or try to adjust the final result to give the right answer. The ratio of the effective mass to the true mass is a correction factor in the final expressions that takes into account the contributions from potential energy. It is a compromise that avoids the complexity of including the potential energy contributions explicitly.

For most applications of defect chemistry, the Boltzmann approximation is adequate, and the mass-action approach is sufficiently accurate. Equations (8.1) and (8.2) or (8.5) are then the most useful relationships.

ELECTRONIC CONDUCTIVITY

In intrinsic semiconductors, the electron and hole concentrations are equal and both make significant contributions to the conductivity. The total conductivity is given by

$$\sigma_T = ne\mu_n + pe\mu_p \tag{8.19}$$

where μ_n and μ_p are the mobilities of the electrons and holes, respectively. Using the relationship in Eq. (8.3), this can be written

$$\sigma_T = (\mu_n + \mu_p)eK_I^{1/2}e^{-E_g/2kT} \tag{8.20}$$

and it is seen that the relative contributions of the two carriers depends only on their relative mobilities. It will be recalled from the earlier discussion of ionic conductivity that the mobility, with units of $cm^2/V \cdot s$, is the velocity per unit field, $(cm/s)/(V/cm)$. This then implies a steady state velocity that is proportional to the field, rather than the steady acceleration expected for a charged body subjected to a field in free space. As in the case of ions, this results from a discontinuous motion, but for a different reason. Electrons and holes can propagate freely through a perfect crystalline lattice, but they will be scattered by any deviation from ideal periodicity. Thus impurities, defects, dislocations, grain boundaries, and even lattice vibrations can interrupt the path. The carriers accelerate for only brief periods between scattering events, and the result is an average constant velocity. Lattice vibrations scatter the charge carriers because a vibrating ion or atom is usually not on its precise crystallographic site. At low temperatures, the scattering is primarily due to impurities, defects, and so on, but

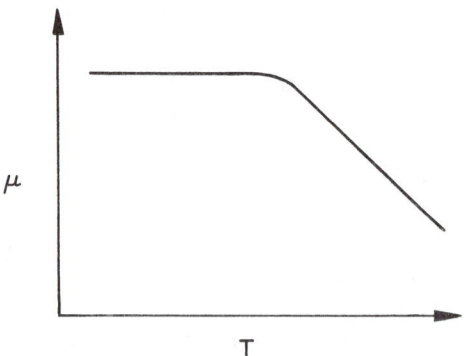

Figure 8.10 Schmatic temperature dependence of the electron mobility in a metal.

with increasing temperature, the vibrations increase and eventually become the most important factor. Thus a schematic plot of mobility versus temperature will appear as in Fig. 8.10. At low temperatures, the mobility is relatively temperature independent, because the impurities and other static sites are doing most of the scattering while the lattice vibrations are minimal. At higher temperatures, the mobility decreases with increasing temperature because of the increasing lattice vibrations. The mobility of a semiconductor typically varies as follows:

$$\mu = \mu_0 T^{-3/2} \tag{8.21}$$

in this region.

In the case of a metal, the current is carried by electrons near the Fermi energy in a partially filled band, and the carrier concentration is essentially independent of temperature. The temperature dependence of the conductivity of metals is then due to the temperature dependence of the electron mobility. The resistivity of samples of a particular metal with different impurity levels will then appear as in Fig. 8.11. At low

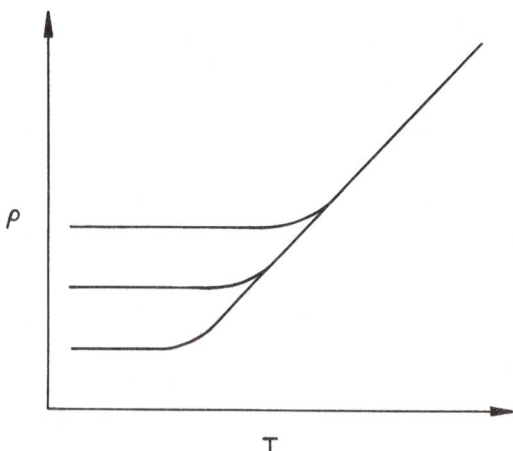

Figure 8.11 The resistivity of three samples of a metal having different levels of impurities. The relative impurity content can be obtained from the "resistivity ratio" between high and low temperatures.

temperatures the resistivity is temperature independent but depends on the impurity level, while all samples have the same resistivity at higher temperatures where lattice vibrations are the controlling factor. The temperature dependence is typically linear in this region. The ratio of the resistivity at some reference temperature in the region where the scattering is primarily vibrational to the resistivity at a temperature in the impurity-controlled region is commonly called the "resistivity ratio" and is used as an indication of the relative impurity content. It does not indicate the chemical identities of the impurities, but it is an easy measurement that gives an integrated picture of the impurity content.

The magnitude of the conductivity of intrinsic semiconductors or insulators is primarily determined by the carrier concentrations, hence by the term $\exp(-E_g^\circ/2kT)$. Values of the band gaps of the elements in group IVA of the periodic table, which includes the important elemental semiconductors silicon and germanium, are given in Table 8.1, along with typical carrier mobilities and the room temperature resistivities. Their melting points are included to show the correlation with the band gaps. Diamond is a very good insulator, while in lead, a metal, the band gap has disappeared and the uppermost band is now only partially filled. Tin has two structural types; below 13°C the structure is the same as that of the lighter elements in the group, the diamond structure (grey tin), and there is a small band gap as shown in the table. Above 13°C the structure is different (tetragonal), and the band gap has collapsed so that the material is a metal (white tin). Note that the band gaps decrease with increasing atomic number; this is an important general trend. The exponential function of the band gaps, $\exp(-E_g^\circ/2kT)$, are plotted in Fig. 8.12 for these materials to demonstrate the power of the band gaps in determining the carrier concentrations. Visual appearance also correlates with the band gap. Diamond is transparent because visible light is not energetic enough to excite electrons across the band gap, hence is not absorbed. Both Si and Ge are opaque to visible light, because their band gaps are

TABLE 8.1
Thermodynamic and transport parameters for the group IV elements

Element	Melting Point (°C)	E_g [eV (kJ/mol)]	Mobilities (cm²/V·s)		ρ_{RT} ($\Omega\cdot$cm)
			μ_n	μ_p	
C	>3550	5.33 (513)	1800	1200	10^{14}
Si	1410	1.14 (110)	1600	400	230,000
Ge	937	0.67 (64.5)	3800	1800	43
Sn (grey)	232	0.08 (8)			10^{-4}
Pb	328	Metallic			10^{-6}

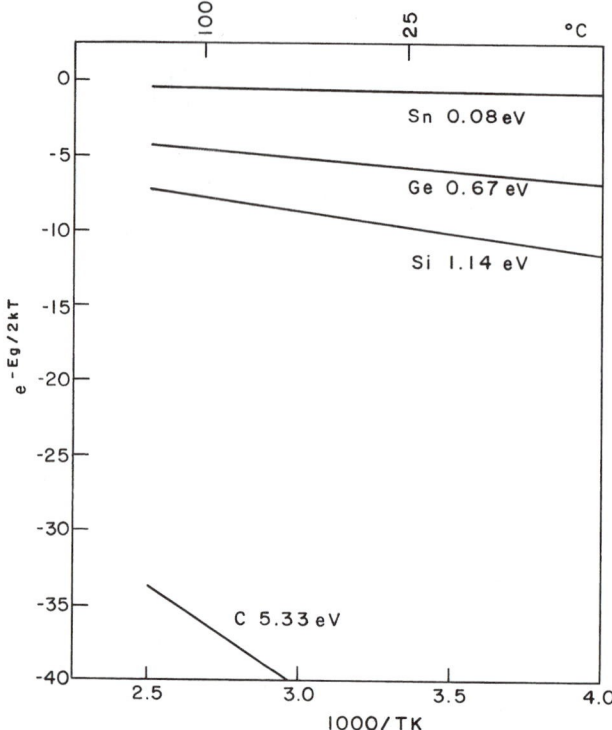

Figure 8.12 Arrhenius plots of $\exp(-E_g/2kT)$ for C (diamond), Si, Ge, and grey Sn.

small enough for optical absorption in the visible range, but both are transparent in regions of the infrared where the optical energy is lower. Lead is a typically opaque metal. Thus insulators tend to be transparent or white, while materials with lower band gaps, as well as metals, are usually opaque or dark in color.

HOPPING MECHANISMS

All the discussion of electronic conduction so far has been limited to band conduction, that is, the free motion of electrons in the conduction band or holes in the valence band, hindered only by scattering due to deviations from the perfect lattice. In many oxides the motion is less uninhibited and takes place by discrete jumps between localized states. Such jumps are thermally activated so that the mobility increases exponentially with temperature. This process is commonly called a hopping mechanism, and the temperature dependence of the mobility is often of the form

$$\mu = \mu_0 T^{-3/2} e^{-E_h/kT} \tag{8.22}$$

where E_h is the activation energy of the hopping process. The localized centers on which the electrons or holes rest between hops may be a host cation or anion, or an impurity center if the impurity ions are close enough together. If the binding is quite strong, there is substantial polarization of the surrounding ions, and this polarization cloud must travel with the carrier. Such centers are called small polarons. In the case of weaker binding, denoted as large polarons, the polarization is less significant and the movement gradually merges into behavior typical of band conduction. It is difficult to predict which systems will conduct by which mechanism, and one must look for experimental evidence. If the activation energy for the concentration of charge carriers is less than that for the conductivity, the difference may be due to a thermally activated mobility (such a discrepancy could also result from significant trapping of the charge carriers with the ionized electrons traveling in the conduction band).

THE BAND STRUCTURE OF COMPOUNDS

The preceding discussion has been rather abstract, and the only real semiconducting materials mentioned were the group IVA elements. They are conveniently simple model materials for describing some of the basic concepts, but in this book we are interested in the behavior of inorganic compounds. Therefore, it is necessary to see how the band model applies to such materials.

In semiconducting and insulating materials, the conduction band is the lowest-lying collection of normally unoccupied energy states, while the valence band is made up of the highest energy states that are normally occupied. How does that apply to compounds? We will again use NaCl as a simple example. Let us assume that we can make a material that consists of equal numbers of sodium atoms and chlorine atoms (not ions), with the atoms arranged in location and spacing exactly where the cations and anions are located in NaCl. Never mind that this is a wildly unstable situation. The energy diagram might look like that in Fig. 8.13a. There is an upper band of electron states that are derived from the Na 3s states and are therefore half-filled. Then there is a lower band derived from Cl 3p states that is also partially filled. If the electrons are now released to find their equilibrium status, all the Na 3s electrons will drop down into the empty Cl 3p states; and the conduction band, derived from the Na 3s states, will become nominally empty, while the valence band, derived from the Cl 3p states, will become nominally filled. This is shown schematically in Fig. 8.13b. Ionization of electrons across the band gap is related to the transfer of an electron from the 3p states of a Cl^- anion to the 3s states of a Na^+ cation. This is highly unfavorable because the transfer of the electron from the Na to the Cl to make the ions in the crystalline solid is extremely favorable. Thus the ionization enthalpy is large, which means that the band gap is large. There is an obvious relationship between the band gap and the ionization potential of the Na atom and the electron affinity of the Cl atom. Another way to look at this is that the ionization reaction is, in effect, breaking chemical bonds. Thus one might expect a correlation between the lattice energy and the band gap. This is illustrated in Fig. 8.14 where the band gaps of a wide variety of materials are plotted against the heat of atomization per chemical equivalent (to normalize for the number of bonds in the formula unit). The excellent correlation indicates that there is a strong relationship between the chemistry of the constituent elements of a solid compound and its electronic properties.

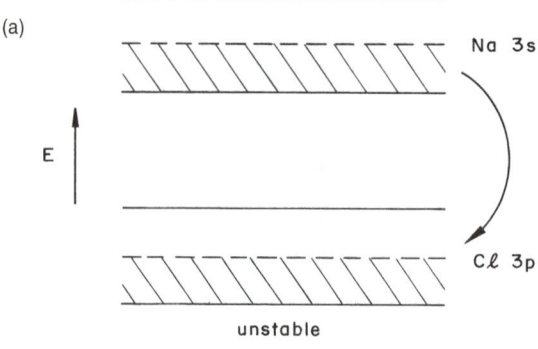

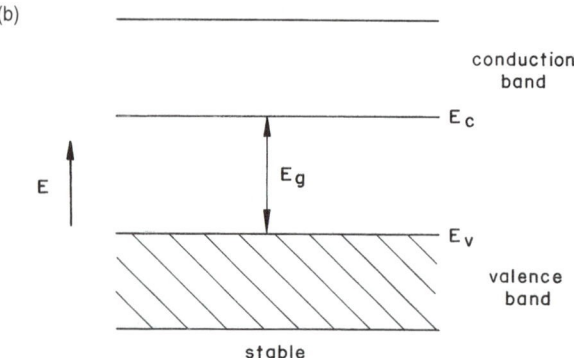

Figure 8.13 The development of valence and conduction bands in NaCl: (a) the fictitious lattice of Na and Cl atoms and (b) the band structure after the shift of the valence electrons from Na to Cl.

In general, the band gap increases with the increasing differences between the electronegativities of the elements in the compound, because this adds an increasing ionic contribution to the bonding and increases the total bond energy.

CHEMISTRY AND THE BAND GAP

Our description of the band gaps of compounds is very qualitative and lacks theoretical rigor. Its major advantage is that it works very well in spite of its simplicity. In most compounds of interest (i.e., halides and oxides), when the cation has the electronic structure of the preceding rare gas, the valence band is derived from the uppermost filled states in the very stable rare gas electronic structure of the anions. This corresponds to the example of NaCl described in the preceding section. In the case of oxides, these are the filled $O\,2p$ states. While the absolute values of the energies of the valence and conduction band edges are less important than their separation in

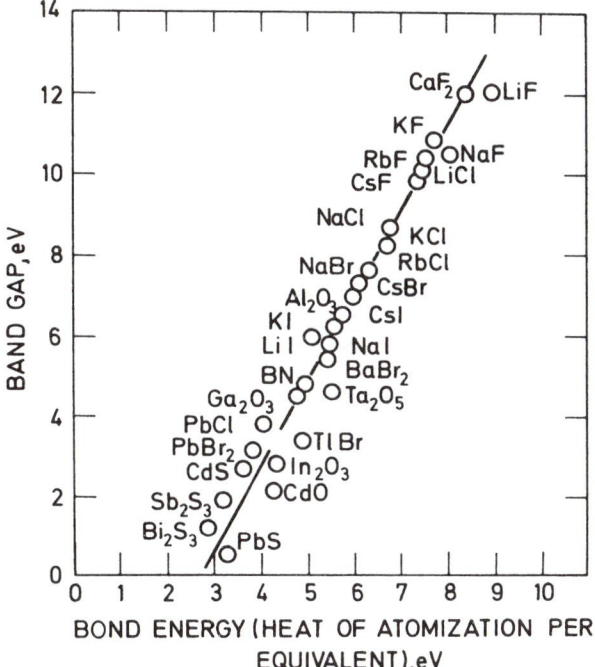

Figure 8.14 The band gaps of various materials as a function of the heat of atomization per chemical equivalent. (Reproduced from Kofstad, 1972, by permission of the editors.)

determining electronic behavior, for the purpose of comparing different materials it is useful to have an absolute energy scale. It is convenient to reference all electron energies to the vacuum level: that is, the energy required to remove an electron from a given state to a remote location in free space. For the uppermost occupied states, this corresponds to the work function of the compound. While there may be small differences between different oxides, in general the electronic states derived from O 2p atomic states should lie at similar distances below the vacuum level, while the energy levels in the valence band of chlorides, derived from the Cl 3p levels, should also form a family with similar energies at the valence band edge. If the ionization energy of the atom that creates the cation in the compound is small [e.g., Na with an ionization energy of 5.14 eV (495 kJ/mol)], then the energy to remove an electron from the bottom of the conduction band to the vacuum level should be modest. As a result, the bottom of the conduction band should lie relatively high above the top of the valence band, and the band gap should be large. For NaCl the band gap is somewhat greater than 8 eV (770 kJ/mol). In the case of Al_2O_3, where the third ionization energy for the aluminum is 28.4 eV (2730 kJ/mol), the band gap is about 6.5 eV (630 kJ/mol). However, for TiO_2, where the fourth ionization energy of the Ti atom is 43.2 eV (4160 kJ/mol), the conduction band derived from these states will lie deeper below the vacuum level and closer to the valence band, giving a band gap

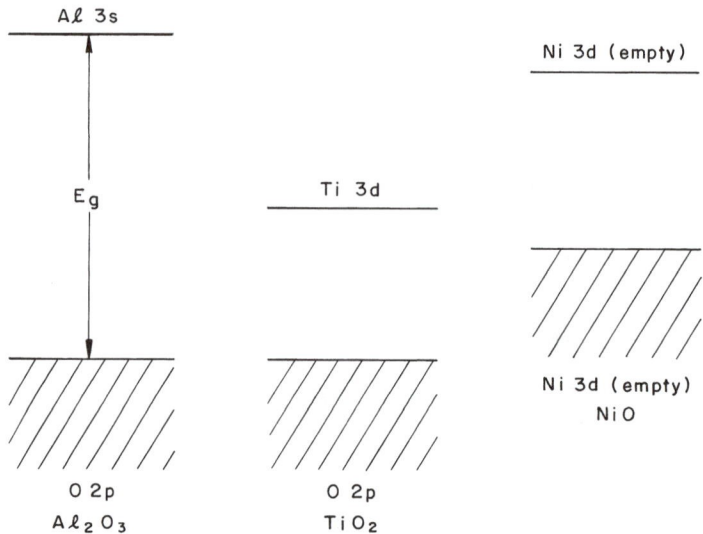

Figure 8.15 Comparison of schematic band gaps for Al_2O_3, TiO_2, and NiO.

of about 3 eV (300 kJ/mol). Figure 8.15 gives a schematic comparison of the energy levels in Al_2O_3 and TiO_2.

This approximation to the relative energies of the valence and conduction bands, and the magnitude of the resulting band gaps, is rather crude, and there will certainly be matrix effects for different compounds. However, insofar that the matrix effects are similar for both bands, and as long as the variations are small compared with the magnitude of the band gap, the approximation is adequate for our purposes. It has one strong argument in its favor: *it works!*

Some of the later transition metal cations do not have the rare gas electronic structure, but retain some electrons in the d shells (e.g., Fe^{2+}, $3d^6$; Co^{2+}, $3d^7$; and Ni^{2+}, $3d^8$). These levels lie above the O 2p levels, and the valence band is then derived primarily from the filled d states of the cations, thus decreasing the band gap. The conduction band is made up primarily of empty d states from the cations. The second ionization energy of Ni is 18.2 eV (1750 kJ/mol) and the conduction band is thus relatively high in energy. But because the valence band is also relatively high, the band gap of NiO is still only of the order of 3 eV (300 kJ/mol), and this is also shown schematically in Fig. 8.15.

Since ionization energies for the metallic elements tend to decrease with increasing atomic number as the atoms get bigger, for equivalent electronic structures the band gaps tend to increase with increasing atomic number; for example, the band gap of Ta_2O_5 is higher than that of Nb_2O_5. Another way of looking at these correlations is the following. If the cation whose empty electronic states make up the conduction band has an accessible lower oxidation state, and thus is receptive to accepting another electron, then the conduction band is relatively low-lying (e.g., Ti^{4+}, Nb^{5+}, but not Al^{3+} or Mg^{2+}). If the ion whose filled electron levels make up the valence band has an accessible more positive oxidation state, and is thus able to lose an electron easily, the valence band lies relatively high (e.g., Ni^{2+}, Fe^{2+}, but not O^{-2} or Cl^{-1}. It is then clear that when all the elements contained in a compound have only a single stable oxidation state, the band gap will be large, and the material will be an insulator. This correlation of the band gaps with the chemical properties of the constituent elements is useful for predicting the electrical properties of their compounds. A little chemical knowledge can be very useful.

SUMMARY

The description of intrinsic electronic disorder in terms of the band model of solids is somewhat different from the usual mass-action treatment of intrinsic ionic disorder. However, the end results are very similar, and the mass-action approach is equally useful for electronic disorder. Whereas there are several possible types of intrinsic ionic disorder, there is only one kind of intrinsic electronic disorder, and it is described by the equilibrium ionization reaction, Eq. (8.1), and its mass-action expression, Eq. (8.2).

The band gap tends to increase as:

the energy of the last ionization step that creates the cation decreases

the ionization energy of the uppermost filled atomic levels increases

the bond energy of the compound increases

the electronegativity difference between the cations and anions increases

the lower oxidation states of the cation that makes up the conduction band become less stable

the more positive oxidation states of the ion whose filled states make up the valence become less stable

It will be seen that differences in the magnitude of the band gaps are extremely important in determining the levels of nonstoichiometry and the resulting electronic conductivities of compounds. It is convenient to use the known chemical properties of the constituent elements as a guide for predicting such behavior.

REFERENCES

Kittel, C. *Introduction to Solid State Physics*, 3rd ed. New York: John Wiley & Sons, 1966.

Kofstad, P. *Nonstoichiometry, Diffusion, and Electrical Conductivity in Binary Metal Oxides*, New York: Wiley-Interscience, 1972, Chapter 4, Fig. 8.

Smith, W. F. *Principles of Materials Science and Engineering*, 2nd ed., New York: McGraw-Hill, 1986, Fig. 3.28.

Extrinsic Electronic Disorder

9

INTRODUCTION

The discussion of extrinsic ionic disorder in Chapter 5 was limited to stoichiometric compositions in which the doped material consists of a solid solution of stoichiometric binary compounds. Under those circumstances electrons and holes arise only by ionization across the band gap; that is, only intrinsic electronic disorder is involved. Up to this point our world of defect chemistry has been limited to the interior of crystalline solids. We have ignored the fact that the crystal is surrounded by some kind of ambient with which it must be in equilibrium if it is to reach a truly defined thermodynamic state. We now consider the result of interactions of doped crystals with their gaseous ambient, which we will assume to contain some activity of the nonmetallic constituent (e.g., oxygen for oxides or chlorine for chlorides). This will introduce the important concept of nonstoichiometry.

The treatment so far has implicitly assumed that the equilibrium state involves only the nonmetal activity at each temperature that is in equilibrium with the stoichiometric composition. In that case, bulk charge neutrality is maintained by the aliovalent impurity centers and the compensating ionic defects. We now deal with the other limiting case, in which interaction with the gaseous atmosphere surrounding the crystal causes the nonmetallic constituent to be either lost or gained until the compensating ionic defect has been eliminated and replaced by electrons or holes of equivalent charge. Chapter 11 analyzes the important transition between these limiting cases in a detailed discussion of nonstoichiometry in doped crystals.

INTERACTIONS WITH THE GASEOUS AMBIENT

An aliovalent dopant compound contains more or less nonmetal per cation than does the component oxide it replaces in the crystal. In the case of purely extrinsic ionic disorder, that relative excess or deficiency of nonmetal is maintained in the crystal and results in the formation of ionic defects. Thus NaCl doped with $CaCl_2$ becomes $Na_{1-2x}Ca_xCl$, and the extra positive charge of the Ca^{2+} is compensated by an equal number of cation vacancies. Nb_2O_5 doped with TiO_2 becomes $Nb_{2-x}Ti_xO_{5-\frac{x}{2}}$, and the lesser charge of the Ti^{4+} is compensated by the presence of half as many oxygen vacancies. In the limiting case of purely extrinsic electronic disorder, the excess nonmetal associated with the dopant compound is expelled from the crystal, leaving behind an equivalent charge of electrons, or the deficiency of nonmetal is filled from the ambient, creating an equivalent charge of holes. This is best demonstrated by some specific examples.

In Chapter 5, we noted two alternative ways of incorporating Nb_2O_5 into TiO_2, Eqs. (5.13) and (5.14), which are repeated here for convenience

$$Nb_2O_5 \xrightarrow{(2TiO_2)} 2Nb_{Ti}^{\cdot} + 4O_O + O_I^{//} \tag{9.1}$$

$$2Nb_2O_5 \xrightarrow{(5TiO_2)} 4Nb_{Ti}^{\cdot} + 10O_O + V_{Ti}^{4/} \tag{9.2}$$

To shift these cases of extrinsic ionic disorder to extrinsic electronic disorder, it is always simplest to start with the case that involves an anion defect: Eq. (9.1) in this case. If the excess interstitial oxygen is expelled from the crystal as neutral oxygen molecules, it leaves behind electrons of equivalent charge:

$$O_I^{//} \to \frac{1}{2}O_2 + V_I + 2e^/ \tag{9.3}$$

The overall incorporation reaction can then be expressed by the sum of Eqs. (9.1) and (9.3):

$$Nb_2O_5 \xrightarrow{(2TiO_2)} 2Nb_{Ti}^{\cdot} + 4O_O + \frac{1}{2}O_2 + 2e^/ \tag{9.4}$$

The material is now nonstoichiometric because it no longer corresponds to the combination of stoichiometric binary constituents. It also contains more electrons than would result from ionization across the band gap. In effect, the oxygen ions dissociate into oxygen atoms, which leave the crystal, and electrons, which are left behind. To maintain bulk charge neutrality, only electrically neutral species can enter or leave the crystal. The same result can be obtained from Eq. (9.2) by the loss of a stoichiometric set of TiO_2 lattice sites

$$V_{Ti}^{4/} + 2O_O \to O_2 + 4e^/ \tag{9.5}$$

Addition of Eqs. (9.2) and (9.5), and division by 2, again gives Eq. (9.4). The advantage of viewing the process as the loss or gain of oxygen from the case involving the anion defect is that there is no need for a loss or gain of lattice sites.

For the case of an acceptor impurity, consider the solution of Al_2O_3 in TiO_2. The acceptor oxide has less oxygen per cation than the binary constituent oxide that it replaces. The two alternative modes of incorporation with the formation of extrinsic ionic disorder are

$$Al_2O_3 \xrightarrow{(2TiO_2)} 2Al_{Ti}^/ + 3O_O + V_O^{\cdot\cdot} \tag{9.6}$$

$$2Al_2O_3 \xrightarrow{\text{(3TiO}_2\text{)}} 3Al'_{Ti} + 6O_O + Al_i^{\cdots} \tag{9.7}$$

We will ignore the possibility that Ti^{4+} may replace Al^{3+} in the interstitial sites because of the much higher concentration of the former. The oxygen vacancy can be eliminated by being filled with oxygen atoms from the surrounding atmosphere

$$V_O^{\cdots} + \tfrac{1}{2}O_2 \rightarrow O_O + 2h^{\cdot} \tag{9.8}$$

In this case, neutral oxygen atoms enter the crystal and become oxygen ions by stealing two electrons from the top of the valence band, the most energetically accessible electrons in the crystal. This leaves two holes in the valence band that replace the positive charge of the lost oxygen vacancies. The incorporation of the acceptor oxide with the formation of compensating electronic disorder can be obtained from the sum of Eqs. (9.6) and (9.8):

$$Al_2O_3 + \tfrac{1}{2}O_2 \xrightarrow{\text{(2TiO}_2\text{)}} 2Al'_{Ti} + 4O_O + 2h^{\cdot} \tag{9.9}$$

The same result could have been obtained from Eq. (9.7) by the addition of a stoichiometric set of TiO_2 lattice sites:

$$Al_i^{\cdots} + O_2 \rightarrow Al'_{Ti} + 2O_O + 4h^{\cdot} \tag{9.10}$$

Addition of Eqs. (9.7) and (9.10), and division by 2, again gives Eq.(9.9).

The addition of a donor oxide with the formation of compensating electrons results from loss of the oxygen that is excess relative to the host oxide. The addition of an acceptor oxide with the formation of compensating holes results from the gain of the deficiency of oxygen relative to the host oxide. In both cases the product is nonstoichiometric because it no longer corresponds to the combination of stoichiometric binary constituents. In the resulting solid solutions, both the cation and anion sublattices have become perfect. These statements can be generalized to any metal compound (e.g., halides, sulfides, etc.).

The reason for designating aliovalent impurities as acceptor or donor impurities should now be obvious. Acceptor impurities give an opportunity for the solid solution to gain nonmetal with creation of the corresponding amount of holes, analogous to the creation of holes by solution of an acceptor dopant, such as boron, into silicon. Donor impurities give the opportunity for the solid solution to lose nonmetal with the creation of the corresponding amount of electrons, just as occurs from the solution of a donor dopant, such as arsenic, into silicon.

It should now be clear that there are three alternative modes of incorporation of aliovalent oxides, involving three optional choices of compensating defect. There may be a cationic defect, an anionic defect, or an electronic defect. Since a donor impurity represents a positively charged center, the choices are between:

cation vacancies

anion interstitials

electrons

For an acceptor impurity, the options are:

cation interstitials

anion vacancies

holes

As always, the system will choose whichever of these options is energetically most favorable. The choice of compensating defect will have obviously important implications for the transport properties of the doped material. In particular, the electrical conductivity may differ by many orders of magnitude, depending on whether ionic or electronic defects are the preferred choice.

THE CHOICE OF COMPENSATING DEFECT

We have already discussed the choice between the two alternative ionic compensating defects and have emphasized the necessary continuity with the most favored type of intrinsic ionic disorder. The issue now is the choice between the most favorable ionic defect and the alternative electronic defect. In NaCl doped with $CaCl_2$, will the positive charge of the donor centers be compensated by cation vacancies or by electrons? In Nb_2O_5 doped with TiO_2, will the negative charge of the acceptor be compensated by oxygen vacancies or by holes? This is another place in which some knowledge of the chemistry of the component elements and its relationship to the electronic states in the solid compound can be extremely helpful.

When Na reacts with Cl_2 to form crystalline NaCl, electrons are given up by the Na atoms to the Cl atoms to form the constituent ions. In the crystalline NaCl, the valence band is made up of the filled anion states, derived from the atomic Cl 3p states, and the conduction band is made up of the emptied cation states, derived from the atomic Na 3s states. Putting an electron in the conduction band of NaCl is akin to changing a Na^+ to a neutral Na, and that is not favorable in the crystalline environment. One can thus perform a simple test: Does putting an electron in the atomic equivalent of a conduction band state yield an achievable oxidation state of that element? Since sodium is always Na^+ in its compounds, the answer in this case is No. Therefore, one does not expect to find electrons in the conduction band of such a compound, especially if there is an energetically feasible alternative. In the case of NaCl, singly charged cation vacancies are quite favorable, as indicated by the modest enthalpy for the formation of Schottky disorder. Another way of looking at this is to consider the relative magnitudes of the intrinsic ionic and electronic types of disorder. With a band gap that is several times larger than the formation enthalpy for Schottky disorder, it is not likely that electronic compensation of aliovalent impurities would be favored over ionic compensation in NaCl. This is slightly hazardous, because each disorder enthalpy combines the energetics of two oppositely charged defects, and one may be much more favorable than its counterpart, but the comparison is usually helpful.

A similar argument can be made for the valence band. Putting a hole in the valence band of NaCl is equivalent to changing a Cl^- to a neutral Cl, and this is also energetically very unfavorable. Again, Cl^- is the only stable ionic state of chlorine, and therefore one does not expect to find holes in a valence band made up of Cl 3p states.

These arguments can be put on a somewhat more quantitative thermodynamic basis. Equation (9.3) represents the reaction to change from compensation by interstitial oxygen to compensation by electrons due to the loss of oxygen from donor-doped TiO_2. It can be assigned a characteristic enthalpy, ΔH_e; small values of ΔH_e favor electronic compensation. Likewise, Eq. (9.8) represents the reaction to change from compensation by oxygen vacancies to compensation by holes due to the addition of oxygen to acceptor-doped TiO_2. It will be assigned the enthalpy ΔH_h. The sum of these two reactions gives

$$V_O^{\cdot\cdot} + O_I^{//} \rightarrow O_O + V_I + 2e' + 2h' \tag{9.11}$$

for which the total enthalpy is $\Delta H_e + \Delta H_h$. But Eq. (9.11) is also the difference between twice the intrinsic electronic ionization reaction, with enthalpy E_g°, and the reaction to create intrinsic anion Frenkel disorder, with enthalpy ΔH_{AF}. Therefore

$$\Delta H_e + \Delta H_h = 2E_g^\circ - \Delta H_{AF} \tag{9.12}$$

For ease of derivation, this result is based on the assumption that anion Frenkel disorder is the preferred type of intrinsic disorder in TiO_2. Experimental results discussed in Chapter 12 suggest that cation Frenkel disorder is the more likely choice. This is one example of a general relationship between the enthalpies of transfer from ionic to electronic compensation and the enthalpies of intrinsic electronic and ionic disorder. For a monovalent anion, one anionic defect would be replaced by a single electronic defect, and the band gap would appear only once. This relationship emphasizes the importance of the band gap in oxides. One might compare NiO and MgO, for example. They have the same crystal structure, and the cations are about the same size. Thus the enthalpies of formation for the preferred Schottky disorder should be similar. However, NiO has a band gap characteristic of transition metal oxides, 3–4 eV (300–400 kJ/mol), while that of MgO is perhaps 7–8 eV (700–800 kJ/mol). Thus the sum of the enthalpies to transfer from ionic to electronic compensation is about 6–10 eV (600–1000 kJ/mol) less for NiO than for MgO. While we cannot separate the enthalpies for the acceptor-doped and donor-doped cases, it is clear that electronic compensation, on the average, is much more favorable in NiO, and this is primarily due to the difference in band gaps. If we make the (unjustified) assumption that the enthalpy difference is evenly divided between the acceptor and donor-doped cases, electronic compensation in NiO is favored over that in MgO by a factor of 10^{12}–10^{20} [obtained from exp $(6-10 \text{ eV})/2kT$] at 1000°C. It should be recalled that in Chapter 8, the magnitude of the band gap was correlated with the availability of adjacent stable oxidation states of the cation. To make a more fair comparison of the type just described for compounds of different kinds, it is advantageous to normalize the treatment to the replacement of single charges. In other words, formulate the relationships on the basis of the enthalpy necessary to replace a single charge of compensating ionic defect with a single electron or hole. Thus for the example just cited, Eqs. (9.3), (9.8), (9.11), and (9.12) can all be divided by 2 to give

$$\Delta H_e^/ + \Delta H_h^\cdot = E_g^\circ - \frac{\Delta H_{AF}}{2} \tag{9.13}$$

where the superscripts on the enthalpy terms indicate that they are normalized to a single charge. For the comparison of NiO with MgO, there is approximately 3–5 eV (300–500 kJ/mol) to divide up between $\Delta H_e^/$ and $\Delta H_h^\cdot$. It is left to the reader to confirm that if cation Frenkel disorder is favored in TiO_2, the right-hand side of Eq. (9.13) is $E_g^\circ - \Delta H_{CF}/4$, where ΔH_{CF} is the enthalpy of formation of a pair of cation interstitials and vacancies. It will be suggested in Chapter 12 that ΔH_{CF} in TiO_2 is about 5 eV (500 kJ/mol). With a band gap of about 3 eV (300 kJ/mol), this means that only about 2 eV (200 kJ/mol) is available to be divided between $\Delta H_e^/$ and $\Delta H_h^\cdot$. Thus electronic compensation of aliovalent impurities should be quite favorable in TiO_2; given the easy reducibility of Ti^{4+}, compensation of donor impurities by electrons should be particularly favored.

The so-called main group elements, those atoms that do not have partially filled d or f shells, almost always have only one stable oxidation state (e.g., the alkali metals, alkaline earth metals, aluminum, etc.). The conduction band is derived from empty cation s and p states, and the placement of an electron in the atomic equivalent of the conduction band does not result in an achievable oxidation state for those elements. In their compounds, electrons are never favored over one of the negatively charged ionic defects. [A notable and peculiar exception to the prevalence of a single oxidation state is the case of the "inert pair" elements that typically have $(n - 1)d^{10}ns^2np^x$ outer electron configurations, where n is usually 5 or 6 and x can be 1, 2, or 3. These elements can have oxidation states that involve all the outer valence electrons (e.g., Tl^{3+}, Pb^{4+}, and Bi^{5+} for $n = 6$, and Sn^{4+} for $n = 5$), but they also have stable states in which the 6s ($n = 6$) or 5s ($n = 5$) electrons do not participate, but act as an "inert pair" (e.g., Tl^+, Pb^{2+}, Bi^{3+}, and Sn^{2+}).]

The anions of oxygen and the halogens also have only a single stable oxidation state. Thus in compounds in which the valence band is derived from the filled outer electronic states of these anions, holes are not energetically favored. This is the case of the oxides and halides of the main group elements described above. Thus in Al_2O_3 doped with MgO, one would expect either oxygen vacancies or cation interstitials to be favored over holes as the compensating defects.

The situation is somewhat more complicated for compounds of the transition metals, in which the d shells are gradually filled. However, the same chemical test is quite useful. This group can be divided into two subgroups. In one, all the valence electrons participate in the bonding, including those in the d shell, and the element has the formal oxidation state corresponding to its group number in the periodic table (e.g., Ti^{4+}, Nb^{5+}, W^{6+}, etc.), and the ions have the electronic structure of the preceding rare gas. In the corresponding oxides, TiO_2, Nb_2O_5, and WO_3, the uppermost filled electronic states are the anion states, and they make up the valence band. As described above, holes are then not favored over an alternative ionic defect. Thus in TiO_2 doped with Al_2O_3, the compensating defects are not holes, but either oxygen vacancies or cation interstitials; experimental evidence favors the latter, as will be seen in Chapter 12. All these cations can be chemically reduced to lower oxidation states, as in Ti_2O_3, TiO, Nb_2O_4, and WO_2, and this corresponds to putting electrons back into the d shells. The emptied d shells of the cations make up the conduction bands in these compounds. Therefore, putting electrons into the conduction bands of the oxides of these elements is quite feasible and may be preferable to the alternative compensating ionic defects. Thus in TiO_2 doped with Nb_2O_5, under suitable conditions (to be described), electrons can be the favored compensating defect for the positively charged donor centers. Compounds in this class are sometimes referred to as reduction-type semiconductors because they contain a reducible cation, and the electronic compensation of donor impurities, which is accomplished by a loss of oxygen or other nonmetal, leads to substantial conductivities.

In the other subgroup of the transition metals, the cations are in their lowest stable oxidation states and the d shells still contain electrons [e.g., Cr^{3+} ($3d^3$), Mn^{2+} ($3d^5$), Fe^{2+} ($3d^6$), and Ni^{2+} ($3d^8$)]. These elements are typically those that have large numbers of d electrons, so that it would be energetically prohibitive to use them all in relatively ionic bonds. As in the other cases already described, the conduction bands in the compounds of these cations are made up of the emptied electronic states, primarily the emptied d states. However, the highest filled states are the remaining occupied d states, and these make up most of the valence band, rather than filled anion states as in the classes of compounds described earlier. The cations cannot be further reduced, so the conduction bands are not receptive to electrons. Thus when NiO is doped with Cr_2O_3 (a situation that corresponds to the oxide scale formed by the oxidation of Ni–Cr alloys), cation vacancies are favored over electrons as the compensating defect. Since, however, all these cations can be oxidized by the loss of electrons (e.g., Mn^{2+} to Mn^{3+} and Fe^{2+} to Fe^{3+}), the occupied d states that make up the valence band are receptive to holes. Thus in NiO doped with Li_2O, holes are the favored compensating defect. Compounds in this class are sometimes referred to as oxidation-type semiconductors, since they contain an oxidizable cation, and the holes that compensate acceptor impurities, as the result of a gain of oxygen or other nonmetal, result in substantial conductivities.

When a metal has two prominent oxidation states, as in the case of Cu, two distinct behaviors can be anticipated. Thus for Cu_2O, in which Cu^+ can be oxidized to Cu^{2+} but cannot be reduced, electrons are not expected to be favored over ionic defects for the compensation of donor impurities such as Zn^{2+}. Note that in this case, acceptor doping is not pertinent because there are no cations with charge less than $+1$, nor are there convenient anions with charge greater than -2. For CuO, in which Cu^{2+} can be reduced to Cu^+, electrons would probably be favored over ionic defects for the compensation of donor impurities such as Cr^{3+}. Cu^{2+} can also be oxidized to Cu^{3+}, albeit with some difficulty. In the high temperature superconductor $YBa_2Cu_3O_7$, holes are the compensating defect at high oxygen activities, and this is an essential feature of its superconducting behavior. Cu^{2+} is an example of a cation that can be both oxidized and reduced, as is also the case for Mn^{3+}, and one expects compensation by electronic defects for both acceptor and donor dopants for compounds of these cations.

In summary, one does not expect electrons in the conduction band to be a major defect species in solid solutions that contain no reducible chemical species, and one does not expect holes in the valence band to be a major defect species in solid solutions that contain no oxidizable chemical species. General expectations for the oxides and halides of the major cations are summarized below in tabular form:

No Electronic Compensation by Electrons or Holes
(Can be neither oxidized nor reduced)

Main group: Alkali metals Li^+, Na^+, K^+, Rb^+, and Cs^+

Alkaline earths: Mg^{2+}, Ca^{2+}, Sr^{2+}, and Ba^{2+}

Others: Al^{3+}, Sc^{3+}, Y^{3+}, La^{3+}, and Si^{4+}

Compensation of Donors by Electrons
(Can be reduced, but not oxidized)

Transition metals: Ti^{4+}, Zr^{4+}, Nb^{5+}, Ta^{5+}, Mo^{6+}, and W^{6+}

Compensation of Acceptors by Holes
(Can be oxidized, but not reduced)

Transition metals: Cr^{3+}, Mn^{2+}, Fe^{2+}, Fe^{3+}, Co^{2+}, and Ni^{2+}

Electronic Compensation of Both Acceptors and Donors
(Can be both oxidized and reduced)

Transition metals: Cu^{2+}, Mn^{3+}, Fe^{3+}

No Compensation of Donors by Electrons
(Cannot be reduced)

Transition metals: Cr^{3+}, Mn^{2+}, Fe^{2+}, Co^{2+}, Ni^{2+}, Cu^+, and Ag^+

No Compensation of Acceptors by Holes
(Cannot be oxidized)

Transition metals: Ti^{4+}, Zr^{4+}, Nb^{5+}, Ta^{5+}, Mo^{6+}, W^{6+}, and Zn^{2+}

THE CHEMICAL CONSEQUENCES OF ELECTRONIC COMPENSATION

Holes are the compensating defects in NiO doped with Li_2O, and this is achieved by the addition of oxygen to the combination of the two stoichiometric oxides:

$$Li_2O + \tfrac{1}{2}O_2 \xrightarrow{\text{(2NiO)}} 2Li'_{Ni} + 2O_O + 2h^{\cdot} \qquad (9.14)$$

The addition of oxygen represents a real oxidation of the system, and the presence of holes in the valence band that is derived from Ni 3d states is the solid state equivalent of oxidizing some of the Ni^{2+} to Ni^{3+}. This is a fairly difficult oxidation, so the latter is quite a strong oxidizing agent. Correspondingly, the solid solution of Li_2O in NiO, with its added oxygen, also has oxidizing power. For example, it will dissolve in hydrochloric acid with the liberation of chlorine

$$2Ni^{3+} + 2Cl^- \rightarrow 2Ni^{2+} + Cl_2 \qquad (9.15)$$

or, more accurately for the solid solution case,

$$2h^{\cdot} + 2Cl^- \rightarrow Cl_2 \qquad (9.16)$$

Analogously, TiO_2 that is doped with Nb_2O_5, with compensation by electrons due to the loss of the crystallographic excess of oxygen, has chemical reducing power similar to that of Ti^{3+}. Loss and gain of nonmetal in these solid solutions are real reductions and oxidations in the chemical sense of the terms. They also represent a real loss or gain in weight relative to the stoichiometric binary constituents.

THE INTERACTIONS OF IMPURITY CENTERS WITH ELECTRONS AND HOLES

In the case of TiO_2 doped with Nb_2O_5, as discussed earlier, the $Nb^{\cdot}_{Ti}$ centers can be compensated by electrons. Since the role of the compensating electronic defect is to balance the abnormal charge of the impurity center, they will always have opposite charges, and there is always the possibility of electrostatic attraction between them. If the electron becomes strongly bound to the donor impurity, or if a hole becomes strongly bound to an acceptor center, neither is available to contribute to the electrical conductivity of the doped material. Thus the question of whether the electronic defects are bound to an impurity center is important in determining the electrical properties of the system. Before attempting to apply similar reasoning to compounds, it is helpful to examine donor and acceptor states in a simple, elemental semiconductor such as Si. Again, the author is indebted to the clear treatment of this subject in Kittel's *Introduction to Solid State Physics* (1966).

Donor and Acceptor States in Elemental Semiconductors

The electronic structure of Si is

Si: $1s^2/2s^22p^6/3s^22p^2$

In solid silicon, the 3s and 3p states hybridize into four equivalent sp^3 orbitals that point toward the corners of a regular tetrahedron, and each contains one electron that is shared with the electron from an equivalent, adjacent Si atom, giving the diamond structure of the crystal, as shown schematically in Fig. 9.1a. The electronic structure of P, the next element after Si, is

P: $1s^2/2s^22p^6/3s^23p^3$

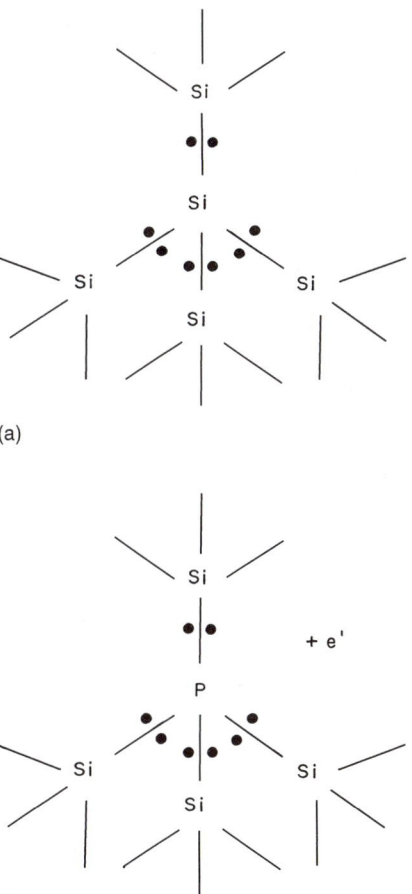

Figure 9.1 (a) The tetrahedral, sp³ bonding configuration in the diamond structure of silicon. (b) The bonding around a substitutional phosphorus in silicon showing the extra electron.

(a)

(b)

If we substitute a neutral P atom for a neutral Si atom, the P atom can duplicate the hybridized outer electronic structure of the Si atom with one electron left over. The local arrangement is shown in Fig. 9.1b. If the extra electron moves away from the P atom, it leaves behind a positively charged center. When remote from that center, the electron must find an allowed electronic state, and the lowest available state will be at the bottom of the conduction band, as shown in Fig. 9.2a. If the electron is attracted to the positively charged P center, and is bound to it, that center must represent a localized lower energy state in the band gap below the conduction band as shown in Fig. 9.2b. The energy difference between the bound state and the bottom of the conduction band is then the ionization energy of the bound state and is the parameter of prime interest. This can be represented by the ionization reaction

$$P_{Si} \rightleftharpoons P_{Si}^{\cdot} + e^{\prime} \tag{9.17}$$

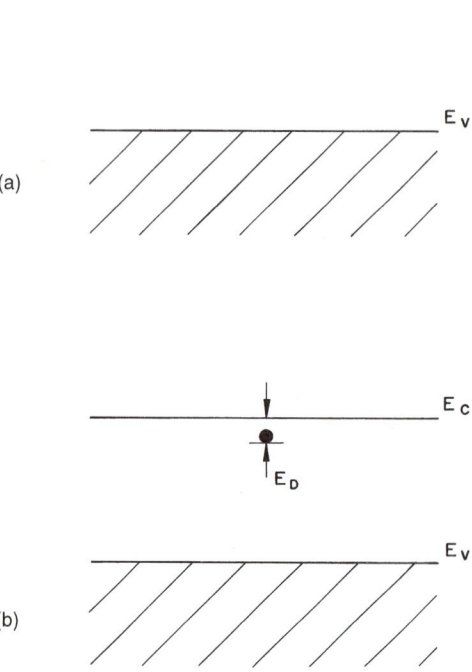

(a)

(b)

Figure 9.2 Primitive band diagrams for phosphorus-doped silicon. (a) Diagram showing the extra electron in the conduction band. (b) Diagram showing the shallow donor level of the impurity center in the band gap.

which has the mass-action expression

$$\frac{[P_{Si}^{\cdot}]n}{[P_{Si}]} = K_D e^{-E_D/kT} \tag{9.18}$$

where E_D is the ionization energy of the donor center, as shown in Fig. 9.2b.

The bound state, with the electron orbiting around the central positive charge of the impurity center, is analogous to a hydrogen atom, and such states are often referred to as hydrogenic states. The major difference is that the electron and proton of the hydrogen atom are separated by completely empty space, while in the case of P-doped Si, if the bound electron is orbiting the central P center at some distance, then the electrostatic attraction is modified by the polarizable material between them. Once again, we can use the dielectric constant of the matrix, Si in this case, as a measure of its polarizability. Since the electron orbits at very high speed, the high frequency dielectric constant is the pertinent parameter. The ionization energy of the hydrogen atom, the energy necessary to remove the electron to an effectively infinite distance from the proton, is 13.6 eV (1310 kJ/mol). To account for the polarizability of the Si lattice, this must be divided by the square of the high frequency dielectric constant of the matrix, $k = 11.7$ for Si, and there is also a correction for the effective mass of the electron. Thus the ionization energy for the donor center in Si is given by

$$E_D \sim \frac{13.6}{k^2} \frac{m^*}{m} \text{ eV} \tag{9.19}$$

where m^* and m are the effective mass and the true rest mass of the electron, respectively. Since the electron is not as tightly bound to the donor center as it is to the proton in the hydrogen atom, its orbital path is much larger. In the hydrogen atom, the radius of the 1s electron orbital is 0.0526 nm. For the donor centers in Si, this is modified by the dielectric constant and effective mass of the electron by the expression

$$r_D \sim 0.0526k \frac{m}{m^*} \text{ nm} \tag{9.20}$$

Approximate values of E_D and r_D for several donor impurities (group V elements) in Si and Ge are given in Table 9.1, and it is seen that the orbitals of the bound electrons are many interatomic distances away from the central positive charge of the donors. For this reason, the specific identity of the donor center is relatively unimportant; it is merely a positive charge. As a result, the ionization energies are similar for all substitutional donors. The large orbitals also justify the use of the bulk dielectric constant to modify the attraction between the oppositely charged species.

It is seen that the ionization energies of donors in Si and Ge are rather small and are close to the average thermal energy available at room temperature, $kT = 0.025$ eV at 300 K. As a result, the donors are essentially all ionized at room temperature, and the free electron concentration is approximately equal to the donor concentration.

The effect of acceptor impurities is similar. Acceptor impurities are those elements that have fewer valence electrons than Si, that is, the group III elements such as B, Al, Ga, and In. Boron has the electronic structure $1s^2/2s^22p^1$, so that when it fits into the Si lattice with sp^3 hybrid orbitals, it is one electron short of completing the bonding, as shown in Fig. 9.3a. If a nearby electron fills this empty bonding state, it leaves a hole that represents a positive defect charge that is attracted to the negative charge of the central acceptor atom, as shown in Fig. 9.3b. If the hole is bound to the impurity center, it orbits the acceptor at some extended distance, as in the case of the electrons and donor centers just described. The bound holes can become mobile by the ionization of an electron from the valence band into the bound hole state, leaving a mobile hole in the valence band. This can be viewed alternatively as the ionization of the bound hole downward into the valence band

TABLE 9.1
Ionization energies and orbital radii for the bound state for donor impurities in Si and Ge

	Ionization energies (eV)			r_D (nm)
	P	As	Sb	
Si	0.045	0.049	0.039	3.0
Ge	0.012	0.013	0.0096	8.0

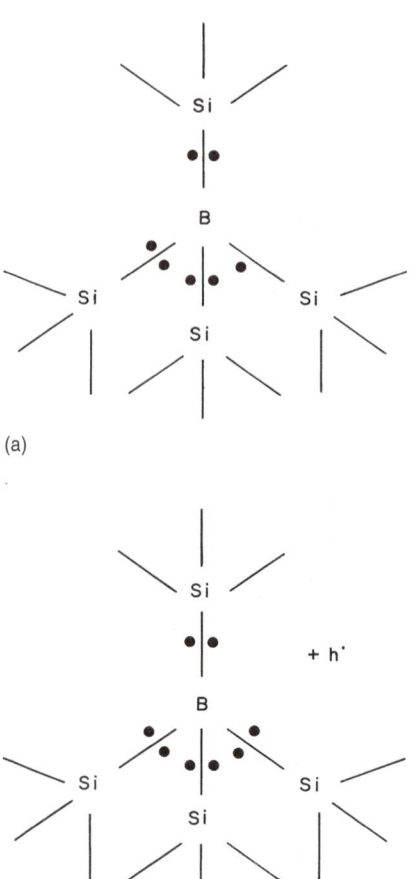

Figure 9.3 (a) The bonding around a substitutional boron atom in silicon, indicating the availability of one less valence electron. (b) Completion of the tetrahedral bonding around the substitutional boron by stealing an electron from the valence band, leaving a hole.

$$B_{Si} \rightleftharpoons B'_{Si} + h^{\cdot} \qquad (9.21)$$

with the mass-action expression

$$\frac{[B'_{Si}]p}{[B_{Si}]} = K_A e^{-E_A/kT} \qquad (9.22)$$

where E_A is the acceptor ionization energy. Typical values are shown in Table 9.2, and are seen to be similar in magnitude to those of the donor impurities. Schematic band diagrams for the bound and ionized states are shown in Fig. 9.4.

Carrier Concentrations and Conductivities

The mass-action treatment of electronic disorder is closely analogous to that of ionic disorder (e.g., cation Frenkel disorder in AgBr). Thus a plot of the electron and hole

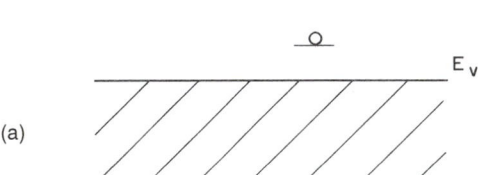

(a)

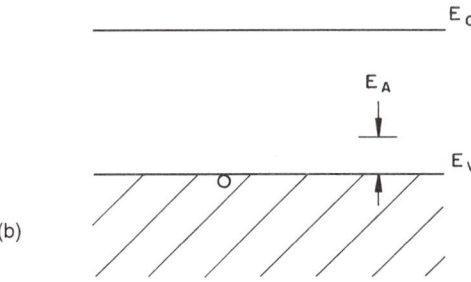

(b)

Figure 9.4 Primitive band diagrams for acceptor-doped Si. (a) Diagram showing the hole trapped in the shallow acceptor level in the band gap. (b) Diagram showing the ionization of the trapped hole downward into the valence band (or elevation of an electron from the valence band into the trapped hole state).

TABLE 9.2
Ionization energies (eV) of acceptor impurities in Si and Ge

	B	Al	Ga	In
Si	0.045	0.057	0.065	0.16
Ge	0.010	0.010	0.011	0.011

concentrations as a function of donor concentration in Si should look just like that of the cation vacancies and interstitials in cadmium-doped AgBr, Fig. 7.11. An Arrhenius plot of the concentrations should be similar to Fig. 7.14, with its regions of intrinsic and extrinsic disorder and of complex formation. A schematic version of such a plot for electrons and holes in a donor-doped material is shown in Fig. 9.5. Some of the terminology is traditionally different from that used for ionic defects. Thus the lowest temperature region, where most of the electrons are bound to the donor centers, is often referred to as the ionization region. The condition of charge neutrality and the carrier concentrations are determined by the mass-action expression for the ionization reaction, Eq. (9.18), where D is a generalized donor center:

$$n \approx [D^{\cdot}] \tag{9.23}$$

$$n \approx \left(K_D[D]_T\right)^{1/2} e^{-E_D/2kT} \tag{9.24}$$

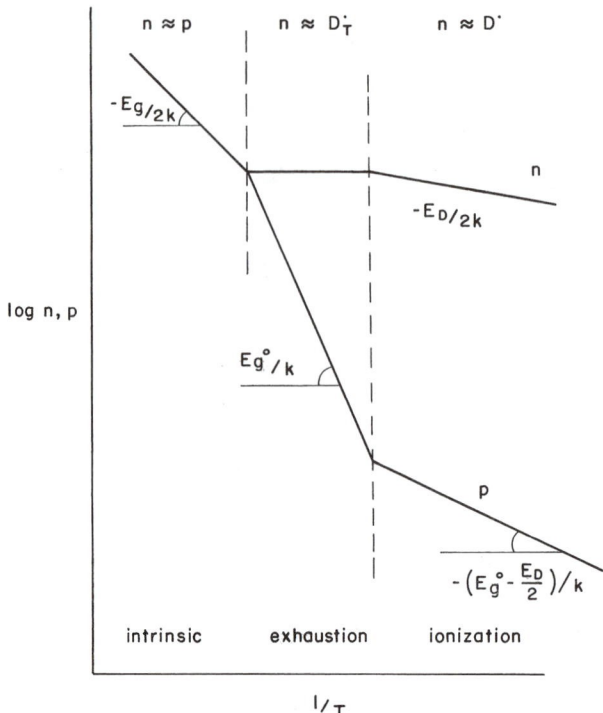

Figure 9.5   A schematic Arrhenius plot for the electron and hole concentrations in donor-doped silicon showing the ionization, exhaustion, and intrinsic regions. The slopes are indicated in terms of the band gap and the donor ionization enthalpy. $[D^{\cdot}]_T$ denotes the total donor concentration.

$$p \approx \frac{K_i}{\left(K_D[D]_T\right)^{1/2}} e^{-\left(E_g^{\circ} - \frac{E_D}{2}\right)/kT} \tag{9.25}$$

where the mass-action expression for intrinsic electronic disorder is given by:

$$np \approx K_i e^{-E_g^{\circ}/kT} \tag{9.26}$$

As the temperature increases, the electron concentration reaches a plateau at the donor concentration; this is called the exhaustion region, because the source of electrons, the neutral donor centers, has been depleted or exhausted. The carrier concentrations are given by

$$n \approx [D^{\cdot}]_T \tag{9.27}$$

$$p \approx \frac{K_i}{[D^{\cdot}]_T} e^{-E_g^{\circ}/kT} \tag{9.28}$$

where the subscript T, for total, indicates that all the donor centers are in the charged, ionized state. Finally, at even higher temperatures, intrinsic behavior is reached and the carrier concentrations are

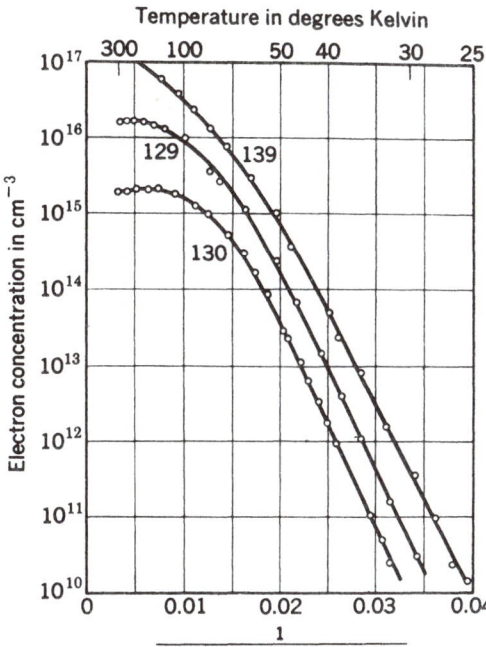

Temperature in degrees Kelvin

Figure 9.6 Arrhenius plot of the electron concentration in silicon containing three different concentrations of arsenic as a donor-dopant. The two lower curves show both the ionization and exhaustion regions. (Reproduced from Kittel, 1966, by permission of John Wiley & Sons.)

$$\mathrm{n} \approx \mathrm{p} \approx K_i^{1/2} e^{-E_g^\circ/2kT} \tag{9.29}$$

Semiconductor devices are usually operated in the exhaustion region, where the carrier concentration is fixed by the dopant level and is temperature independent.

Where is the Fermi level in these various regions? In the ionization region, near 0 K, the donor levels are nominally all filled and the conduction band is nominally empty, so the Fermi level must be halfway between the donor levels and the conduction band. In the intrinsic region, where the impurity effects are negligible, it must drop back to the midgap position.

Figure 9.6 shows an Arrhenius plot of the electron concentration in silicon doped with three different levels of arsenic, a donor dopant. At the lowest temperatures, the electron concentration is increasing with increasing temperature as the electrons are ionized from their bound state with the donor centers. The two lesser-doped samples have reached the exhaustion region just below room temperature. None of the samples has reached the intrinsic region at the maximum temperatures. Log–log plots of the electron mobilities for the same samples are shown in Fig. 9.7. In the high temperature region, increased scattering of the electrons by lattice vibrations is causing the mobilities to decrease with increasing temperature approximately as $T^{-3/2}$. At low temperatures, this effect has decreased to the point that scattering by impurities becomes dominant, and the mobilities are relatively temperature insensitive.

Figure 9.8 shows an Arrhenius plot of the resistivity of germanium doped with three different concentrations of gallium, an acceptor impurity. All three samples

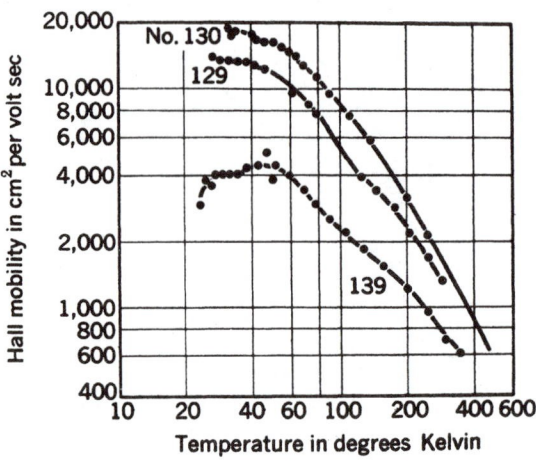

Figure 9.7 Log–log plot of electron mobility as a function of temperature for the three samples shown in Fig. 9.6. The slopes at the higher temperatures indicate a temperature dependence of approximately $T^{-3/2}$; at the lower temperatures, the mobilities are primarily determined by impurity scattering. (Reproduced from Kittel, 1966, by permission of John Wiley & Sons.)

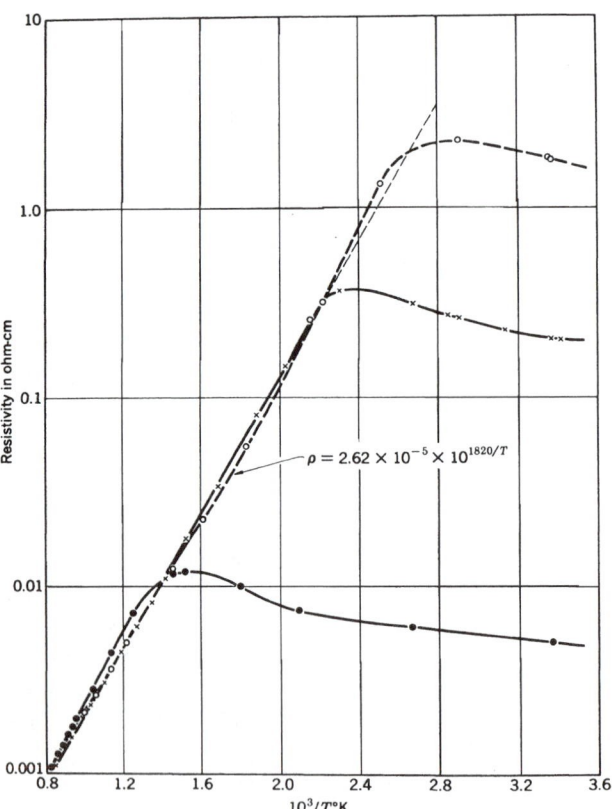

$$\rho = 2.62 \times 10^{-5} \times 10^{1820/T}$$

Figure 9.8 Arrhenius plot of the resistivity of germanium doped with three different levels of gallium as an acceptor impurity. The two lower curves show ionization, exhaustion, and intrinsic regions. The intrinsic slope corresponds to a band gap of 0.72 eV (69 kJ/mol). (Reproduced from Kittel, 1966, by permission of John Wiley & Sons.)

show the same intrinsic behavior at the highest temperatures, and the slope gives a band gap of 0.72 eV (69 kJ/mol). On cooling, this is followed by the exhaustion region in which the resistivity decreases with decreasing temperature (conductivity increases with decreasing temperature). In this region, the carrier concentration is independent of temperature, but the mobility increases with decreasing temperature because of the decreased scattering by lattice vibrations. At the lowest temperatures, the resistivity becomes less temperature dependent, because the gradual binding of the holes to the acceptor centers with decreasing temperature counters some of the temperature dependence of the mobility.

Compensation

So far we have assumed that the semiconductor contains only one type of dopant, either donor or acceptor. In practice, the level of deliberately added dopants is usually large enough to allow us to ignore trace amounts of naturally occurring impurities. However, in the presence of significant amounts of both donor and acceptor impurities, compensation will occur, and the behavior will be determined by the net excess of whichever type is present in the larger amount. It is not possible to maintain electrons in the conduction band, or trapped on donor centers, in the presence of a large concentration of holes in the valence band, or holes trapped on acceptor centers, at lower energy levels. A hole is an empty electron state, and it will be filled by

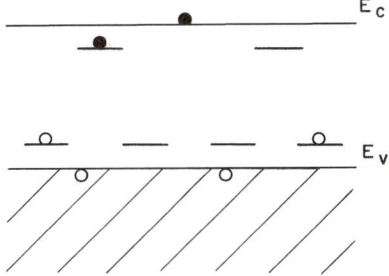

Figure 9.9 Band diagrams for a semiconductor containing twice as many acceptor centers as donor centers. (a) Primitive diagram showing the semiconductor without recombination of electrons and holes (a nonequilibrium situation). (b) After recombination of electrons and holes, resulting in partial compensation.

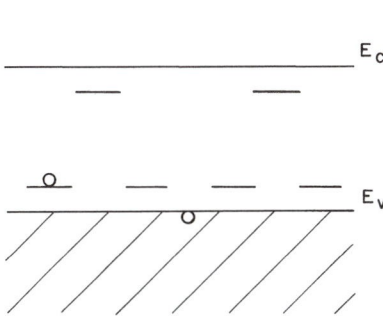

electrons from any higher energy level. Thus the situation shown in Fig. 9.9a is unstable and will decay to that shown in Fig. 9.9b. Note that while electrons tend to drop down the energy scale, holes "drop up," because a hole increasing in energy is equivalent to an electron decreasing in energy.

THE SITUATION FOR COMPOUNDS

The interaction of electrons and holes with charged impurity centers in compounds is very similar to that just described for the elemental semiconductors. Since the impurity centers and the compensating electronic defects must have opposite charges to maintain charge neutrality, they have the potential to attract each other so that the electronic defects may be immobilized. Once again, chemical properties can be used to predict the extent of the binding due to electrostatic attraction.

In MgO, for example, both the cations and the anions have stable rare gas electronic structures. The valence band is made up primarily of filled O 2p states and the conduction band is made up of empty Mg 3s states. The addition of electrons to the conduction band is equivalent to forming Mg^+ ions, while the addition of holes to the valence band is related to the formation of O^- ions, and neither is a stable oxidation state. Thus if MgO could be donor-doped with compensation by electrons, for example, via

$$D_2O_3 \xrightarrow{\text{(2MgO)}} 2D^{\cdot}_{Mg} + 2O_O + \tfrac{1}{2}O_2 + 2e' \qquad (9.30)$$

the conduction band would not be receptive to these electrons, and it would take quite a lot of energy to put them there. Then where will they go? They must be in states far below the conduction band, where they are bound to the donor centers:

$$D^{\cdot}_{Mg} + e' \rightleftharpoons D^{x}_{Mg} \qquad (9.31)$$

where the relative lengths of the equilibrium arrows indicate that this reaction proceeds strongly toward the right. The situation is then as shown in Fig. 9.10a. The donor centers create electron energy levels far down in the band gap, and thermal energy is generally insufficient to excite them up to the conduction band. Of course, we have already seen that in the absence of a reducible cation, there is little likelihood that electrons will be favored over an ionic defect as a compensating defect in the first place. So the chemical test has double power. In this case, since Mg^{2+} is not reducible, ionic defects will be favored as the compensating defects for donor centers, and, even if electrons were present, they would be frozen out by strong attraction to the donors. Donor-doped MgO is then expected to be an electronic insulator, with perhaps a trace of ionic conduction.

The same argument holds for acceptor doping of MgO. If compensation were by holes:

$$A_2O + \tfrac{1}{2}O_2 \xrightarrow{\text{(2MgO)}} 2A'_{Mg} + 2O_O + 2h^{\cdot} \qquad (9.32)$$

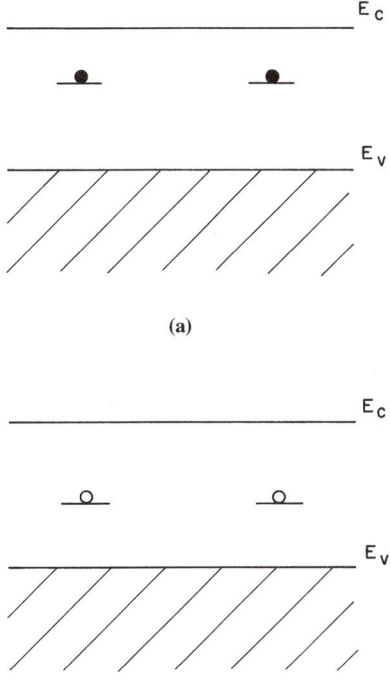

E_c

E_v

(a)

E_c

E_v

(b)

Figure 9.10 Primitive band diagrams for doped MgO: (a) donor-doped with electron compensation, showing the electrons trapped in deep donor levels, and (b) acceptor-doped with hole compensation showing the holes trapped in deep acceptor levels.

the presence of these holes in the valence band would be the equivalent of forming O^- ions, an unstable species. Therefore, the holes must be at energy levels considerably higher than the valence band, where they are trapped by the acceptor centers:

$$A'_{Mg} + h^{\cdot} \rightleftharpoons A^x_{mg} \qquad (9.33)$$

Again, it would take a large amount of energy to excite the trapped holes down into the valence band, and acceptor-doped MgO is expected to be an electronic insulator with perhaps a trace of ionic conduction. This situation is depicted in Fig. 9.10b. These expectations are in accord with the observed properties of these materials. Once again, the chemical test gives two votes against hole conduction in acceptor-doped MgO. Since the valence band states do not represent an oxidizable chemical species, ionic compensation of the acceptor centers is favored over compensation by holes. Even if there were a significant concentration of holes, they would be tightly bound to the acceptor centers.

Let us consider donor-doped TiO_2 as another example. In this material the conduction band is made up of empty Ti 3d states, and since Ti^{4+} is reduced fairly easily to Ti^{3+}, this band should be receptive to electrons. Indeed, in donor-doped TiO_2 electrons are the major compensating defect for the impurity centers under most conditions:

$$D_2O_5 \xrightarrow{(2TiO_2)} 2D^{\cdot}_{Ti} + 4O_O + \frac{1}{2}O_2 + 2e' \qquad (9.34)$$

E_c

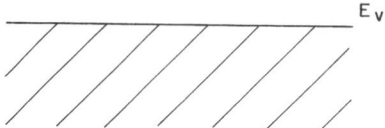
E_v

Figure 9.11 Primitive band diagram for donor-doped TiO$_2$ with electron compensation showing the electrons ionized into the conduction band from the shallow donor levels.

Because of the presence of a reducible cation, the conduction band is at a lower level than when the cation has the rare gas electronic structure, and it moves down closer to the donor levels in the band gap. Since the conduction band is receptive to electrons, it must not require much energy to place them there, and this is consistent with the donor trapping levels being quite close to the conduction band. In other words, these levels are shallow, and even at room temperature there is enough thermal energy to excite the electrons into the conduction band. This situation is depicted in Fig. 9.11. In this case the trapping reaction moves strongly toward the left:

$$D_{Ti}^{\cdot} + e' \rightleftharpoons D_{Ti}^{x} \tag{9.35}$$

The chemical test again gives double evidence. Because of the reducible cation, electrons are favored over ionic defects as compensating defects, and they are not strongly trapped. As expected, donor-doped TiO$_2$ is a dark-colored, electronic conductor (under appropriate equilibration conditions).

The situation for acceptor-doped TiO$_2$ is unchanged from that of acceptor-doped MgO, since, in both cases, the valence band is made up of filled O 2p states and is not receptive to holes. Thus ionic compensation is favored over hole compensation, and any holes that might be present will be strongly trapped by the acceptor centers, whose energy levels thus must lie far above the valence band. This is similar to the situation shown in Fig. 9.10b. Acceptor-doped TiO$_2$ is then expected to be an electronic insulator, as is indeed observed.

Thus we see that the electrical properties of TiO$_2$ are strongly asymmetric with regard to doping. Donor-doped TiO$_2$ is a black semiconductor, while the acceptor-doped material is a white or transparent insulator. This set of properties is extremely important for electrical applications of TiO$_2$ and the titanates. Since the most abundant naturally occurring metallic impurities have smaller ionic charges than Ti^{4+} (e.g., Fe^{3+}, Mg^{2+}, Al^{3+}, etc.), TiO$_2$ is almost always furnished to us by nature already doped with a net acceptor excess. Elements that would either act as acceptor impurities, be a neutral dopant, or have no solubility in TiO$_2$ make up about 99.5% of the earth's crust. Potential donor dopants (e.g., Nb^{5+} or W^{6+}), are very scarce indeed. Thus we consider TiO$_2$ and most titanates to be insulating materials and design their applications accordingly.

Ti^{4+} belongs to the family of transition metal cations that have the stable rare gas electronic structure and are thus not oxidizable. Let us now look at those transition metal cations that retain electrons in their d shells and are thus oxidizable (e.g., Mn^{2+}, Fe^{2+}, Co^{2+}, and Ni^{2+}). These are the lowest achievable oxidation states for these cations, and they are therefore not reducible. We will select NiO as a specific example with behavior typical of this group. If NiO could be donor-doped with compensation by electrons, the incorporation reaction would be:

$$D_2O_3 \xrightarrow{(2NiO)} 2D_{Ni}^{\cdot} + 2O_O + {}^{1}\!/{}_{2}O_2 + 2e' \tag{9.36}$$

However, the conduction band is made up of empty Ni 3d states, and, since the cation is not reducible, it will not be receptive to electrons. As a result, the conduction band remains at a high energy level and is substantially above the donor states, which thus act as effective trapping levels for the electrons, similar to the situation shown in Fig. 9.10a. The trapping reaction is:

$$D_{Ni}^{\cdot} + e' \rightleftharpoons D_{Ni}^{x} \tag{9.37}$$

and the equilibrium lies far to the right. In the absence of a reducible chemical species, it is unlikely that electrons would be preferred over cation vacancies in the first place, and in addition, any electrons that might be in the system would be effectively trapped by the deep donor levels. As expected, donor-doped NiO is an insulator.

For the incorporation of an acceptor impurity with compensation by holes, the reaction is:

$$A_2O + {}^{1}\!/{}_{2}O_2 \xrightarrow{(2NiO)} 2A_{Ni}' + 2O_O + 2h^{\cdot} \tag{9.38}$$

In this family of cations, the valence band is made up of the filled Ni 3d states. The cations can be oxidized to higher oxidation states with varying degrees of ease: Fe^{2+} extremely easily, Co^{2+} not quite so easily, and Ni^{2+} with even less ease. This means that the valence band is receptive to holes and it therefore lies above the O 2p levels, thereby reducing the band gap, and moving closer to the acceptor levels. Thus the acceptor levels are shallow and are easily ionized, as shown in Fig. 9.12. The equilibrium for the trapping reaction

$$A_{Ni}' + h^{\cdot} \rightleftharpoons A_{Ni}^{x} \tag{9.39}$$

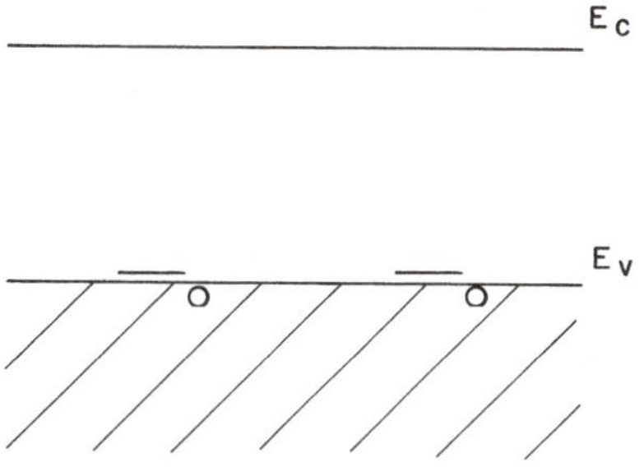

Figure 9.12 Primitive band diagram for acceptor-doped NiO with hole compensation showing the holes ionized into the valence band from shallow acceptor levels.

lies far to the left. As a result, compensation of acceptor dopants by holes is favored, and the holes remain free even at room temperature. As expected, acceptor-doped NiO is a black semiconductor. Li_2O-doped NiO is a well-known example of this behavior.

As in the case of TiO_2, with a reducible cation, we again see a strong asymmetry in the electrical properties of NiO, which has an oxidizable cation, but in the reverse sense. In the case of NiO, the acceptor-doped material is semiconducting, while the donor-doped material is not.

It should now be apparent that for compounds in which the cation can be both oxidized and reduced (e.g., Fe^{3+} or Mn^{3+}), the band gap will be quite narrow, and both donor and acceptor states will be shallow. Compensation by electrons and holes will be favored over ionic defects, and these electronic carriers will tend not to be trapped, so that both donor-doped and acceptor-doped materials will be conductors. Schematic band diagrams for these four classes of material are compared in Fig. 9.13, which shows the relative widths of the band gaps and the relative depths of the donor and acceptor levels.

This correlation of chemical behavior with the relative positions of the conduction and valence bands, and the depths of donor and acceptor levels, is extremely simplistic and probably will cause major cases of indigestion in band theorists. However, it has one strong point in its favor: IT WORKS!

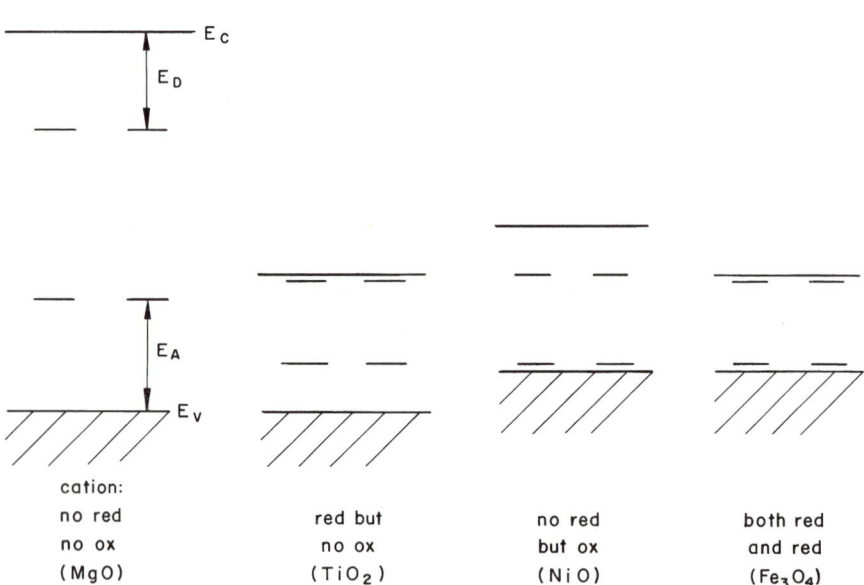

Figure 9.13 Very approximate schematic showing the relative positions of the valence and conduction bands and the depths of donor and acceptor levels for various oxides based on the relative ease of reduction (red) or oxidation (ox) of the cation.

SUMMARY

We have now dealt with the two limiting cases of charge compensation in doped oxides: compensation by ionic defects when the solid solution remains stoichiometric and retains the same oxygen content as the binary constituents, and compensation by electronic defects when the solid solution has had its ideal crystal lattice restored by the gain or loss of oxygen from the ambient. We are now prepared to study the intermediate ranges of nonstoichiometry as a system shifts from one limiting case to the other as a function of oxygen activity.

REFERENCES

Kittel, C. *Introduction to Solid State Physics*, 3rd ed. New York: John Wiley & Sons, 1966.

Mehta, A., and D. M. Smyth. Defect model for nonstoichiometry in $YBa_2Cu_3O_{6+y}$. *Phys. Rev. B* 51:15382–15387, 1995.

Intrinsic Nonstoichiometry

10

INTRODUCTION

The concept of nonstoichiometry was introduced in Chapter 9 as the limiting case of compensation of aliovalent impurities by electronic defects, electrons in the case of donor impurities and holes in the case of acceptor impurities. Actually there is a gradual transition between the limiting cases of electronic and ionic compensation, and this entire range of mixed compensation corresponds to nonstoichiometric compositions. As soon as the solid solution of the host compound and the aliovalent impurity loses or gains a significant amount of nonmetal in order to achieve some amount of electronic compensation, the material is nonstoichiometric. We can now bring together all the concepts of defect chemistry covered thus far into a systematic treatment of the concentrations of various defects as functions of the temperature and nonmetal activity of equilibration, and of the impurity content. But the presence of aliovalent impurities is not necessary to obtain nonstoichiometric compositions; pure compounds can also exhibit nonstoichiometry. After all, if we can dissolve Cr_2O_3 into NiO, we certainly should be able to dissolve Ni_2O_3 into NiO. The resulting solid solution can be generalized as $Ni^{2+}_{1-x}Ni^{3+}_xO_{1+\frac{x}{2}}$, and it is nonstoichiometric for $x > 0$ up to the solubility limit for trivalent Ni. This chapter deals with the phenomenon of nonstoichiometry in pure compounds, while Chapter 11 will add the effects of aliovalent dopants.

Nonstoichiometry was a very controversial subject in the nineteenth century. Those who accepted the universality of Dalton's laws emphasized the stoichiometric nature of simple halides, oxides, and sulfides, and of gaseous species such as CO and CO_2. Others noted the complex atomic ratios in such compounds as the paraffin hydrocarbons, [CH_4 (methane), CH_3 (C_2H_6, ethane), $CH_{2.67}$ (C_3H_8, propane), $CH_{2.50}$ (C_4H_{10}, butane), $CH_{2.40}$ (C_5H_{12}, pentane), etc.]. We now understand that simple valence rules do not apply to these highly covalent materials. Further complicating matters was the later discovery of such anomalies as the tungsten bronzes, Na_xWO_3 with $0 < x < 1$. Two seemingly irreconcilable schools of thought developed, but the situation is now well understood, and both schools are in full accordance with our basic concept of the atomic structure of matter.

Much of the early work on the combining ratios of elements was carried out on gases because they are easy to work with in a quantitative manner. For example, exactly two volumes of hydrogen react with one volume of oxygen, at the same temperature and pressure, to form two volumes of gaseous water. Also, there are two common gaseous oxides of carbon, CO and CO_2, and they differ by exactly one oxygen per carbon. In these cases, the compounds are all discrete molecular species, made up of only a few atoms, and their compositions can differ only by integral numbers of atoms. We can have CO or CO_2, but no molecular species of intermediate composition. Stoichiometry is quite clear, and is inviolate. However, consider a perfect, stoichiometric crystal of NiO that contains exactly 10^{21} cations and 10^{21} anions. The entire crystal is now the unitary structural unit, not a small molecule of two or three atoms. The composition of this crystal can also be changed by the addition of a single oxygen atom, giving a composition of 10^{21} cations and $10^{21} + 1$ anions. This infinitesimal change in composition will not be enough to affect any observable properties. However, if we add 10^{18} oxygen atoms, the composition becomes $Ni(10^{21})O(1.001 \times 10^{21})$, or $NiO_{1.001}$. This is, in fact, the approximate composition of NiO that has been heated to high temperatures in air. It is black and semiconducting, whereas the stoichiometric NiO is light green and insulating. In a chemical sense, it can be viewed as a solid solution of Ni_2O_3 in NiO as described above, in this case with $x = 0.002$. Much of the argument between the competing advocates of stoichiometry and nonstoichiometry can be resolved by comparing the relative sizes of the chemical ensemble, that is, small discrete molecules versus large crystals. The consequences of changing the compositions of the two extreme cases by the addition or subtraction of individual atoms is enormously different for both composition and properties.

It is possible to predict the direction and magnitude of nonstoichiometry in crystalline compounds based on the chemical properties of the constituent atoms. This capability is directly related to the prediction of the nature of the compensating defect for aliovalent impurities described in the previous chapter. Thus for NiO, it is possible to oxidize Ni^{2+} to Ni^{3+} or even to Ni^{4+}; such higher oxides of Ni are the basis of the Ni–Cd and Ni–Fe batteries. However, there are no attainable lower oxidation states for Ni. Therefore, one expects nonstoichiometry in NiO to lie in the direction of excess oxygen, which corresponds to partial oxidation of the Ni^{2+} content toward Ni^{3+}. The amount of nonstoichiometry is modest, because rather strong oxidizing conditions are required to oxidize Ni^{2+}. Co^{2+} is more easily oxidized, and the composition of CoO heated in air is approximately $CoO_{1.01}$ (i.e., about 1% excess oxygen, 10 times that of NiO treated under the same conditions). The amount of nonstoichiometry can have profound effects on the properties of a given compound, especially the electrical conductivity. Thus within the same phase, only a minute variation in the cation/anion ratio may cause a compound to change from being an excellent insulator to being a highly semiconducting material.

NONSTOICHIOMETRY IN PURE CRYSTALLINE COMPOUNDS

Nonstoichiometric compositions are obtained by the partial oxidation or reduction of a stoichiometric compound without a change of crystal structure. This means that there is a finite width to the phase field of that structure in terms of the ratio of metal to nonmetal. In principle, this is true of all compounds; it is required by thermodynamic equilibrium, in that the nonmetal activity in the compound must be equal to that in the ambient. In practice, the resulting variation in composition may be so small that it is experimentally unobservable (e.g., in MgO and other compounds containing only single-valent ions). At the other extreme, in a compound that contains an easily oxidized cation, such as FeO, there can be as much as 15% excess oxygen.

Since nonstoichiometry represents a variable composition within a given crystal structure, it must be accommodated by defects, in fact by a combination of ionic and electronic defects. When the solubility limit of these defects is reached, further change in the nonmetal activity in the same direction will result in the separation of an additional phase; that is, the phase boundary for the nonstoichiometric compound will have been reached. Thus oxidation of FeO will ultimately result in the separation of a second phase of Fe_3O_4, while excessive reduction will result in the formation of metallic Fe.

It is convenient to describe nonstoichiometry by a "thought experiment" that involves a process that would be completely impractical in the laboratory. Thermodynamics allows us to indulge ourselves in such a procedure, since it only cares about the initial and final states. How we get from one to the other is immaterial. Consider an ideally pure, stoichiometric crystal of NiO. Carefully remove a small corner from the crystal and heat it in a high oxygen activity such that it is oxidized to Ni_2O_3:

$$2Ni_{Ni} + 2O_O + \tfrac{1}{2}O_2 \rightleftharpoons Ni_2O_3 \tag{10.1}$$

Now dissolve the small fragment of Ni_2O_3 into the remaining crystal of NiO; there are two alternative modes of incorporation:

$$Ni_2O_3 \xrightarrow{(2NiO)} 2Ni_{Ni}^{\cdot} + 2O_O + O_I^{//} \tag{10.2}$$

$$Ni_2O_3 \xrightarrow{(3NiO)} 2Ni_{Ni}^{\cdot} + V_{Ni}^{//} + 3O_O \tag{10.3}$$

Since NiO has the NaCl structure, interstitial oxygen is unfavorable, and excess oxygen is accommodated by cation vacancies as indicated in Eq. (10.3). $Ni_{Ni}^{\cdot}$ represents trivalent Ni on a site normally occupied by divalent Ni and is formally equivalent to Ni^{2+} plus a hole:

$$Ni_{Ni}^{\cdot} \rightleftharpoons Ni_{Ni}^{x} + h^{\cdot} \tag{10.4}$$

For the time being, we shall assume that this dissociation reaction is complete and that all cations in the crystal have the same formal oxidation state. The net result of the process just described is obtained by the addition of Eqs. (10.1) and (10.3) to twice Eq. (10.4):

$$\tfrac{1}{2}O_2 \rightleftharpoons O_O + V_{Ni}^{//} + 2h^{\cdot} \tag{10.5}$$

Thus the total process amounts to the addition of a neutral oxygen atom to the NiO lattice, creating a new pair of lattice sites, of which the cation site remains vacant.

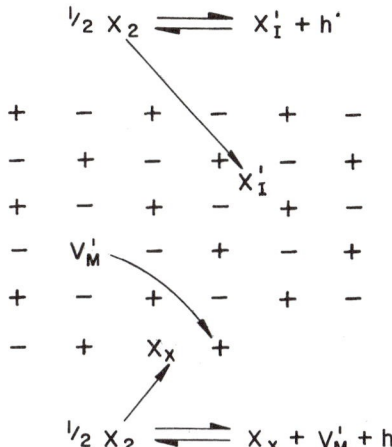

Figure 10.1 Schematic diagram illustrating the two optional modes for adding excess nonmetal to a lattice, forming either interstitial anions or cation vacancies.

The oxygen atom accepts two electrons from the top of the valence band to become an oxygen ion, leaving two holes. Equation (10.5) then is the equilibrium reaction that describes the equilibration of NiO with oxygen activities that lie above that in equilibrium with the stoichiometric composition and that result in oxygen-excess compositions. This is, in fact, the mode of incorporation of excess oxygen into MnO, FeO, CoO, and NiO, all of which have the NaCl structure and contain an oxidizable cation. Note that excess oxygen can be accommodated as a lattice excess of oxygen, Eq. (10.2), or as a lattice deficiency of the cation, Eq. (10.3). The latter is the choice of compounds having the NaCl structure, and their compositions are better represented as $M_{1-x}O$ than as MO_{1+x}. The optional modes of incorporation of excess anion are shown schematically in Fig. 10.1. Note that in oxides having the fluorite structure, with the availability of large, unoccupied octahedral sites surrounded by cations, excess nonmetal can be expected to occupy these interstitial sites, and the incorporation reaction is represented by:

$$\tfrac{1}{2}O_2 + V_I \rightleftharpoons O_I'' + 2h^\cdot \tag{10.6}$$

If the amount of excess nonmetal is small, the empty interstitial site is often not included in the equation explicitly, but then site balance is not clearly indicated.

The same kind of "thought experiment" can be carried out for the loss of nonmetal. Thus a small fragment taken from an ideally pure, stoichiometric crystal of Nb_2O_5 can be heated in a reducing atmosphere to form NbO_2:

$$2Nb_{Nb} + 5O_O \rightleftharpoons 2NbO_2 + \tfrac{1}{2}O_2 \tag{10.7}$$

The resulting NbO_2 can then be dissolved back into the remaining Nb_2O_5 crystal according to one of the following options:

$$2NbO_2 \xrightarrow{(Nb_2O_5)} 2Nb_{Nb}' + 4O_O + V_O^{\cdot\cdot} \tag{10.8}$$

$$5NbO_2 \xrightarrow{(2Nb_2O_5)} 4Nb'_{Nb} + Nb_I^{4\cdot} + 10O_O \tag{10.9}$$

It will be assumed that all the Nb^{4+} ions dissociate into the normal lattice constituent Nb^{5+} plus electrons. Oxygen vacancies have proved to be favorable defects in many oxides, and this appears to be the case for Nb_2O_5, that is, Eq. (10.8). The net result of the overall process can then be obtained by adding Eqs. (10.7) and (10.8) and changing each Nb^{4+} into $Nb^{5+} + e'$:

$$O_O \rightleftharpoons \tfrac{1}{2}O_2 + V_O^{\cdot\cdot} + 2e' \tag{10.10}$$

An oxygen ion in the lattice dissociates into a neutral oxygen atom, which leaves the crystal, and two electrons, which are left behind. Equation (10.10) will become very familiar because it describes the reduction reaction for many transition metal oxides that contain reducible cations.

In the case of the rutile structure of TiO_2, which also contains a reducible cation, only half the octahedral sites are filled by cations in an approximately hcp lattice. Thus the option of interstitial cations, rather than oxygen vacancies, cannot be ignored. In fact, these two options seem to be energetically similar for TiO_2, and both cases have been reported. Since, however, the best evidence seems to favor cation interstitials, in this case the net reduction reaction should be written:

$$Ti_{Ti} + 2O_O \rightleftharpoons O_2 + Ti_I^{4\cdot} + 4e' \tag{10.11}$$

The two optional modes for accommodating a loss of nonmetal through reduction are shown schematically in Fig. 10.2.

For both oxidation and reduction, the two optional equilibrium reactions differ by the way in which lattice site ratios are conserved. For oxidation, either lattice sites remain unchanged and nonmetal interstitials are formed, or new sets of lattice sites are created in stoichiometric ratio and cation vacancies appear. For reduction, either lattice sites remain unchanged and nonmetal vacancies are the result, or stoichiometric sets

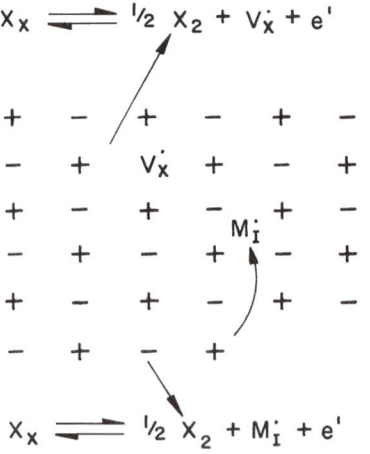

$$X_X \rightleftharpoons \tfrac{1}{2}X_2 + V_X^{\cdot} + e'$$

$$X_X \rightleftharpoons \tfrac{1}{2}X_2 + M_I^{\cdot} + e'$$

Figure 10.2 Schematic diagram illustrating the two optional modes for removing nonmetal from a lattice, leaving either anion vacancies or interstitial cations.

of lattice sites disappear and cation interstitials are formed. Electrons are invariably the product of reduction and a deficiency of nonmetal, while holes always result from oxidation and an excess of nonmetal.

NONSTOICHIOMETRY AND EQUILIBRIUM DEFECT CONCENTRATIONS

If a sufficient number of the pertinent mass-action constants, their enthalpies, and the expression for charge neutrality are known, the concentrations of the various defects can be calculated as a function of temperature and nonmetal activity. Just as it is convenient to represent phase relationships visually in the form of phase diagrams, it is convenient to have schematic representations of defect concentrations as a function of the major experimental variables. The latter are, of course, temperature and nonmetal activity, and since a sheet of paper has only two dimensions, such representations are traditionally presented either as a function of temperature at constant nonmetal activity (i.e., as Arrhenius plots), or as a function of nonmetal activity at constant temperature. The latter, in the form of log–log plots, have proved to be particularly useful. These are generally known as Kröger–Vink diagrams in honor of two scientists at the Philips Research Laboratories in the Netherlands who pioneered their use in the 1950s. While these diagrams can be confusing to the casual observer, with a little practice they are extremely useful. They are a particularly valuable check for the consistency of any proposed defect model derived from experimental observation. Defect models have been proposed in the literature for which it is impossible to construct self-consistent Kröger–Vink diagrams, and this excludes them from serious consideration.

The usual procedure in real life is to obtain clues to the functional dependences of some of the defects by experimental measurements of such things as the equilibrium electrical conductivity or diffusion constants as functions of temperature and nonmetal activity. An attempt is then made to fit the results to a self-consistent defect model that is usually summarized in the form of a Kröger–Vink diagram. The model often suggests further experiments that may or may not support the model. An iterative process may then be pursued to obtain better and better models. While it is virtually impossible to find a model that is proved beyond doubt to be uniquely correct, the evidence can become extremely convincing if extensive and precise experimental data are available. For pedagogical reasons, we initially proceed with an artificial approach that assumes that certain thermodynamic parameters (i.e., mass-action constants), are known and then derive the appropriate Kröger–Vink diagram. The analysis of experimental data for real systems will be dealt with in detail in later chapters.

THE HYPOTHETICAL COMPOUND MX WITH SCHOTTKY DISORDER

We now derive the functional dependences of the equilibrium defect concentrations on nonmetal activity at constant temperature for the simple example of a hypothetical compound MX that contains monovalent cations and anions. Let us assume that Schottky disorder is the preferred type of intrinsic ionic disorder and that the concentrations of intrinsic ionic defects substantially exceed those of the intrinsic electronic defects. The thermodynamic state of the system is completely described by the equilibrium reactions and mass-action expressions for intrinsic ionic and electronic disorder, the oxidation reaction, the reduction reaction, and a complete expression for bulk electrical neutrality. (Actually, the thermodynamic state is completely described by any three of the equilibrium reactions plus the expression for electrical neutrality, since the fourth equilibrium reaction can always be derived from the other three. However, it is convenient to include both the oxidation and reduction reactions explicitly to describe the deviation from stoichiometry above and below the stoichiometric composition, respectively, to show the increase in the concentrations of the species that dominate the properties in those regions.) Since Schottky disorder has been chosen as the preferred type of lattice disorder, these reactions and relationships for intrinsic disorder are:

$$\text{nil} \rightleftharpoons V'_M + V^{\cdot}_X \tag{10.12}$$

$$[V'_M][V^{\cdot}_X] = K_S(T) = K'_S e^{-\Delta H_S/kT} \tag{10.13}$$

where K_S is the gross mass-action constant and K'_S is all of the mass-action constant except for the enthalpy term, which is given explicitly:

$$\text{nil} \rightleftharpoons e' + h^{\cdot} \tag{10.14}$$

$$np = K_I(T) = K'_I e^{-E^{\circ}_g/kT} \tag{10.15}$$

Charge neutrality is expressed by

$$n + [V'_M] = p + [V^{\cdot}_X] \tag{10.16}$$

It will be assumed that the values of these mass-action constants are known at the temperature selected for our analysis:

$$
\begin{aligned}
K_S &= 10^{34} \ (\text{defects/cm}^3)^2 \\
K_I &= 10^{26} \ (\text{defects/cm}^3)^2
\end{aligned}
$$

It will be further assumed that MX is precisely stoichiometric at $P(X_2) = 10^{-10}$ atm. The defect concentrations at this nonmetal activity are then:

$$[V'_M]_0 = [V^{\cdot}_X]_0 = K_S^{1/2} = 10^{17} \text{defect/cm}^3 \tag{10.17}$$

$$n_0 = p_0 = K_I^{1/2} = 10^{13} \text{defect/cm}^3 \tag{10.18}$$

These values can be placed on a log–log plot of the defect concentrations as a function of nonmetal activity as shown in Fig. 10.3.

If $P(X_2)$ is decreased below the value at the stoichiometric composition, the compound will lose a certain amount of nonmetal in order to stay in equilibrium.

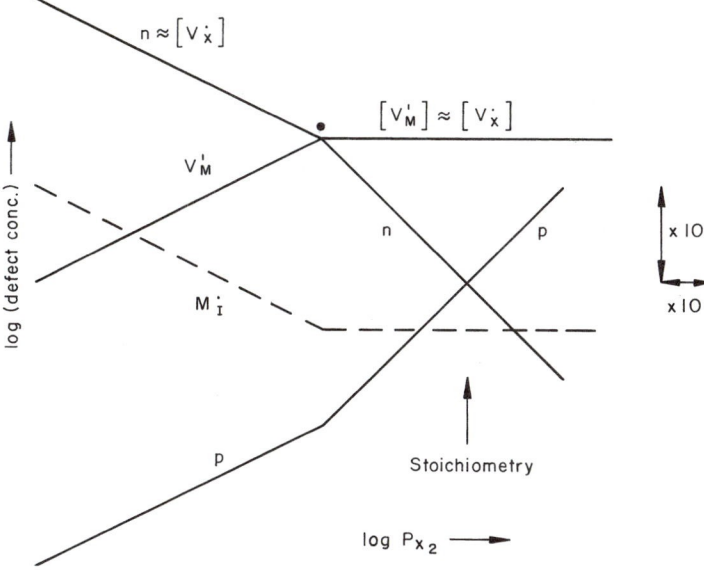

Figure 10.3 The reduction side of the Kröger–Vink diagram for MX as defined in the text.

This should result in the creation of anion vacancies, one of the preferred intrinsic defects, and electrons:

$$X_X \rightleftharpoons \tfrac{1}{2}X_2 + V_X^{\cdot} + e' \tag{10.19}$$

with the mass-action expression:

$$\frac{[V_X^{\cdot}]n}{[X_X]} = K_N P(X_2)^{-1/2} = K_N' e^{-\Delta H_N/kT} P(X_2)^{-1/2} \tag{10.20}$$

where the subscript N refers to the nonmetal-deficient, n-type region, and ΔH_N is the enthalpy of the reduction reaction per nonmetal atom lost. It has been assumed that there is no association between electronic and ionic defects. Since a Kröger–Vink diagram is an isothermal plot, only the dependence on $P(X_2)$ is needed to construct the diagram. However, it is often useful to include the exponential enthalpy term explicitly, since experiments are usually carried out at several temperatures. For small deviations from stoichiometry, the normal lattice components, such as $[X_X]$, are hardly affected, and their concentrations are often subsumed into the preexponential constant. Thus it is convenient to define a new mass-action constant, designated K_n in this case, that includes the concentrations of normal lattice components and the exponential enthalpy term. Thus Eq. (10.20) can be simplified to:

$$[V_X^{\cdot}]n \approx K_n P(X_2)^{-1/2} \tag{10.21}$$

For nonmetal activities above that at the stoichiometric composition, MX must take up an excess of nonmetal to maintain equilibrium. Since cation vacancies are one of the preferred intrinsic ionic defects, they are expected to be a product of the oxidation reaction

$$\tfrac{1}{2}X_2 \rightleftharpoons X_X + V_M' + h^{\cdot} \tag{10.22}$$

with the mass-action expression

$$[X_X][V_M']p = K_p P(X_2)^{1/2} = K_p' e^{-\Delta H_P/kT} P(X_2)^{1/2} \tag{10.23}$$

The subscript p refers to the oxygen-excess, p-type region, and ΔH_p is the enthalpy of the oxidation reaction per added nonmetal atom. As in the reduction region, it is convenient to define a new mass-action constant that contains the exponential enthalpy term as well as the concentrations of normal lattice components, which will be nearly constant for small deviations from stoichiometry:

$$[V_M']p \approx K_p P(X_2)^{1/2} \tag{10.24}$$

The numerical values of K_n and K_p are easily obtained from Eqs. (10.21) and (10.24), the defect concentrations at the stoichiometric composition, and the nonmetal activity at that composition, $P(X_2)_0$. It is possible to obtain the value of the mass-action constants for the nonstoichiometric reactions from data obtained at the stoichiometric composition because all valid mass-action expressions must be obeyed simultaneously everywhere in the single-phase region:

$$\begin{aligned} K_n &= [V_X^{\cdot}]_0 n_0 P(X_2)_0^{1/2} \\ &= 10^{17} \times 10^{13}(10^{-10})^{1/2} = 10^{25} \end{aligned} \tag{10.25}$$

$$\begin{aligned} K_p &= [V_M']_0 p_0 P(X_2)_0^{-1/2} \\ &= \frac{10^{17} \times 10^{13}}{(10^{-10})^{1/2}} = 10^{35} \end{aligned} \tag{10.26}$$

We now have four mass-action expressions, Eqs. (10.13), (10.15), (10.21), and (10.24), and four unknowns, the four defect concentrations. However, the mass-action expressions represent only three independent relationships, since any one of them can be obtained from the other three. The expression of charge neutrality, Eq. (10.16), serves as the fourth relationship required to solve for the four defect concentrations. These four simultaneous equations in four unknowns can be explicitly solved to give the defect concentrations as a function of the nonmetal activity. However, this is a rather sterile mathematical exercise that does not give much insight into the underlying defect chemistry. We will use a more approximate approach, in which the behavior is divided into regions in which the expression for charge neutrality is dominated by only two oppositely charged defects, while the contributions of the other two defects to charge neutrality are ignored. This approach was originally suggested by Brouwer (1954), another scientist at the Philips Research Laboratories in the Netherlands, and has been extensively developed by Kröger and Vink.

The Near-Stoichiometric Region: Reduction

Suppose that the nonmetal activity is reduced from that at the stoichiometric composition to a value at which 10^{14} additional anion vacancies and electrons are formed, according to Eq. (10.19). At this point, the new total electron concentration is 1.1×10^{14}, an increase by a factor of 11, and the new anion vacancy concentration is 1.001×10^{17}, an increase of only 0.1%. Thus the electron concentration can initially be increased substantially by reduction, while the anion vacancy concentration is hardly affected. As long as this condition persists, the condition of charge neutrality can be approximated by

$$[V_M']_0 \approx [V_X^{\cdot}]_0 \approx K_S^{1/2} \approx 10^{17} \tag{10.27}$$

The units of defect concentration (defects/cm^3), are not explicitly included in the following discussion. This region adjacent to the stoichiometric composition is conveniently referred to as the near-stoichiometric region. An approximate expression for the electron concentration can be obtained by combination of Eqs. (10.21) and (10.27) to give:

$$n \approx \frac{K_n}{K_S^{1/2}} P(X_2)^{-1/2} \tag{10.28}$$

Thus the electron concentration can be extended toward lower $P(X_2)$ from its value at stoichiometry as a line with a slope of $-1/2$ on the log–log plot of Fig. 10.3. The concentrations of cation and anion vacancies can be extended as a horizontal line as long as Eq. (10.27) is valid. There are two alternative ways to obtain an expression for the hole concentration in this region. The electron concentration, Eq. (10.28), can be substituted into the mass-action expression for intrinsic electronic disorder, Eq. (10.15), to give

$$p \approx \frac{K_1 K_S^{1/2}}{K_n} P(X_2)^{1/2} \tag{10.29}$$

or, perhaps more directly, the approximate value of $[V_M']$ in the near-stoichiometric region, $K_S^{1/2}$, can be substituted into the mass-action expression for the oxidation reaction, Eq. (10.22)

$$p \approx \frac{K_p}{K_S^{1/2}} P(X_2)^{1/2} \tag{10.30}$$

This gives a expression that is more symmetrical with that for the electron concentration in this region, Eq. (10.28). Even though this expression corresponds to the region of reduction from the stoichiometric composition, it is permissible to use the mass-action expression for the oxidation reaction, since all mass-action expressions must be valid everywhere on the diagram. It is easily shown that Eqs. (10.29) and (10.30) are equivalent. Since the product of the electron and hole concentrations must be independent of $P(X_2)$ according to the mass-action expression for intrinsic electronic disorder, Eq. (10.15), they must have complementary dependences on $P(X_2)$: that is, a log–log slope of $-1/2$ for the electrons and $+1/2$ for the holes.

Lines for n and p can be added to our diagram and extended until the electron concentration reaches that of the intrinsic ionic defects. We will then enter into a region having a different approximation to charge neutrality

The Highly Nonstoichiometric Region: Reduction

If $P(X_2)$ is decreased until the reduction reaction becomes the major source of both anion vacancies and electrons, there will be approximately the same number of each, and the approximate condition of charge neutrality changes from Eq. (10.27) to:

$$n \approx [V_X^{\cdot\cdot}] \tag{10.31}$$

Substitution of this relationship into Eq. (10.21) gives:

$$n \approx [V_X^{\cdot\cdot}] \approx K_n^{1/2} P(X_2)^{-1/4} \tag{10.32}$$

The dependence of n on the oxygen activity changes to $P(X_2)^{-1/4}$, and $[V_X^{\cdot\cdot}]$ must follow along on the same line.

It is usually an adequate approximation to neglect the curvature of the defect concentrations that occur as one slope is changing to another. The defect concentrations are then represented as intersecting straight lines. But where on the diagram do they intersect? There are several ways to answer this question. It is possible to calculate a value for n at some low value of $P(X_2)$ [e.g., 10^{-28} atm] from Eq. (10.32) and the value of K_n given in Eq. (10.25). The electron concentration at 10^{-28} atm can then be calculated to be 3.16×10^{19}, whose log is 19.5. This point can be placed on Fig. 10.4, and a line of slope $-1/4$ drawn to the point of intersection with the line drawn for the electron concentration in the near-stoichiometric region. The lines meet at $P(X_2) = 10^{-18}$ atm, where they both intersect the line for intrinsic ionic disorder at 10^{17} defects/cm^3. It will be shown in a later section that the value of $P(X_2)$ at the intersection point can be calculated directly from the appropriate mass-action constants.

Expressions for $[V_M^{/}]$ and p can be obtained for this highly nonstoichiometric region by combination of Eq. (10.32) with the mass-action expressions for intrinsic ionic and electronic disorder, Eqs. (10.13) and (10.15):

$$[V_M^{/}] \approx \frac{K_S}{K_n^{1/2}} P(X_2)^{1/4} \tag{10.33}$$

$$p \approx \frac{K_I}{K_n^{1/2}} P(X_2)^{1/4} \tag{10.34}$$

The slopes of $[V_M^{/}]$ and $[V_X^{\cdot\cdot}]$ as well as those of n and p are complementary, as required by Eqs. (10.13) and (10.15). Note that the lines representing all the defect concentrations must change slope at exactly the same value of $P(X_2)$. The results of this analysis are shown in Fig. 10.3 in a log–log plot of the defect concentrations in the reduction region as a function of $P(X_2)$ at constant temperature.

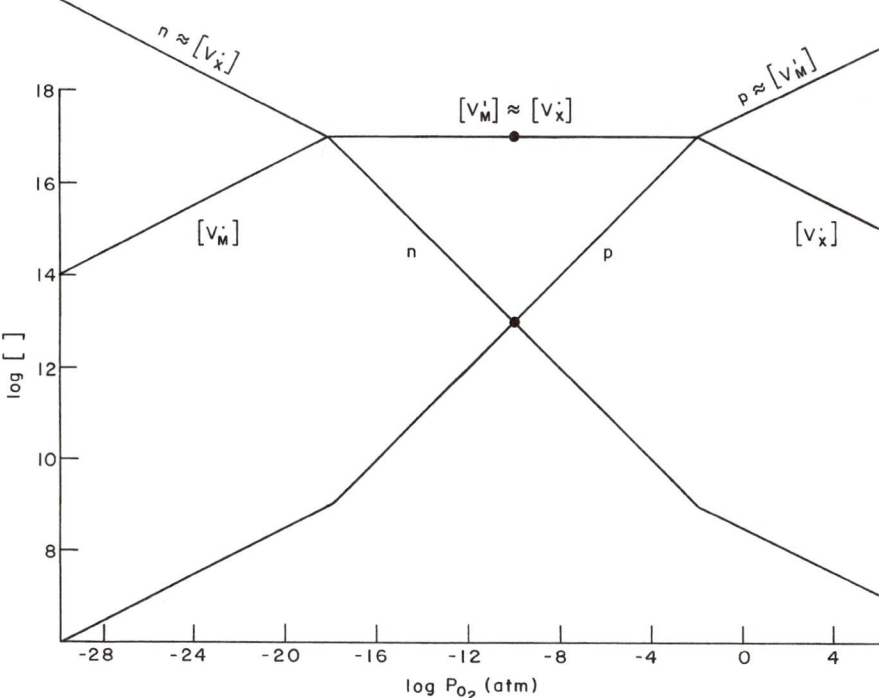

Figure 10.4 The complete Kröger–Vink diagram for MX.

In the preceding discussion, the defect chemistry of anion-deficient nonstoichiometry was treated as two limiting approximations to the condition of bulk charge neutrality. A more rigorous analysis is obviously possible. For most purposes, it is adequate to consider only the two-term approximations and to draw the defect dependences as intersecting straight lines, as in Fig. 10.3, while understanding that there is a transitional region in between where one dependence on $P(X_2)$ curves smoothly into the next dependence. This breakdown of the rigorous treatment into regions that involve only two dominant defects in an approximation to charge neutrality remains the standard way of depicting nonstoichiometry in the form of Kröger–Vink diagrams. A later section presents an example of a more rigorous treatment of the transition region.

The Near-Stoichiometric Region: Oxidation

As in the reduction region, the oxidation reaction for our hypothetical compound MX for values of $P(X_2)$ near that at the stoichiometric composition can create significantly increased numbers of holes while the concentration of intrinsic ionic defects remains essentially unchanged. Thus the approximate condition of charge neutrality is still represented by Eq. (10.27), just as it is on the other side of stoichiometry. Combination of Eqs. (10.24) and (10.27) then gives:

$$p \approx \frac{K_p}{K_S^{1/2}} P(X_2)^{1/2} \tag{10.35}$$

This is the same dependence on $P(X_2)$ for holes as was found on the reduction side of the near-stoichiometric region, Eq. (10.30). Thus the electrons must obey Eq. (10.28) across the entire near-stoichiometric region. All defects must always have constant dependences on the nonmetal activity across this region. Thus the dependences of n and p on $P(X_2)$ in this region are an extension of those found on the reduction side of stoichiometry.

The Highly Nonstoichiometric Region: Oxidation

When $P(X_2)$ is sufficiently high, the oxidation reaction will become the major source of defects, and the approximate condition of charge neutrality will shift to

$$p \approx [V_M'] \tag{10.36}$$

Combination of Eqs. (10.24) and (10.36) gives

$$p \approx [V_M'] \approx K_p^{1/2} P(X_2)^{1/4} \tag{10.37}$$

$[V_X^{\cdot\cdot}]$ and n can be obtained from Eq. (10.37) and the mass-action expressions for intrinsic ionic and electronic disorder, Eqs. (10.13) and (10.15):

$$[V_X^{\cdot\cdot}] \simeq \frac{K_S}{K_p^{1/2}} P(X_2)^{-1/4} \tag{10.38}$$

$$n \approx \frac{K_I}{K_p^{1/2}} P(X_2)^{-1/4} \tag{10.39}$$

The entire Kröger–Vink diagram for MX can now be completed as shown in Fig. 10.4. Table 10.1 summarizes the exponents of the $P(X_2)$ dependences, $d \log [\ ]/d \log P(X_2)$, and the components of the apparent enthalpies for each of the four defects for MX in the three regions of different approximations to charge neutrality.

SUMMARY OF THE KRÖGER–VINK DIAGRAM FOR MX

General Considerations

Surprisingly little information is needed to construct a defect diagram, since it must obey the laws of plane geometry as well as of defect chemistry. If we know either the point at which the slope of the electron concentration changes from $-1/2$ to $-1/4$, or the point at which the slope of the anion vacancy (or cation vacancy) concentration changes from $-1/4(+1/4)$ to horizontal, all the oxygen-deficient side of the diagram,

TABLE 10.1
d log [ ]/d log $P(X_2)$, the exponents of the dependences of the defect concentrations on $P(X_2)$, and the apparent activation enthalpies for ionic and electronic defects in M^+X^-, with Schottky disorder, and $K_S \gg K_I$.

	Highly reduced $n = [V_X^{\bullet}]$	Near-stoichiometric $[V_M'] = [V_X^{\bullet}]$	Highly oxidized $p = [V_M']$
$[V_M']$	$1/4$ $\Delta H_S - \Delta H_N/2$	0 $\Delta H_S/2$	$1/4$ $\Delta H_P/2$
$[V_X^{\bullet}]$	$-1/4$ $\Delta H_N/2$	0 $\Delta H_S/2$	$-1/4$ $\Delta H_S - \Delta H_P/2$
n	$-1/4$ $\Delta H_N/2$	$-1/2$ $\Delta H_N - \Delta H_S/2$	$-1/4$ $E_g^{\circ} - \Delta H_P/2$
p	$1/4$ $E_g^{\circ} - \Delta H_N/2$	$1/2$ $\Delta H_P - \Delta H_S/2$	$1/4$ $\Delta H_P/2$

as shown in Fig. 10.3, can be constructed, except for the hole concentration. The latter would require knowledge of the $P(X_2)$ at which the compound is stoichiometric, or of the mass-action constant for intrinsic electronic disorder. With that information, the entire diagram as shown in Fig. 10.4 can be constructed. This ability to extend the diagram far beyond the region over which experimental results are available is extremely helpful in developing defect models for real systems.

There are several ways to check a diagram such as that shown in Fig. 10.4 for consistency. There must be two dominant defects of opposite charge that satisfy charge neutrality in each region, and these must lie on the same line or on parallel lines (no crossover or convergence of the lines for these dominant defects!). There must be a common dominant defect in adjacent regions. Finally, all the mass-action expressions must be satisfied everywhere in the diagram. For example, the product of the expressions for the electron and hole concentrations must satisfy the mass-action expression for intrinsic electronic disorder, Eq. (10.15), and, for the example just described, the product of the expressions for the anion vacancy and cation vacancy concentrations must satisfy the mass-action expression for intrinsic ionic disorder, Eq. (10.13). Thus the product of Eqs. (10.28) and (10.30), the expressions for n and p in what we have called the near-stoichiometric region, gives

$$np = \frac{K_n K_P}{K_S} \tag{10.40}$$

which agrees with the mass-action expression for intrinsic electronic disorder. The implied relationship between the mass-action constants on the right-hand side of Eq. (10.40) is explained in the next section.

Numerous checks for self-consistency are available in Table 10.1. Since both $[V_M'][V_X^{\bullet}]$ and np must be independent of $P(X_2)$, the sum of the exponents on $P(X_2)$ for $[V_M']$ and $[V_X^{\bullet}]$ must equal zero for each of the three regions, and the same is true for the exponents on $P(X_2)$ for n and p. This is seen to be the case. Also, the sum of the exponents for $[V_X^{\cdot}]$ and n must always equal $-1/2$, Eq. (10.21), while that for $[V_M']$ and p must always equal $+1/2$, Eq. (10.24). Similarly, the sum of the enthalpies for $[V_M']$ and $[V_X^{\cdot}]$ must always equal ΔH_S, and the sum of the enthalpies for n and p must always equal E_g°. Also the sum of the enthalpies for $[V_X^{\cdot}]$ and n must be ΔH_N, Eq. (10.20), while that for $[V_M']$ and p must equal ΔH_P, Eq. (10.23). All these conditions are confirmed by the values listed in Table 10.1.

In the construction of Fig. 10.4 it has been assumed that the phase in question remains stable over the entire width of the diagram: that is, that the crystal structure of our hypothetical example is maintained and that no second phases are formed. When a phase boundary is crossed, a two-phase region will be entered, and the original phase will be gradually transformed into the new phase as the nonmetal activity is changed further, until the phase boundary of the new phase is reached. At that point the original phase will have completely disappeared. The new phase (e.g., FeO that results from the reduction of Fe_3O_4), will then have its own characteristic Kröger–Vink diagram that will have no relationship to that of the original compound. It has also been assumed that only the four defects shown on the diagram have significant concentrations. In principle, all possible kinds of defect are present at some concentration. It is mathematically possible that some minority defect that has a steeper dependence on the nonmetal activity than any of the majority defects could become significant if the nonmetal activity can be changed sufficiently.

To sum up the preceding discussion, the guiding principles for developing Kröger–Vink diagrams are:

1. All valid mass-action expressions must be satisfied simultaneously everywhere in the single-phase field.

2. Bulk charge neutrality must always be obeyed.

3. There must be continuity of the various defect concentrations between adjacent regions.

4. All defects have specific dependences on the nonmetal activity, and do not suddenly appear or disappear on the diagram.

The development of a self-consistent Kröger–Vink diagram is a useful check on the validity of any proposed defect model.

Enthalpy Relationships

The sum of the oxidation and reduction reactions gives

$$
\begin{array}{ll}
\tfrac{1}{2}X_2 \rightleftharpoons X_X + V_M' + h^{\cdot} & \Delta H_P \\[4pt]
X_X \rightleftharpoons \tfrac{1}{2}X_2 + V_X^{\cdot} + e' & \Delta H_N \\[4pt]
\hline
\text{nil} \rightleftharpoons V_M' + V_X^{\cdot} + e' + h^{\cdot} & \overline{\Delta H_P + \Delta H_N}
\end{array}
\qquad (10.41)
$$

This is also the sum of the formation reactions for intrinsic ionic and electronic disorder, and their enthalpies, $\Delta H_S + E_g^{\circ}$. Therefore, we write:

$$\Delta H_N + \Delta H_P = \Delta H_S + E_g^\circ \tag{10.42}$$

Since the addition of enthalpies is mathematically equivalent to the multiplication of mass-action constants, this is equivalent to the statement that the product of the mass-action expressions for oxidation and reduction is equal to the product of those for intrinsic ionic and electronic disorder:

$$K_p K_n = K_S K_I \tag{10.43}$$

This is the relationship used in Eq. (10.40), and it supplies confirmation that the four mass-action expressions represent only three independent equations. These relationships emphasize the point that both ionic and electronic defects are created by nonstoichiometry and that the energetics of nonstoichiometry are therefore related to the energetics of intrinsic defect formation. Once again, these effects are lumped together and do not separate the effects of the different directions of nonstoichiometry. There will be a similar relationship between the enthalpies of the oxidation and reduction reactions and those of intrinsic ionic and electronic disorder for any compound with any kind of intrinsic ionic disorder.

The importance of the band gap in setting the scale of nonstoichiometry is apparent from Eq. (10.42). Transition metal compounds typically have band gaps of the order of 2–4 eV (200–400 kJ/mol), reflecting the stability of adjacent oxidation states. Main group compounds, with only one stable oxidation state, have much larger band gaps, perhaps 8–10 eV (800–1000 kJ/mol). Thus in the latter case, there is 4–8 eV (400–800 kJ/mol) more to be divided up between the enthalpies of oxidation and reduction than is found in the case of transition metal compounds. As a result, the main group compounds may have only minute or even undetectable deviations from stoichiometry, while the amount of nonstoichiometry in transition metal compounds may be substantial, exerting a very pronounced effect on physical properties.

It is seen from Table 10.1 that the temperature dependences of the n and $[V_X^{\cdot\cdot}]$ concentrations in the highly reduced region are both characterized by one-half of the enthalpy of the reduction reaction (i.e., $\Delta H_N/2$). This does not mean that the energetics of forming these two defects are equal, merely that there is a certain total enthalpy of formation for the two defects, and the experiment gives no information on how to partition it between them. Since the defects must have the same temperature dependence in this region, they must appear to share the enthalpy equally. The same argument holds for $[V_M']$ and p in the highly oxidized region.

The temperature dependence of n in the near-stoichiometric region is characterized by the enthalpy of the reduction reaction, ΔH_N, minus half the enthalpy of formation of Schottky defects, $\Delta H_S/2$. This reduction of the reaction enthalpy can be attributed to the fact that in this region the reduction reaction does not contribute significantly to the concentration of $V_X^{\cdot\cdot}$, and therefore their formation energy as part of the Schottky disorder can be subtracted from the reaction enthalpy. The remaining enthalpy is then borne by the electrons, which are the only significant product of the reaction. Equivalent enthalpy behavior is noted for the holes in the near-stoichiometric region.

Calculation of Critical Points

There are three definable points on the Kröger–Vink diagram just derived for MX, the intersection points between the near-stoichiometric region and the highly oxidized and reduced regions, and the stoichiometric composition defined by n = p. The concentration levels at which these occur are simply determined by the levels of intrinsic disorder, but the values of nonmetal activity at which they occur are often of interest. The value of $P(X_2)$ at the stoichiometric composition was given in this example, but its value can be determined in general by relating the concentrations of the pertinent defects at that composition. In this case, that involves equating the expressions for the electron and hole concentrations in the near-stoichiometric region, Eqs. (10.28) and (10.30). and solving for $P(X_2)$

$$n_0 = \frac{K_n}{K_S^{1/2}} P(X_2)_0^{-1/2} = \frac{K_p}{K_S^{1/2}} P(X_2)_0^{1/2} = p_0$$

(10.44)

$$P(X_2)_0 = \frac{K_n}{K_p} = \frac{10^{25}}{10^{35}} = 10^{-10} \text{ atm}$$

where the subscripts on n, p, and $P(X_2)$ refer to the value at the stoichiometric composition. The answer is in accord with the assigned value.

The value of the nonmetal activity at the boundary between the near-stoichiometric region and the region of highly reduced nonstoichiometry can be obtained by equating the expressions for the electron concentrations in those two regions, Eqs. (10.28) and (10.32):

$$n_n = \frac{K_n}{K_S^{1/2}} P(X_2)_n^{-1/2} = K_n^{1/2} P(X_2)_n^{-1/4}$$

(10.45)

$$P(X_2)_n = \left(\frac{K_n}{K_S} \right)^2 = \left(\frac{10^{25}}{10^{34}} \right)^2 = 10^{-18} \text{ atm}$$

where the subscript n refers to this boundary on the reduction side. The calculated value agrees with that obtained by the construction of the diagram.

The value of $P(X_2)$ at the corresponding boundary on the nonmetal-rich side is obtained similarly:

$$p_p = \frac{K_p}{K_S^{1/2}} P(X_2)_p^{1/2} = K_p^{1/2} P(X_2)_p^{1/4}$$

(10.46)

$$P(X_2)_p = \left(\frac{K_S}{K_p} \right)^2 = \left(\frac{10^{34}}{10^{35}} \right)^2 = 10^{-2} \text{ atm}$$

This value also agrees with that found by construction of Fig. 10.4.

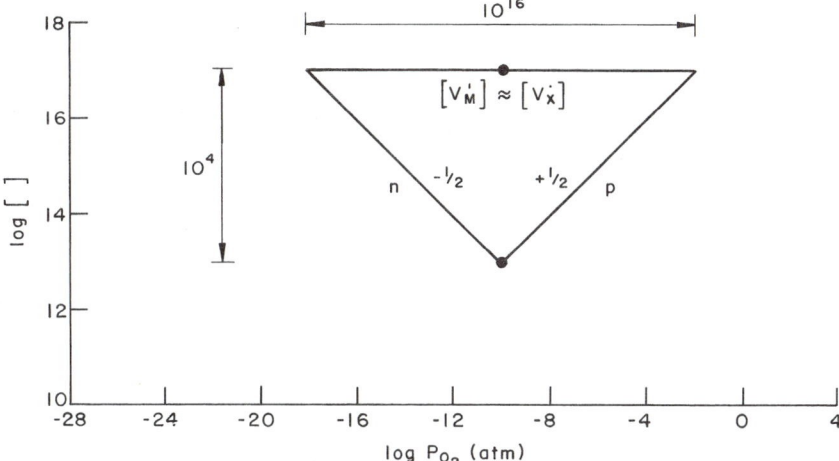

Figure 10.5 The width of the near-stoichiometric region.

The Width of the Near-Stoichiometric Region

The width of the near-stoichiometric region is of general importance in determining the detailed Kröger–Vink diagram. It is easily obtained if the concentrations of the intrinsic defects, and their functional dependences on the nonmetal activity, are known. For the hypothetical MX, we determined that the electrons and holes vary as $P(X_2)^{-1/2}$ and $P(X_2)^{+1/2}$, respectively, in the near-stoichiometric region, and that the concentrations of intrinsic ionic defects are 10^4 times greater than those of intrinsic electrons and holes. Thus, the concentrations of both intrinsic electrons and holes must increase by a factor of 10^4 from the stoichiometric composition to the boundary with the highly nonstoichiometric regions. As shown in Fig. 10.5, this requires 8 orders of magnitude for each, and thus a total of 16 orders of magnitude of $P(X_2)$ must be traversed across the near-stoichiometric region between the boundaries for highly reduced and highly oxidized compositions. Once the numbers are known, this is a trivial exercise in plane geometry.

Defect Concentrations in the Transition Regions

Figure 10.4 was constructed from a series of intersecting straight lines in adjacent regions in which only two oppositely charged defects were included in approximations to charge neutrality. This ignored the gradual change in slope of the defect concentrations in the transitions between the regions. It is instructive to see how much inaccuracy results from these approximations. Figure 10.6 represents an enlarged view of the transition region between the near-stoichiometric region and the highly reduced region for our hypothetical MX. Charge neutrality in this region is more accurately given by:

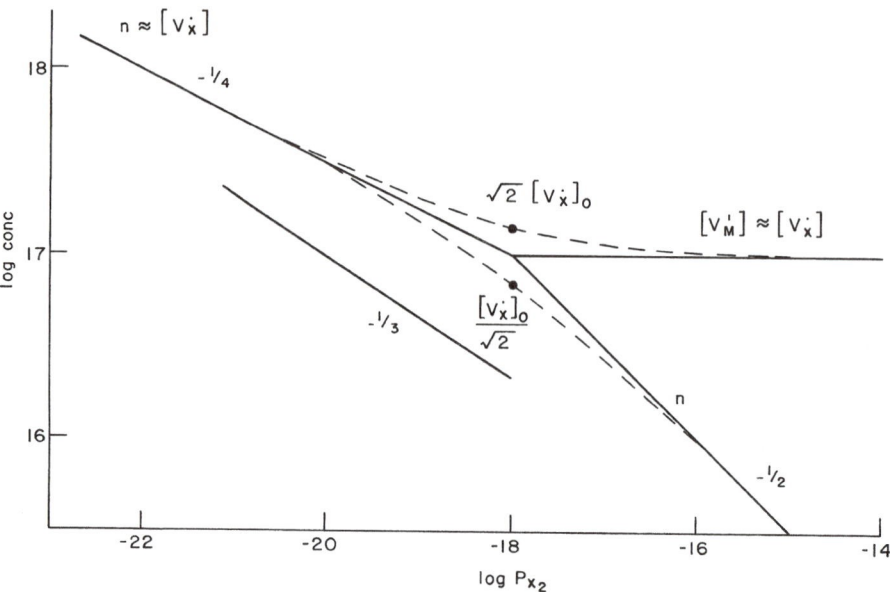

Figure 10.6 True defect concentrations in the transition region between the near-stoichiometric and highly reduced regions (dashed lines). A line with slope $-1/3$ is shown for comparison.

$$n + [V_M'] \approx [V_X^{\cdot}] \tag{10.47}$$

In this three-term approximation, only the hole concentration has been neglected, and, since their concentration is lower by about eight orders of magnitude, the resulting error will be truly insignificant. Combination of this expression with Eq. (10.21), the mass-action expression for the reduction reaction, gives a more accurate expression for $[V_X^{\cdot}]$ in this transition region:

$$[V_X^{\cdot}] \approx [K_n P(X_2)^{-1/2} + K_S]^{1/2} \tag{10.48}$$

and the electron concentration can be obtained by substituting the resulting values into Eq. (10.21). The results are shown in Fig. 10.6. The difference between the straight-line approximation and the more exact expression is modest. The maximum difference occurs at the intersection of the straight lines and amounts to 40%. This point was also included in Fig. 10.3 to put it in perspective with the entire diagram. The discrepancy has almost entirely disappeared after two decades of $P(X_2)$ on either side. While modest in magnitude, the difference can be a subtle trap. A line with a slope of $-1/3$ is included in Fig. 10.6 for comparison. It is a fairly good approximation to the apparent slope in the transition region, but, of course, it has no real meaning of its own. There are unfortunate examples in the literature where data have been unknowingly taken only in such a transition region, and defect models have been proposed based on the apparent slope that actually lies between two defined regions.

The Composition at the Near-Stoichiometric Extremes

In the development of Fig. 10.6, the defect concentrations near the boundary between the near-stoichiometric and highly reduced regions were calculated accurately for the hypothetical compound MX. In a similar way, it is possible to calculate the composition at such boundaries. First of all, it is necessary to have a mathematical definition for the boundary. This is best accomplished by defining the boundary as the point of crossing of two defect lines, one of which is increasing while the other is decreasing with nonmetal activity. Thus on the reduction side of MX, as shown in Fig. 10.4, the boundary can be defined as the point where the cation vacancy concentration crosses the electron concentration. For this simple example, we have shown that this intersection takes place at $P(X_2) = 10^{-18}$ atm. This definition is expressed as follows:

$$n = [V_M']\qquad(10.49)$$

One might think that the boundary could also be defined as the point at which $n = [V_X^{\cdot}]$, but this does not work because these two concentrations approach each other asymptotically and are truly equal only at infinitely low $P(X_2)$. Combination of Eq. (10.49) with the three-term approximation to charge neutrality, Eq. (10.47), and multiplication of both sides by $[V_X^{\cdot}]$, gives:

$$[V_X^{\cdot}]^2 = 2[V_M'][V_X^{\cdot}] = 2K_S\qquad(10.50)$$

Thus we have:

$$[V_X^{\cdot}] = (2K_S)^{1/2}$$

$$[V_M'] = \left(\frac{K_S}{2}\right)^{1/2}\qquad(10.51)$$

The concentrations of the ionic defects at the boundary are some near-unity factor of their concentrations at the stoichiometric composition. If N_0 is the concentration of cation and anion sites in the crystal, and if the cation sublattice remains ideally filled, the composition of MX at this boundary is given by:

$$MX_{[N_0-(2K_S)^{1/2}]/\left[N_0-\left(\frac{K_S}{2}\right)^{1/2}\right]}\qquad(10.52)$$

If N_0 is 10^{22} sites/cm³, and $K_S = 10^{34}$, as in our example, the composition at the boundary is:

$$MX_{0.999993}$$

Clearly, the compound is still very, very close to the stoichiometric composition at this boundary.

It will be seen that the defect concentrations, hence the composition, at these boundaries are always scaled to the level of the dominant intrinsic disorder, whether it is ionic or electronic. Note that if the concentrations of the dominant intrinsic defects are of the order of 1%, a very high value, the amount of nonstoichiometry at the ends of the near-stoichiometric region will also be of the order of 1%.

The Effect of Temperature

Since a Kröger–Vink diagram is an isothermal representation of the defect concentrations as a function of nonmetal activity, there will be a different diagram at each temperature for a given compound. The change in the diagrams with temperature will obviously be related to the enthalpies of the various defect reactions. Let us assume that the Kröger–Vink diagram just derived for MX referred to 600°C, and construct a new diagram for 1000°C. To do this, we must assign enthalpies for the various equilibrium reactions. Each mass-action expression has the general form:

$$K = K'e^{-\Delta H/kT} \tag{10.53}$$

Given the values of K used earlier at 600°C, and the assigned enthalpies, values of K' can be calculated for each reaction. Assume that the enthalpies of formation for intrinsic ionic and electronic disorder, Eqs. (10.12) and (10.14), are 1.5 and 3.0 eV (140 and 290 kJ/mol), respectively. This is in accord with the original assumption that the concentrations of Schottky defects substantially exceed those of electrons and holes at the stoichiometric composition. The equilibrium constants for these two reactions are then:

$$K_S = 4.54 \times 10^{42}e^{-1.5/kT} \tag{10.54}$$

$$K_I = 2.06 \times 10^{43}e^{-3.0/kT} \tag{10.55}$$

From Eq. (10.42) it is seen that we have a total of 4.5 eV (430 kJ/mol) to divide up between ΔH_N and ΔH_P. Let us assume that MX prefers oxidation over reduction, and assign the values $\Delta H_N = 3.0$ eV (290 kJ/mol) and $\Delta H_P = 1.5$ eV (140 kJ/mol). The mass-action expressions for reduction and oxidation are then:

$$K_n = 2.06 \times 10^{42}e^{-3.0/kT} \tag{10.56}$$

$$K_p = 4.54 \times 10^{43}e^{-1.5/kT} \tag{10.57}$$

The mass-action constants at 1000°C are then given by

$$
\begin{aligned}
K_S &= 5.25 \times 10^{36} \\
K_I &= 2.76 \times 10^{31} \\
K_n &= 2.76 \times 10^{30} \\
K_p &= 5.25 \times 10^{37}
\end{aligned}
$$

Note that these values satisfy Eq. (10.43). At the stoichiometric composition:

$$[V'_M]_0 = [V^\bullet_X]_0 = K_S^{1/2} = 2.293 \; 10^{18} \; (\log = 18.36)$$

$$n_0 = p_0 = K_I^{1/2} = 5.253 \; 10^{15} \; (\log = 15.72)$$

The value of $P(X_2)$ at the stoichiometric composition is given by Eq. (10.44):

$$P(X_2)_0 = \frac{K_n}{K_p} = 5.25 \times 10^{-8} \text{ atm (log} = -7.28)$$ (10.58)

The Kröger–Vink diagram for MX at 1000°C is now easily drawn, as was done previously for the lower temperature. The results are shown as the solid lines in Fig. 10.7, superimposed on the diagram for 600°C, shown in dashed lines. Several specific points are worth noting.

The value of $P(X_2)$ at stoichiometry moves to higher values with increasing temperature. The reason for this can be seen from an expansion of Eq. (10.39)

$$P(X_2)_0 = \frac{K_n}{K_p} = \frac{K'_n}{K'_p} e^{[(\Delta H_p - \Delta H_n)/kT]}$$ (10.59)

Since ΔH_N is larger than ΔH_P, the exponent is negative, and the diagram is pushed toward higher $P(X_2)$ with increasing temperature. Obviously, if ΔH_P had been greater than ΔH_N, the stoichiometric composition would have moved to lower $P(X_2)$ with increasing temperature.

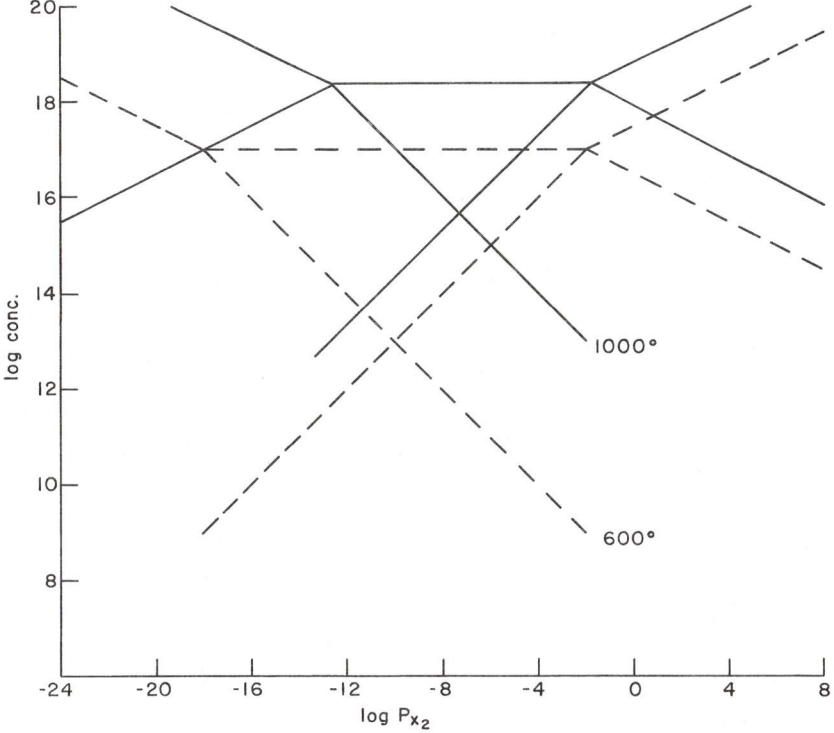

Figure 10.7 Kröger–Vink diagrams for MX at both 600°C (dashed lines) and 1000°C (solid lines).

On the log–log plot, temperature differences scale linearly with the enthalpies. Thus the vertical distance between the lines for $[V_X^{\bullet}]$ and n at the two temperatures in the highly reduced region is exactly twice that between those for $[V_M']$ and p in the highly oxidized region, because the enthalpy of reduction is twice that of oxidation.

The inverted central triangle defined by the near-stoichiometric region decreases in size with increasing temperature. This is because the concentrations of intrinsic electrons and holes, with a higher enthalpy of formation, are increasing faster than the concentrations of Schottky defects. This is to be expected, since the relative concentrations of the intrinsic defects are determined primarily by their enthalpies of formation. In principle, this central triangle could converge with increasing temperature to a single point and then open up again, with the electrons and holes becoming the dominant form of intrinsic disorder. No such extreme behavior has been reported, because the required temperatures exceed the melting or decomposition temperatures of the compounds.

There is no temperature dependence for $[V_M']$ in the highly reduced region. This is an accident caused by the specific enthalpies chosen for the system. $[V_M']$ is given in this region as follows:

$$[V_M'] \approx \frac{K_S}{K_n^{1/2}} P(X_2)^{1/4}$$

$$\approx \frac{K_S'}{(K_n')^{1/2}} e^{\left(\frac{\Delta H_n}{2} - \Delta H_S\right)/kT} P(X_2)^{1/4}$$

(10.60)

Since $\Delta H_N/2 - \Delta H_S = 3/2 - 1.5 = 0$, there is no temperature dependence. This emphasizes the fact that the concentration of a minority defect may increase, not change, or even decrease with increasing temperature in a nonstoichiometric material. Thus one cannot always assume that an increase in temperature will result in an increase in the diffusion rate of the ion of a minority defect in a highly nonstoichiometric compound. This will depend on the balance between the change in the defect concentration and the change in its mobility.

The lack of a temperature dependence for the value of $P(X_2)$ at the transition from the near-stoichiometric region to the highly oxidized region results from a similar coincidental cancellation of enthalpy terms: see Eq. (10.41).

CONCLUSION OF THE DISCUSSION OF MX

This completes the discussion of the hypothetical compound MX. Our example, and its specific properties, were deliberately chosen to be as simple as possible, to present the principles of nonstoichiometry and the development of a Kröger–Vink diagram with a minimum of complexity. However, it is now time to move on to a more complex example that is closer to what one may expect to find in a real material.

A MORE COMPLEX KRÖGER–VINK DIAGRAM

Definition of the Problem

The construction and reading of Kröger–Vink diagrams is so essential to the understanding of defect chemistry that it is a worthwhile exercise to develop a more complex example in detail. For our second hypothetical example, consider an oxide of generic formula M_2O_3 (e.g., Cr_2O_3 or Fe_2O_3), with Schottky disorder as the predominant form of intrinsic ionic disorder. In contrast with the previous example of the hypothetical MX, let us assume that in M_2O_3 the concentrations of intrinsic electronic defects considerably exceed those of intrinsic ionic defects at the stoichiometric composition. Once again, we shall assume that there are no significant interactions between any of the defects. Our persistent use of hypothetical compounds is convenient because there are hardly any real examples that extend across several regions with different approximations to the expression for charge neutrality. We will look at some real examples later to see where they fit into the overall range of behavior.

The following equilibrium reactions and their mass-action expressions are needed for derivation of the diagram:

Intrinsic Ionic Disorder: Schottky Disorder

$$\text{nil} \rightleftharpoons 2V_M''' + 3V_O^{\cdot\cdot} \tag{10.61}$$

$$[V_M''']^2[V_O^{\cdot\cdot}]^3 = K_S = K_S'e^{-\Delta H_S/kT} \tag{10.62}$$

Intrinsic Electronic Disorder

$$\text{nil} \rightleftharpoons e' + h^{\cdot} \tag{10.63}$$

$$np = K_I = K_I'e^{-E_g^{\circ}/kT} \tag{10.64}$$

Oxidation

$$\tfrac{3}{2}O_2 \rightleftharpoons 3O_O + 2V_M''' + 6h^{\cdot} \tag{10.65}$$

$$[V_M''']^2p^6 = K_pP(O_2)^{3/2} = K_p'e^{-\Delta H_P/kT}P(O_2)^{3/2} \tag{10.66}$$

Reduction

$$O_O \rightleftharpoons \tfrac{1}{2}O_2 + V_O^{\cdot\cdot} + 2e' \tag{10.67}$$

$$[V_O^{\cdot\cdot}]n^2 = K_n P(O_2)^{-1/2} = K_n'e^{-\Delta H_n/kT} P(O_2)^{-1/2} \tag{10.68}$$

It has been assumed that the concentration of $[O_O]$ is not significantly affected by the deviations from stoichiometry, so it is included in the preexponential coefficients. The oxidation reaction, Eq. (10.65), has been written to avoid fractional coefficients for the defects. This is an arbitrary decision. If it had been written for the addition of a single oxygen atom, to make it symmetrical with the reduction reaction, the products would have included $2/3[V_M''']$, and the enthalpy would have been only one-third of that defined by Eq. (10.66).

The mass-action expressions give us three independent equations with four unknowns, the four defect concentrations. The fourth necessary relationship is the condition of charge neutrality:

$$p + 2[V_O^{\cdot\cdot}] = n + 3[V_M'''] \qquad (10.69)$$

The complete Kröger–Vink diagram can now be derived by solution of the four simultaneous equations, but once again we will break it down into separate regions having different approximations to charge neutrality according to the Brouwer approximation.

A Kröger–Vink diagram refers to a specific equilibrium temperature, so let us assume that the intrinsic mass-action constants at that temperature are known:

$$K_I \quad = \quad 10^{36} \text{ (defects/cm}^3)^2$$
$$K_S \quad = \quad 3.37 \times 10^{70} \text{ (defects/cm}^3)^5$$

and that the stoichiometric composition occurs at

$$P(O_2)_0 \quad = \quad 10^{-10} \text{ atm}$$

where the subscript denotes the situation at the stoichiometric composition. As in the previous example, this derivation must be understood as an educational exercise, rather than as a realistic solution to a real system.

The mass-action constants for the oxidation and reduction reactions, K_p and K_n, can be calculated from the data at the stoichiometric composition and Eqs. (10.61) and (10.63):

$$K_p = \frac{\left(10^{14}\right)^2 \left(10^{18}\right)^6}{\left(10^{-10}\right)^{3/2}} = 10^{151} \qquad (10.70)$$

$$K_n = \left(1.5 \times 10^{14}\right) \left(10^{18}\right)^2 \left(10^{-10}\right)^{1/2} = 1.5 \times 10^{45} \qquad (10.71)$$

The stoichiometric composition is defined by the point at which

$$n_0 = p_0 = K_I^{1/2} = 10^{18} \qquad (10.72)$$

and from Eq. (10.69) it is then apparent that the concentrations of the intrinsic ionic defects at that composition are:

$$[V_M''']_0 = \tfrac{2}{3}[V_O^{\cdot\cdot}]_0 = 10^{14} \qquad (10.73)$$

These points can be placed on the Kröger–Vink diagram. We can now proceed to construct the diagram shown in Fig. 10.8.

The Near-Stoichiometric Region: Oxidation

At the stoichiometric composition, $P(O_2)_0 = 10^{-10}$ atm, the defect concentrations are given by Eqs. (10.72) and (10.73). In this region, the expression of charge neutrality is dominated by the intrinsic electronic defects, and Eq. (10.72) is an adequate approximation to charge neutrality. An expression for $[V_M''']$ can then be obtained by combination of Eqs. (10.66) and (10.72)

$$[V_M'''] \approx \left(\frac{K_p}{K_I^3} \right)^{1/2} P(O_2)^{3/4} \tag{10.74}$$

A line for $[V_M''']$ can be extended from its value at stoichiometry toward higher $P(O_2)$ with a slope of 3/4, while the electron and hole concentrations are nearly unchanged. An expression for the oxygen vacancies can be obtained by substituting Eq. (10.74) into the mass-action expression for Schottky disorder, Eq. (10.62)

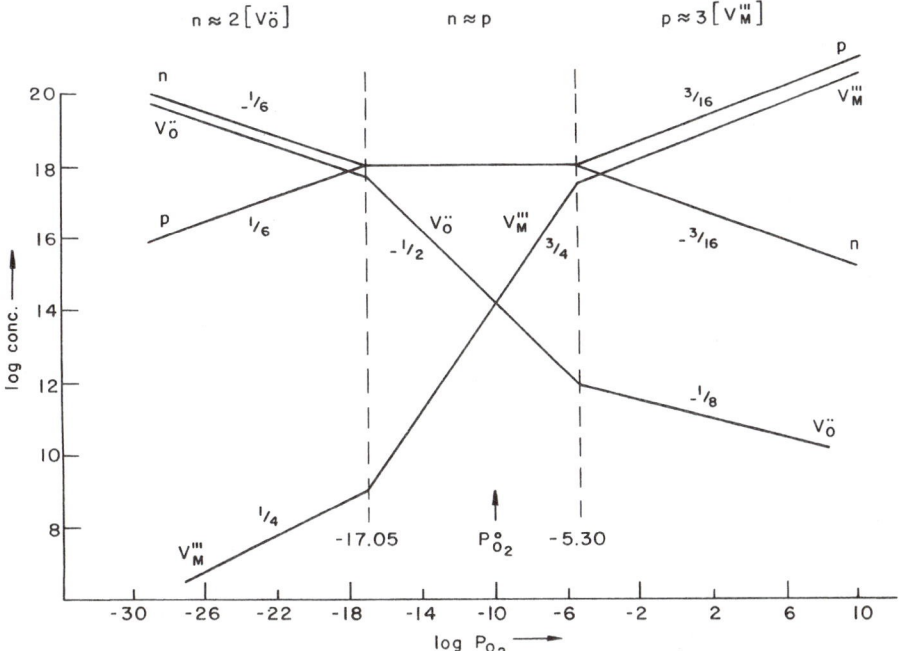

Figure 10.8 The Kröger–Vink diagram for M_2O_3.

$$[V_O^{\cdot\cdot}] \approx K_I \left(\frac{K_S}{K_p}\right)^{1/3} P(O_2)^{-1/2} \tag{10.75}$$

Since all mass-action expressions must be satisfied under all conditions, $[V_O^{\cdot\cdot}]$ can also be obtained by combination of the mass-action expression for the reduction reaction, Eq. (10.68), and the approximate expression for charge neutrality, Eq. (10.72):

$$[V_O^{\cdot\cdot}] \approx \frac{K_n}{K_I} P(O_2)^{-1/2} \tag{10.76}$$

It is easily shown that the last two equations are equivalent. One could also substitute Eq. (10.76) into the mass-action expression for Schottky disorder, Eq. (10.62), and obtain an alternative expression for $[V_M^{///}]$ that is equivalent to Eq. (10.75). In any case, a line for $[V_O^{\cdot\cdot}]$ can be extended from the value at stoichiometry toward higher $P(O_2)$ with a slope of $-1/2$.

The Highly Nonstoichiometric Region: Oxidation

With increasing oxidation, the cation vacancy concentration will reach and ultimately surpass the level of intrinsic electronic disorder. The hole concentration will then start to increase with $P(O_2)$, and there will be a transition region as the slopes of the defect concentrations adjust to a new situation. That new situation occurs when the oxidation reaction becomes the major source of defects. At that point, the condition of charge neutrality can be approximated by including only cation vacancies and holes

$$p \approx 3[V_M^{///}] \tag{10.77}$$

When this is substituted into the mass-action expression for the oxidation reaction, we obtain:

$$p \approx 3[V_M^{///}] \simeq (9K_p)^{1/8} P(O_2)^{3/16} \tag{10.78}$$

The electron concentration can be obtained by substituting this into the mass-action expression for intrinsic electronic disorder, Eq. (10.64):

$$n \approx \frac{K_I}{(9K_p)^{1/8}} P(O_2)^{-3/16} \tag{10.79}$$

and the oxygen vacancy concentration can be obtained from Eqs. (10.62) and (10.78):

$$[V_O^{\cdot\cdot}] \approx \frac{(9K_S)^{1/3}}{(9K_p)^{1/12}} P(O_2)^{-1/8} \tag{10.80}$$

An equivalent expression for $[V_O^{\cdot\cdot}]$ in this region can be obtained from Eqs. (10.68) and (10.79):

$$[V_O^{\cdot\cdot}] \approx \frac{K_n}{K_I^2} (9K_p)^{1/4} P(O_2)^{-1/8} \tag{10.81}$$

With the Brouwer approximation, which develops a diagram of intersecting straight lines and ignores the curvature in the transitional region, the change of slopes in this case will occur when $[V_M^{///}]$ reaches p/3. This satisfies the approximate condition of charge neutrality in the new region, and this condition persists in the highly oxidized region. It will be left to the reader to show that the intersection point between these two regions occurs at the proper values of $P(O_2)$ and $[V_M^{///}]$. The oxidized side of the Kröger–Vink diagram can now be completed. Note once again that the slopes of the lines for all defects must change at exactly the same value of $P(O_2)$.

The Near-Stoichiometric Region: Reduction

We can now return to the near-stoichiometric region and look at the reduction side. The procedure is exactly the same in mirror image. Charge neutrality is still approximated by Eq. (10.72), and this can be combined with Eq. (10.68) to give:

$$[V_O^{\cdot\cdot}] \approx \frac{K_n}{K_I} P(O_2)^{-1/2} \tag{10.82}$$

and this can be combined with the mass-action expression for Schottky disorder, Eq. (10.62), to give:

$$[V_M^{///}] \approx \left(\frac{K_S K_I^3}{K_n^3} \right)^{1/2} P(O_2)^{3/4} \tag{10.83}$$

Or, the approximate condition of charge neutrality can be combined with the mass-action expression for the oxidation reaction to give the equivalent expression:

$$[V_M^{///}] \approx \left(\frac{K_p}{K_I^3} \right)^{1/2} P(O_2)^{3/4} \tag{10.84}$$

Note that Eqs. (10.82) and (10.84) are identical to Eqs. (10.76) and (10.74), which were derived for the oxidation side of the near-stoichiometric region. This is a necessary consequence of continuity, since both relations can be derived from the mass-action expression for intrinsic electronic disorder and that of either the oxidation or the reduction reaction.

The Highly Nonstoichiometric Region: Reduction

As the $P(O_2)$ is lowered further, the concentration of oxygen vacancies will eventually reach n/2. Further reduction will push the process into a new region, where the reduction reaction becomes the major source of defects. In that region, charge neutrality can be approximated by:

$$n \approx 2[V_O^{\cdot\cdot}] \tag{10.85}$$

This can be combined with the mass-action expression for the reduction reaction, Eq. (10.68), to give:

$$n \approx 2[V_O^{\cdot\cdot}] \approx (2K_n)^{1/3} P(O_2)^{-1/6} \tag{10.86}$$

This can be combined with the mass-action expression for intrinsic electronic disorder, Eq. (10.64), to give:

$$p \approx \frac{K_I}{(2K_n)^{1/3}} P(O_2)^{1/6} \tag{10.87}$$

and with the mass-action expression for Schottky disorder, Eq. (10.62), to give

$$[V_M^{///}] \approx 2 \left(\frac{K_S}{K_n}\right)^{1/2} P(O_2)^{1/4} \tag{10.88}$$

The diagram can be extended into the heavily reduced region and is now complete, assuming that no other defects appear with further oxidation or reduction.

Summary of the Kröger–Vink Diagram for M_2O_3

The specific shape of a Kröger–Vink diagram depends on the nature of the compound, that is, on the type of intrinsic ionic disorder and on whether intrinsic ionic disorder is greater than intrinsic electronic disorder, or vice versa. The Kröger–Vink diagram for our example of M_2O_3 (Fig. 10.8), is less symmetrical than that for the hypothetical MX. This is due to the different charges on the ionic defects in the case of M_2O_3. If the preferred type of ionic disorder had been cation or anion Frenkel, the diagram would have been symmetrical.

The relationship between the various enthalpies can be seen by adding three times the reduction reaction to the oxidation reaction

$$
\begin{array}{ll}
3O_O \rightleftharpoons {}^3\!/_2 O_2 + 3V_O^{\cdot\cdot} + 6e' & 3\Delta H_N \\
{}^3\!/_2 O_2 \rightleftharpoons 3O_O + 2V_M^{///} + 6h^{\cdot} & \Delta H_P \\
\hline
\text{nil} \rightleftharpoons 2V_M^{///} + 3V_O^{\cdot\cdot} + 6e' + 6h^{\cdot} & 3\Delta H_N + \Delta H_P
\end{array}
\tag{10.89}
$$

The result is also the sum of the reaction to form Schottky disorder and six times the reaction to form intrinsic electronic disorder. Therefore, the net enthalpy must also equal $\Delta H_S + 6E_g^\circ$:

$$3\Delta H_N + \Delta H_P = \Delta H_S + 6E_g^\circ \tag{10.90}$$

The tremendous importance of the band gap in setting the scale of nonstoichiometry in oxides, where the addition or loss of each oxygen atom yields two electronic defects, is readily apparent. For two similar oxides with significantly different band gaps (e.g., Al_2O_3 and Fe_2O_3), the amount of nonstoichiometry is vastly different. If the band gaps differ by only 2 eV (200 kJ/mol), there will be an extra 12 eV (1200 kJ/mol) to be divided up between $3\Delta H_N$ and ΔH_P, and that is the difference between a highly stoichiometric compound and one that can be significantly nonstoichiometric.

The exponents of the dependence of the defect concentrations on $P(O_2)$ for M_2O_3, and the components of the apparent activation enthalpies, are summarized in Table 10.2. Once again, it is seen that the combinations of the exponents and of the enthalpies are consistent with the various mass-action expressions. As an example, the exponents for $[V_M^{///}]$ and p in the highly reduced region when inserted in the mass-action expression of the oxidation reaction give

$$2(1/4) + 6(1/6) = 3/2$$

while the enthalpies give

$$2\frac{(\Delta H_S - \Delta H_N)}{2} + 6\left(E_g^\circ - \frac{\Delta H_N}{3}\right) = \Delta H_S + 6E_g^\circ - 3\Delta H_N = \Delta H_P$$

These are the values in the mass-action expression, Eq. (10.66).

The calculation of the precise location of the critical points on the diagram [i.e., the values of $P(O_2)$ and the defect concentrations at which the slopes change between adjacent regions], is left as an exercise for the reader.

The Width of the Near-Stoichiometric Region

Once a compound leaves the near-stoichiometric region and enters a highly nonstoichiometric region, whether on the reduction or the oxidation side, the dependence of the defect concentrations, hence of the composition, becomes well-defined. Within the near-stoichiometric region, however, the situation is less clear, especially for the dominant defects. The complete picture can be obtained by solution of a sufficient number of simultaneous equations, but a more general treatment is enlightening. The width of the near-stoichiometric region is clearly important, because the greater

TABLE 10.2
d log [ ]/d log $P(O_2)$, the exponents of the dependences of the defect concentrations on $P(O_2)$, and the apparent activation enthalpies for ionic and electronic defects in M_2O_3, with Schottky disorder, and $K_I^{1/2} \gg K_S^{1/5}$

	Highly reduced $n = 2[V_O^{\cdot\cdot}]$	Near-stoichiometric $n = p$	Highly oxidized $p = 3[V_M^{///}]$
$[V_M^{///}]$	1/4 $(\Delta H_S - \Delta H_N)/2$	3/4 $(\Delta H_P - 3E_g^\circ)/2$	3/16 $\Delta H_P/8$
$[V_O^{\cdot\cdot}]$	$-1/6$ $\Delta H_N/3$	$-1/2$ $\Delta H_N - E_g^\circ$	$-1/8$ $\Delta H_S/3 - \Delta H_P/12$
n	$-1/6$ $\Delta H_N/3$	0 $E_g^\circ/2$	$-3/16$ $E_g^\circ - \Delta H_P/8$
p	1/6 $E_g^\circ - \Delta H_N/3$	0 $E_g^\circ/2$	3/16 $\Delta H_P/8$

dependence of the defect concentrations on the nonmetal activity begins only when the experimental conditions move into a highly nonstoichiometric region.

For our hypothetical compound M_2O_3, shown in Fig. 10.8, the situation is somewhat more complex, but follows the same pattern. Once again, the ratio of the concentrations of the majority and minority intrinsic defects is about 10^4. Since the oxygen vacancies vary as $P(O_2)^{-1/2}$, they traverse about eight orders of magnitude of $P(O_2)$ between the stoichiometric composition and the lower boundary of the near-stoichiometric region. The cation vacancies vary as $P(O_2)^{3/4}$; hence it takes them about $4 \times 4/3 = 16/3$ orders of magnitude of $P(O_2)$ to reach the upper boundary. Hence the near-stoichiometric region is about 13 orders of magnitude wide in terms of $P(O_2)$. A more exact result, which takes into account the exact ratio of the defect concentrations between the stoichiometric composition and the boundaries of the region, indicates that the width is closer to 12 orders of magnitude, as shown in Fig. 10.8. As an alternative, the values of $P(O_2)$ at the boundaries between adjacent regions can be calculated exactly from the appropriate mass-action expressions, and a precise value of the width of the near-stoichiometric region is then obtained by difference.

The Composition at the Near-Stoichiometric Extremes

For the reduction side of M_2O_3, the boundary between the near-stoichiometric and highly nonstoichiometric regions can be defined as follows:

$$p = [V_O^{\cdot\cdot}] \tag{10.91}$$

and charge neutrality is closely approximated by:

$$p + 2[V_O^{\cdot\cdot}] \approx n \tag{10.92}$$

Combination of these two equations and multiplication by n leads to:

$$n \approx (3K_I)^{1/2}$$

$$p \approx \left(\frac{K_I}{3}\right)^{1/2} \approx [V_O^{\cdot\cdot}] \tag{10.93}$$

The concentration of cation vacancies is negligible at this point. If there are N_0 cations/cm^3, the composition at the boundary is:

$$M_2O_{3-\frac{3}{N_0}\left(\frac{K_I}{3}\right)^{1/2}} \tag{10.94}$$

and if $N_0 = 10^{22}$ cations/cm^3 and $K_I = 10^{36}$ (defects/cm^3)^2, as in our example, the composition is:

$$M_2O_{2.9998}$$

Note that once again, the concentration of the significant ionic defect at the boundary is scaled to the value of the concentration of the majority intrinsic defects, even though in this case these are electrons and holes. This is a general principle:

The amount of nonstoichiometry at the boundary between the near-stoichiometric region and an adjacent highly nonstoichiometric region is scaled to the concentration of the majority intrinsic defects.

This in turn sets the scale of nonstoichiometry, because it is from this level that the defects start to rise in the highly nonstoichiometric regions.

SUMMARY OF KRÖGER–VINK DIAGRAMS FOR INTRINSIC NONSTOICHIOMETRY

The two hypothetical examples used in this chapter, MX and M_2O_3, should suffice to demonstrate the principles involved in displaying the properties of intrinsically nonstoichiometric compounds in the form of Kröger–Vink diagrams. The principles and procedures remain the same, independent of the compound and its formula type. The construction and interpretation of such diagrams requires only simple logic and common sense within the framework of the mass-action approach that depends on the validity of dilute solution thermodynamics.

ENTHALPY RELATIONSHIPS

Comparison of Tables 10.1 and 10.2 discloses some generalities for the enthalpies associated with each defect:

1. In each region, the enthalpy for each defect that participates in the condition of charge neutrality is the enthalpy of the appropriate defect reaction divided by the total number of defects created by that reaction. Thus for the highly oxidized region of M_2O_3, charge neutrality is approximated by $p = 3[V_M''']$, and the product of the oxidation reaction is $2V_M''' + 6h^{\cdot}$, a total of eight defects. Thus the enthalpy associated with each defect is $\Delta H_P/8$.

2. In the near-stoichiometric region, the enthalpy associated with each of the defects not included in the approximation to charge neutrality is the reaction enthalpy per defect minus the enthalpy to form the corresponding number of the reaction product defects that are part of the approximation to charge neutrality. Thus for M_2O_3, where charge neutrality in the near-stoichiometric region is approximated by $n = p$, the enthalpy associated with $[V_M''']$ is the enthalpy of the oxidation reaction per V_M''', $\Delta H_P/2$, minus the enthalpy of formation of three intrinsic holes at $E_g^{\circ}/2$ each, or $3E_g^{\circ}/2$. This reduces the reaction enthalpy per cation vacancy by the amount of enthalpy that would otherwise be needed

to create the holes, because they are already present in abundance, and the oxidation reaction does not appreciably affect their concentration.

These generalities will be valid for any case of intrinsic nonstoichiometry.

CONCLUSION

While an understanding of intrinsic nonstoichiometry is essential as the basis for understanding the effects of impurities, there are not many actual cases in which it is the dominant effect. For oxides in particular, the enthalpies of formation for intrinsic disorder are usually quite high, so that the concentrations of intrinsic defects are overwhelmed by the impurity content of real materials. Thus in the chapter that follows, we will build on this foundation to illustrate nonstoichiometric behavior for the case that an aliovalent impurity is one of the dominant charged defects. This will correspond more closely to reality in most systems.

REFERENCE

Brouwer, G. A general asymptotic solution of reaction equations common in solid-state chemistry. *Philips Res. Rep.* 9:366–376, 1954.

☑ PROBLEMS

10.1. Assume that an oxide M_2O has cation Frenkel defects for the predominant type of intrinsic ionic disorder and that the compound is precisely stoichiometric when in equilibrium with $P(O_2) = 10^{-10}$ atm at 1000°C. Given that the concentration of cation Frenkel defects equals 10^{16} cm^{-3} and n = p = 10^{18} cm^{-3}, both at 10^{-10} atm and 1000°C, construct a defect diagram for pure M_2O at that temperature for oxygen pressures between 10^{-30} and 10^{10} atm. Include all four major defects, identify each line with its slope, and label each region with the appropriate approximation to charge neutrality. Assume that all ionic defects are fully ionized.

What is the width of the near-stoichiometric region?

10.2. Assume that M_2O has Schottky defects for the predominant type of intrinsic ionic disorder and that the compound is precisely stoichiometric when in equilibrium with $P(O_2) = 10^{-20}$ atm at 1000°C. At this temperature and oxygen activity, the concentration of cation vacancies is 2×10^{18} cm^{-3}, and n = p = 10^{16} cm^{-3}.

a. Construct a defect diagram for pure M_2O at 1000°C that clearly shows the near-stoichiometric, highly reduced, and highly oxidized regions.

b. Give numerical values for the mass-action constants for both the reduction and oxidation reactions.

c. Calculate values for $P(O_2)$ at the boundaries between the near-stoichiometric region and the highly reduced and highly oxidized regions.

10.3. For pure AgBr at 550 K, the mass-action constants for cation Frenkel disorder and intrinsic electronic disorder are 10^{-8} and 10^{-26}, respectively. The mass-action constant for the oxidation reaction, in terms of the addition of a single bromine atom, is 10^{-10}. (Concentrations are in site fractions and pressures in atm.)

a. *Derive* the value of $P(Br_2)$ for the condition of exact stoichiometry at 550 K.

b. Sketch a defect diagram for pure AgBr at 550 K and for $10^{-20} < P(Br_2) < 10^{16}$ atm.

c. The mass-action constant for intrinsic Schottky disorder has been estimated to be 10^{-18} for AgBr at 550 K. Add a line for bromine vacancies to your diagram.

10.4. Consider an ideally pure compound M_2O for which Schottky defects represent the major type of intrinsic ionic disorder. Assume that at 800°C it has been determined that $p = [V_M'] = 10^{19}$ cm^{-3} at $P(O_2) = 1$ atm, and that the compound is precisely stoichiometric at $P(O_2) = 10^{-16}$ atm where $n = p = 10^{18}$ cm^{-3}.

a. Draw a complete defect diagram for pure M_2O at 800°C for $10^{-32} < P(O_2) < 10^4$ atm. Show all four major defects in each region and label all lines with the proper $P(O_2)$ dependence.

b. Give numerical values for the mass-action constants for both kinds of intrinsic disorder and for the oxidation and reduction reactions.

10.5. Equilibrium defect diagrams for two very similar pure oxides, MO and NO, are shown on the following graph. Schottky defects are the only intrinsic ionic defects of importance in both cases. Assume that all lines can be extended across the full width of the graph and that all ionic defects are completely ionized.

a. Copy the diagrams on a piece of graph paper. Label all the lines for both oxides with defect symbols.

b. Give values for the mass-action constants for the formation of intrinsic ionic and electronic defects for both compounds.

c. Can the concentration of any defect in NO ever exceed the concentrations of all defects in MO? If so, over what range of experimental conditions?

d. Which compound has the greatest change in composition across the near-stoichiometric region (region II for MO compared with region II' for NO)?

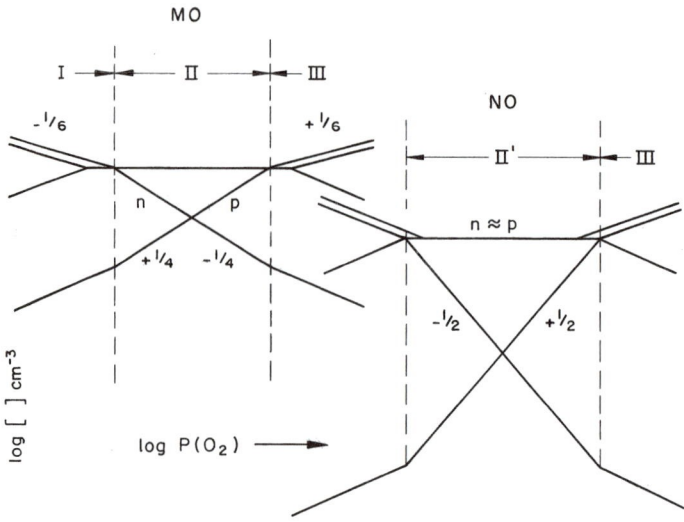

e. In both compounds, if the mobilities for electrons and holes are equal, and are 100 times greater than that of any ionic defect, can either compound ever have a 50% ionic contribution to the total electrical conductivity?

f. Which compound can have the largest fractional ionic contribution to the conductivity, and under what experimental conditions is this a maximum?

Extrinsic Nonstoichiometry

11

INTRODUCTION

Extrinsic nonstoichiometry is defined as the case when the aliovalent impurity content exceeds the concentration of any defect resulting from intrinsic disorder. Thus the net impurity content is one component of the two-term approximation to charge neutrality over some range of nonmetal activity at a given temperature. The near-stoichiometric region described for intrinsic nonstoichiometry, where intrinsic defects control the approximation to charge neutrality, is replaced by an impurity-controlled region.

Intrinsic stoichiometry is limited to two situations:

1. A form of intrinsic disorder, either ionic or electronic, dominates the defect population and the condition of charge neutrality at the stoichiometric composition, and impurity concentrations are negligible by comparison.

2. The concentrations of the defects that result from nonstoichiometry exceed the impurity content.

The common factor in the two cases is that the impurity content does not play a significant role in the condition of charge neutrality, hence has no significant effect on defect concentrations. It is possible to observe both intrinsic and extrinsic nonstoichiometry in a single sample of a specific compound. In this case, an impurity and its compensating defects may dominate the near-stoichiometric region, but the degree of nonstoichiometry may become so great under more extreme oxidizing or reducing conditions that the impurity content is overwhelmed. In this case, neither intrinsic ionic nor intrinsic electronic disorder is dominant under any conditions.

While there are a number of systems that correspond to case 2 above, some examples of which were described in Chapter 10, there are very few examples of case 1. The reasons are the same as those discussed in the earlier chapters on intrinsic and extrinsic disorder. Unless the intrinsic defects carry low effective charges, preferably a single charge, and unless the crystal structure can easily accommodate the two complementary defects, or unless the band gap has a very modest magnitude, then given the realities of purification, impurities will most probably control the material's properties, at least over some range of conditions. It will be seen that in some cases important practical properties of widely used compounds are totally dependent on their impurity content, whether naturally present or deliberately added.

It cannot be overemphasized that an impurity acts on the defect chemistry of a material only through its charge; it will have no significant effect unless it is a major component of the condition of charge neutrality. Since an acceptor impurity is a negatively charged center, it can be compensated by an anion vacancy (or metal interstitial) or a hole. Since an anion vacancy is a product of reduction and a hole is a product of oxidation, we can anticipate that the acceptor centers will be compensated by anion vacancies under reducing conditions and by holes under oxidizing conditions, and that there will be a transition region in which one compensating defect is replaced by the other. Likewise, a donor impurity is a positively charged center that can be compensated by a cation vacancy (or interstitial anion) or an electron. The electron will be preferred under reducing conditions but will be replaced by cation vacancies under more oxidizing conditions. Finally, we can anticipate that when the amount of nonstoichiometry overwhelms the impurity content, the nonstoichiometric behavior will revert to that of the ideally pure material. The two usual conditions must be satisfied: all valid mass-action expressions must be obeyed simultaneously under equilibrium conditions, and bulk charge neutrality must always be satisfied. Those guiding principles and a little common sense will lead to a logical and consistent understanding of the effects of impurities on nonstoichiometry.

As described for extrinsic ionic disorder in Chapter 5, impurities are almost never incorporated into crystalline solids under equilibrium conditions. They are usually incorporated into the lattice during processing (e.g., solid state reactions), and their concentrations are then fixed and are not functions of temperature or nonmetal activity. Thus the concentrations of ionic and electronic defects vary with the experimental parameters against a background of invariant impurity content. The impurity content may affect the concentrations of ionic and electronic defects, but the reverse is not usually the case.

A SIMPLE EXAMPLE: DONOR-DOPED MX

Basic Relationships

Let us build on the simple example of intrinsic nonstoichiometry derived earlier in Chapter 10, the compound MX exemplified by the Kröger–Vink diagram shown in Fig. 10.4. It is convenient to repeat here the essential relationships, the intrinsic disorder reactions, the oxidation and reduction reactions, and their mass-action expressions:

$$\text{nil} \rightleftharpoons V_M' + V_X^{\bullet} \tag{11.1}$$

$$[V_M'][V_X^{\bullet}] = K_S \tag{11.2}$$

$$\text{nil} \rightleftharpoons e' + h^{\cdot} \tag{11.3}$$

$$np = K_I \tag{11.4}$$

$$\tfrac{1}{2}X_2 \rightleftharpoons X_X + V_M' + h^{\cdot} \tag{11.5}$$

$$[V_M']p = K_p P(X_2)^{1/2} \tag{11.6}$$

$$X_X \rightleftharpoons \tfrac{1}{2}X_2 + V_X^{\cdot} + e' \tag{11.7}$$

$$[V_X^{\cdot}]n = K_n P(X_2)^{-1/2} \tag{11.8}$$

We have incorporated $[X_X]$ into the mass-action constants, assuming once again that it is not significantly affected by the nonstoichiometry. The Kröger–Vink diagram for pure MX is reproduced as dashed lines in Fig. 11.1. The basic assumptions remain the same: $K_S = 10^{34}$, $K_I = 10^{26}$, and stoichiometry occurs at $P(X_2) = 10^{-10}$ atm; K_p and K_n were derived to be 10^{35} and 10^{25}, respectively.

The values of all of the mass-action constants will remain unchanged by the addition of an impurity.

The Compensating Defects

Let us assume that a donor impurity in the form of a divalent cation, $D_M^{\cdot}$ is incorporated into MX on normal cation sites at a concentration level of 10^{18} cm^{-3}. This is greater than the concentration of intrinsic defects at what was the stoichiometric composition, so either $[V_M']$ or n must increase to balance the increased positive charge. Since the other possible compensating ionic defect, anion interstitials, played no part in intrinsic disorder, it can be ignored. The two alternative reactions for the incorporation of the donor compound, DX_2, into MX are then

$$DX_2 \xrightarrow{\text{(2MX)}} D_M^{\cdot} + V_M' + 2X_X \tag{11.9}$$

$$DX_2 \xrightarrow{\text{(MX)}} D_M^{\cdot} + X_X + \tfrac{1}{2}X_2 + e' \tag{11.10}$$

Equation (11.9) represents the formation of the stoichiometric composition of the donor-doped compound, since it corresponds to a solid solution of the two stoichiometric binary constituents, while Eq. (11.10) leads to a nonstoichiometric composition through the loss of nonmetal. The addition of an aliovalent impurity requires that it be included in the complete statement of charge neutrality:

$$n + [V_M'] = p + [V_X^{\cdot}] + [D_M^{\cdot}] \tag{11.11}$$

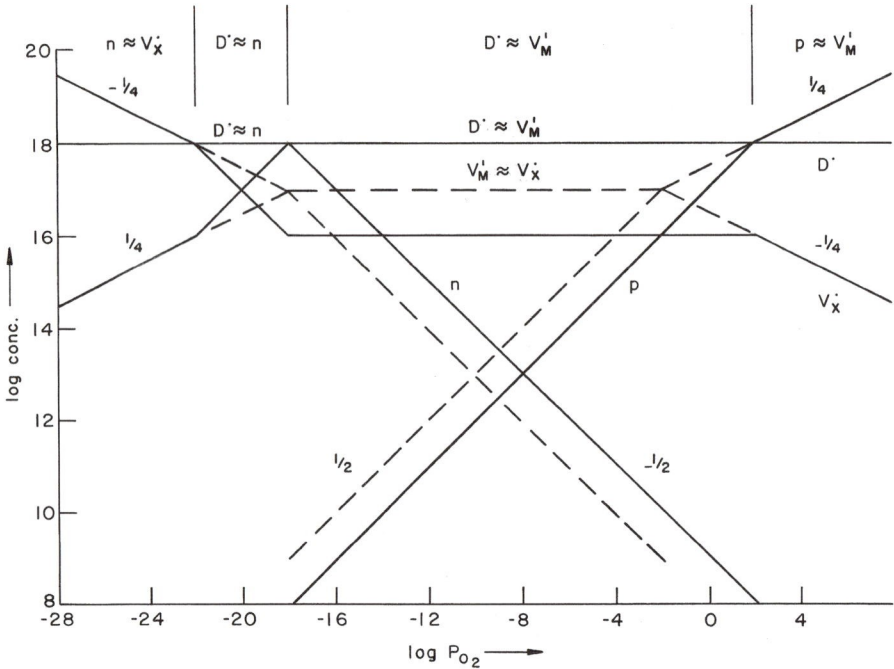

Figure 11.1 The Kröger–Vink diagram for donor-doped MX with the parameters given in the text: dashed lines, diagram for pure MX.

One might suspect that the more favored intrinsic defect $[V'_M]$ would be the preferred choice for the compensating defect, but let us ignore that reasonable assumption and try the electrons as the compensating defects at $P(X_2) = 10^{-10}$ atm. Then n will have to increase from 10^{13}, its value in pure MX, to 10^{18}, or five orders of magnitude, to balance the charge of the donor impurity. According to Eq. (11.8), $[V_X^{\cdot}]$ must then decrease by five orders of magnitude, and that means, according to Eq. (11.2), that $[V'_M]$ must increase by five orders of magnitude, to 10^{22} cm^{-3}. This will place it four orders of magnitude above the donor impurity level without any oppositely charged defect available to compensate its charge, a ludicrous situation. Moreover, in this particular situation, such a change would mean that virtually every cation site in the crystal has become vacant. Compensation by electrons at this nonmetal activity was indeed not a wise choice. So let us try the cation vacancies as the compensating defect for the donor impurities. At $P(X_2) = 10^{-10}$ atm, $[V'_M]$ will then have to increase from 10^{17} cm^{-3} to 10^{18} cm^{-3}, an increase by one order of magnitude. Then, according to Eq. (11.6), p will have to decrease by one order of magnitude, and Eq. (11.4) then requires n to increase by one order of magnitude. There are no inconsistencies here, and $[V'_M]$ is a viable choice as the compensating defect. These new concentrations for all the defects can now be entered at $P(X_2) = 10^{-10}$ atm on Fig. 11.1. This is the general situation: of the two possible compensating defects, one will give an untenable result, while the other will give a reasonable result. Note that the addition of positively charged donor centers causes an increase in the concentration of all negatively charged defects, and a decrease in all positively charged defects. We can now proceed to construct a Kröger–Vink diagram for donor-doped MX. It will contain an impurity-controlled region, but after sufficient oxidation or reduction, when the impurity is no longer part of the approximation to charge neutrality, it must blend back into the diagram for intrinsic nonstoichiometry.

The Impurity-Controlled Region: Ionic Compensation

Near $P(X_2) = 10^{-10}$ atm, $[D_M^\bullet]$ and its compensating defect $[V_M']$ are the major defects, and the approximate condition of charge neutrality in this impurity-controlled region near the stoichiometric composition can be given as follows:

$$[V_M'] \approx [D_M^\bullet] \tag{11.12}$$

MX can be slightly oxidized or reduced without disturbing this approximation, so, over some range of $P(X_2)$ $[V_M']$ will be fixed at the donor level and will be independent of $P(X_2)$. According to Eq. (11.2), $[V_X']$ must then also be independent of $P(X_2)$, and its concentration must have decreased by a factor of 10 from its value in the pure compound:

$$[V_X^\bullet] \approx \frac{K_S}{[D_M^\bullet]} \tag{11.13}$$

As the compound is oxidized, combination of Eqs. (11.6) and (11.12) indicates that the hole concentration must increase as follows:

$$p \approx \frac{K_p}{[D_M^\bullet]} P(X_2)^{1/2} \tag{11.14}$$

and therefore n must decrease as follows:

$$n \approx \frac{K_1[D_M^\bullet]}{K_p} P(X_2)^{-1/2} \tag{11.15}$$

Or, the electron concentration can be obtained from Eqs. (11.8) and (11.13):

$$n \approx \frac{K_n[D_M^\bullet]}{K_S} P(X_2)^{-1/2} \tag{11.16}$$

These are the same dependences on $P(X_2)$ as derived for the pure compound, but the absolute values of the concentrations have been displaced vertically by one order of magnitude, the electrons increasing and the holes decreasing. Thus lines can be added to Fig. 11.1 for these electronic defects, parallel to the lines for the intrinsic case, but displaced accordingly. Note the point at which n = p has moved to higher $P(X_2)$ by one order of magnitude but remains at the same concentration level, since the latter is defined by Eq. (11.4). This point corresponds to the stoichiometric composition of the doped compound, which consists of the solid solution of the stoichiometric binary constituents, $(1 - 2x)MX + xDX_2$, or $M_{1-2x}D_xX$.

The hole concentration will continue to rise along this slope until it intersects the value it had in the pure compound. This occurs exactly as the extended horizontal line for $[D_M^{\cdot}] = [V_M']$ arrives at the same point, $P(X_2) = 10^2$ atm. At this point the hole and cation vacancy concentrations are equal, and this condition continues indefinitely for further oxidation and becomes the approximation to charge neutrality in this region:

$$p \approx [V_M'] \tag{11.17}$$

Since the donors have been replaced in the approximation to charge neutrality, they no longer influence the defect concentrations. With further oxidation, the behavior of the pure MX is identical to that of the doped MX, and the properties thus revert to those of intrinsic nonstoichiometry. Note that the lines for all the defects intersect those for the pure compound at exactly the same value of $P(X_2)$, a requirement of plane geometry as well as of defect chemistry. If all these intersections do not occur at the same $P(X_2)$, the diagram has been drawn incorrectly, because such failure would violate the simultaneous solution of the mass-action expressions.

The Impurity-Controlled Region: Electronic Compensation

On the reduction side of the stoichiometric composition, in the region where the donors are compensated by cation vacancies, the electrons and holes will have the same dependence on $P(X_2)$ as on the oxidation side. With decreasing $P(X_2)$, the rising electron concentration will cross the line for $V_X^{\cdot}$ and will eventually intersect the horizontal line for $[D_M^{\cdot}]$ and $[V_M']$. With further reduction, electrons replace cation vacancies as the compensating defects for the donors. The electron concentration cannot yet rise above the donor level because that would violate charge neutrality, since there are no positively charged defects that can rise with them. Thus the electron concentration is temporarily frozen at the donor level and the approximate condition on charge neutrality becomes:

$$n \approx [D_M^{\cdot}] \tag{11.18}$$

When combined with Eq. (11.8), this requires that

$$[V_X^{\cdot}] \approx \frac{K_n}{[D_M^{\cdot}]} P(X_2)^{-1/2} \tag{11.19}$$

and the substitution of this expression into Eq. (11.2) gives:

$$[V_M'] \approx \frac{K_S[D_M^{\cdot}]}{K_n} P(X_2)^{1/2} \tag{11.20}$$

The hole concentration, of course, must remain independent of $P(X_2)$ along with the electrons. In this "crisscross" region, $[V_X^{\cdot}]$ rises to the level of n, which it intersects just as it meets their values for the pure MX. All the defect concentrations arrive at their value for the pure compound at this same value of $P(X_2)$. With further reduction, the effect of the donor dopant dies away, and the doped compound is indistinguishable from the pure compound with the same approximation to charge neutrality:

$$n \approx [V_X^{\cdot}] \qquad (11.21)$$

This completes the diagram for donor-doped MX. Note that MX cannot be acceptor-doped with cations because that would require a zero-valent cation; it could be acceptor-doped with a higher-valent anion, such as O_X'.

It is now apparent that both limiting cases of charge compensation occur, Eqs. (11.9) and (11.10), but in different ranges of nonmetal activity. Since electrons are a product of reduction, they are the compensating defect in the lower range of nonmetal activities, while cation vacancies, a product of oxidation, take over in the higher range of nonmetal activities.

An Important Aspect of Impurity-Controlled Nonstoichiometry

The alert reader may have noticed something peculiar in the preceding development of the Kröger–Vink diagram for donor-doped MX. As the material is reduced from the stoichiometric composition, the reduction reaction, Eq. (11.7), indicates that electrons and anion vacancies are being produced in equal numbers. Yet the electron concentration increases up to the level of $[V_X^{\cdot}]$, passes right through it, and continues up to the donor level, without any change in $[V_X^{\cdot}]$. This is required by the thermodynamics of the system as expressed by the mass-action expressions. As n rises through $[V_X^{\cdot}]$, $[V_M']$ must still be fixed at the donor level, independent of $P(X_2)$, because there is no other defect available to replace it in the approximation to charge neutrality. The mass-action expression for Schottky disorder, Eq. (11.2), then requires that $[V_X^{\cdot}]$ also remain independent of $P(X_2)$ until n reaches the donor level and $[V_M']$ can begin to decrease with further reduction into the electronically compensated region. But what happens to the additional anion vacancies formed after n has risen above $[V_X^{\cdot}]$? The answer is that they are absorbed by recombination and elimination with cation vacancies in a reversal of the reaction for the formation of Schottky disorder. Since $[V_M']$ is much larger than $[V_X^{\cdot}]$ in this region, $[V_X^{\cdot}]$ can be held constant with reduction without significant reduction in $[V_M']$ until n approaches the donor level.

The net result is that the reduction reaction in this region proceeds by the consumption of cation vacancies rather than by the production of anion vacancies.

$$
\begin{aligned}
X_X &\rightleftharpoons \tfrac{1}{2}X_2 + V_X^{\cdot} + e' & \Delta H_N \\
-(\text{nil} &\rightleftharpoons V_M' + V_X^{\cdot}) & -\Delta H_S \\
\hline
X_X + V_M' &\rightleftharpoons \tfrac{1}{2}X_2 + e' & \Delta H_N - \Delta H_S
\end{aligned}
\qquad (11.22)
$$

As $V_X^{\cdot}$ are formed by reduction, they combine with a corresponding number of V_M' present as compensating defects for the donors, and pairs of vacant cation and anion sites are eliminated from the lattice. The net result is given by Eq. (11.22), with a new apparent enthalpy of reduction, which we will designate as ΔH_{RI}:

$$\Delta H_{RI} = \Delta H_N - \Delta H_S \tag{11.23}$$

where the subscript RI denotes the reduction reaction in the region where the donors are compensated by an ionic defect. Equation (11.16) confirms that the activation enthalpy for electron production in this region corresponds to the right-hand side of Eq. (11.23). Comparison of Eqs. (11.16) and (10.28) indicates that the enthalpy for the production of electrons is less in the impurity-controlled region for the donor-doped case than in the near-stoichiometric region of the pure compound by $\Delta H_S/2$. This is because in the doped case, the reduction reaction is consuming ionic defects rather than creating them, and because these ionic defects result from the donor content and are not created by a thermally activated process.

There is nothing wrong with using Eq. (11.7), the reduction reaction with the production of anion vacancies, and its mass-action expression, Eq. (11.8), because in a consistent derivation of the electron concentration, as was done for Eq. (11.16), the Schottky formation enthalpy will automatically be included. However, Eq. (11.22) is equally valid, since it is obtained by combination of two valid equilibrium reactions and is a more accurate representation of what is happening at the atomic level. Thus Eq. (11.22) is the preferred way of describing the reduction reaction in this region. Its mass-action expression

$$\frac{n}{[V_M']} = K_{RI} P(X_2)^{-1/2} = K_{RI}' e^{-\Delta H_{RI}/kT} P(X_2)^{-1/2} \tag{11.24}$$

can then be used in the description of the system. Note that combination of Eq. (10.42), the relationship between the enthalpies for pure MX, and Eq. (11.23) gives:

$$\Delta H_{RI} + \Delta H_P = E_g^{\circ} \tag{11.25}$$

ΔH_S disappears in this formulation because the concentration of the majority ionic defect in this region is determined by the donor concentration, not by the thermally activated process of creating Schottky disorder. This is a very important relationship because in the evaluation of experimental results, an Arrhenius plot of the electron concentration in this region will yield ΔH_{RI} as the apparent activation enthalpy, and the investigator will have no a priori way of knowing that this contains $\Delta H_N - \Delta H_S$. It also indicates that a nonstoichiometric reaction, the reduction reaction in the near-stoichiometric region in this case, can be considerably enhanced (i.e., its net enthalpy can be reduced) by the presence of an appropriate dopant. Equation (11.25) also indicates that the band gap has become the sole determinant of the amount of enthalpy to be divided up between the enthalpies of oxidation and reduction for the impurity-controlled material.

The oxidation reaction is not subject to this modification because it still results in the creation of additional V_M'. In the later description of acceptor-doped M_2O_3, the reverse will be seen to be true: oxidation will result in the consumption of ionic defects, while reduction still results in the creation of additional ionic defects.

General Summary

As anticipated, the donor impurity was compensated by cation vacancies, an oxidation product, at relatively high $P(X_2)$, but at lower values was compensated by electrons, a reduction product. Thus the introduction of an impurity replaces the near-stoichiometric region of a pure compound, where charge neutrality is dominated by a pair of intrinsic defects, with two impurity-controlled regions in which the impurity is compensated by an ionic defect and an electronic defect. On one side of the diagram, the impurity-controlled region will blend in directly with the diagram for the pure compound, while, on the other side there is a characteristic "crisscross" region, where one product of the pertinent reaction for nonstoichiometry must hold constant at or parallel to the impurity level, while its companion reaction product rises up to its level so that intrinsic nonstoichiometry can take over. The combination of the two impurity-controlled regions, where the impurity is compensated by ionic and electronic defects, is wider in terms of nonmetal activity than the near-stoichiometric region in an ideally pure compound. In the extrinsic region, the approximations to charge neutrality occur at a higher concentration level than those in the near-stoichiometric region of the pure compound, necessarily covering a broader range of nonmetal activity before they blend back into the diagram for intrinsic nonstoichiometry. In this specific example, a near-stoichiometric region with a width of 10^{16} atm in pure MX has been replaced with an impurity-controlled region with a width of 10^{24} atm.

ENTHALPY RELATIONSHIPS

The exponents of $P(X_2)$ in the expressions for the various defects and the components of the apparent activation enthalpies for donor-doped MX are summarized in Table 11.1 for the two impurity-controlled regions. It is seen that the introduction of these exponents and enthalpies into the various mass-action expressions gives consistent results. Comparison of the enthalpy values with those for the near-stoichiometric region of pure MX leads to some important considerations.

First of all, in the region where $[V_M']$ is fixed by the donor content, this defect is dependent on neither $P(X_2)$ nor temperature. Thus instead of splitting the enthalpy of Schottky formation between the intrinsic ionic defects, as is the case for pure MX, the full enthalpy is thrown onto $[V_X^{\cdot}]$, as is the resulting temperature dependence. This was discussed in Chapter 5 with regard to extrinsic ionic disorder. The same situation prevails for the electronic defects in the region where the donor impurity is compensated by electrons: the full band gap is thrown onto the holes, as is the resulting temperature dependence.

A point of greater significance results from a comparison of the enthalpy terms for the electronic defects in the near-stoichiometric regions (defined as the regions on either side of the condition n = p) for the doped and undoped materials. In the expressions for the electron and hole concentrations in pure MX, the reaction

TABLE 11.1
The exponent of the nonmetal dependences of the defect concentrations, and the components of the apparent enthalpy (per defect) for donor-doped MX, for the two impurity-controlled regions

	$[D_M^\bullet] = n$	$[D_M^\bullet] = [V_M']$
V_M'	$1/2$ $\Delta H_S - \Delta H_N = 1.5\,\text{eV}$ $\Delta H_P - E_g^\circ$	0 0
$V_X^\bullet$	$-1/2$ ΔH_N	0 ΔH_S
n	0 0	$-1/2$ $\Delta H_N - \Delta H_S = 1.5\,\text{eV}$ $E_g^\circ - \Delta H_P$
p	0 E_g°	$1/2$ ΔH_P

enthalpies for reduction and oxidation are reduced by half of the Schottky formation enthalpy, as shown in Table 10.1. In the doped case, this reduction is lost by the holes and is doubled for the electrons, as shown in Table 11.1. In other words, relative to pure MX, an enthalpy equivalent to $\Delta H_S/2$ has been shifted from the electrons to the holes. This is directly related to the earlier discussion, in which it was demonstrated that in the donor-doped case, the reduction reaction actually consumes ionic defects rather than producing them.

In the case of the pure compound in the near-stoichiometric region, the oxidation and reduction reactions produce ionic defects in a large reservoir of the same defects and do not significantly change the total concentrations. Thus the reaction enthalpies can be reduced by the enthalpy of formation of the appropriate ionic defect, $\Delta H_S/2$ in this case. In donor-doped MX, V_M' are actually consumed by the reduction reaction, so another unit of $\Delta H_S/2$ can be taken from the enthalpy of formation of electrons. However, since the enthalpies of electron and hole formation must sum up to the enthalpy of formation of intrinsic electronic disorder, the band gap, this unit of $\Delta H_S/2$ taken from the enthalpy of formation of electrons must be added to the enthalpy of formation of holes. The hole concentration in the near-stoichiometric region is then even more temperature dependent than in the pure compound.

A comparison of Eqs. (10.42) and (11.25), the relationships between the enthalpy terms for the pure and donor-doped cases, offers an important distinction. If the phenomenological enthalpies of oxidation and reduction determined experimentally sum up to the band gap, as in Eq. (11.25), it is clear that the system is exhibiting extrinsic nonstoichiometry. This is a clue that the intrinsic defect that is dominating the condition of charge neutrality is not thermally activated and therefore must be controlled extrinsically.

A MORE COMPLEX EXAMPLE: ACCEPTOR-DOPED M_2O_3

Basic Relationships

Once again, we build on an example used in Chapter 10 to illustrate intrinsic non-stoichiometry: pure M_2O_3, with Schottky defects being the dominant type of ionic intrinsic disorder, and with these defects being present in concentrations substantially lower than those of the intrinsic electronic defects at the stoichiometric composition. We will now construct a Kröger–Vink diagram for this material with an added acceptor impurity. For convenience, we repeat the equilibrium reactions and their mass-action expressions for M_2O_3. Remember, to a very close approximation, the addition of an impurity does not change the thermodynamic parameters.

$$\text{nil} \rightleftharpoons 2V_M''' + 3V_O^{\cdot\cdot} \tag{11.26}$$

$$[V_M''']^2 [V_O^{\cdot\cdot}]^3 = K_S \tag{11.27}$$

$$\text{nil} \rightleftharpoons e' + h^{\cdot} \tag{11.28}$$

$$np = K_I \tag{11.29}$$

$$\tfrac{3}{2}O_2 \rightleftharpoons 3O_O + 2V_M''' + 6h^{\cdot} \tag{11.30}$$

$$[V_M''']^2 p^6 = K_p P(O_2)^{3/2} \tag{11.31}$$

$$O_O \rightleftharpoons \tfrac{1}{2}O_2 + V_O^{\cdot\cdot} + 2e' \tag{11.32}$$

$$[V_O^{\cdot\cdot}]n^2 = K_n P(O_2)^{-1/2} \tag{11.33}$$

The values of the various mass-action constants were given as follows:

$$
\begin{aligned}
K_S &= 3.375 \times 10^{70} \\
K_I &= 10^{36} \\
K_p &= 10^{151} \\
K_n &= 1.5 \times 10^{45}
\end{aligned}
$$

Compensating Defects

It will be assumed that our new sample of M_2O_3 contains 10^{19} cm^{-3} substitional cation acceptor impurities A_M', added in the form of its oxide, AO. Given the preferred type of intrinsic ionic disorder, the two possible compensating defects are anion vacancies and holes. The corresponding incorporation reactions are:

$$2AO \xrightarrow{(M_2O_3)} 2A_M' + 2O_O + V_O^{\cdot\cdot} \tag{11.34}$$

$$2AO + \tfrac{1}{2}O_2 \xrightarrow{(M_2O_3)} 2A_M' + 3O_O + 2h^{\cdot} \tag{11.35}$$

The first case corresponds to the stoichiometric solid solution, while the second gives a nonstoichiometric product by virtue of the addition of excess oxygen. The complete statement of charge neutrality is:

$$n + 3\,[V_M'''] + [A_M'] = p + 2[V_O^{\cdot\cdot}] \tag{11.36}$$

The Kröger–Vink diagram for pure M_2O_3 is repeated as dashed lines in Fig. 11.2, and we will superimpose the diagram for the acceptor-doped version on top of it. Since intrinsic electronic disorder is favored over intrinsic ionic disorder in M_2O_3, let us assume that at $P(O_2) = 10^{-10}$ atm, the value at the stoichiometric composition for the pure compound, the primary compensating defects for the acceptor impurity are holes [i.e., Eq. (11.35)]. This will prove to be a smarter choice than the anion vacancies, in this particular case. To compensate the charge of the 10^{19} acceptor centers, the hole concentration will have to increase by one order of magnitude from its value in the stoichiometric, pure compound. According to Eq. (11.29), the electron concentration will then have to decrease by one order of magnitude. From Eq. (11.33), it is seen that $[V_O^{\cdot\cdot}]$ will have to increase by two orders of magnitude, and from either Eq. (11.27) or (11.31), $[V_M''']$ will have to decrease by three orders of magnitude. These changes do not violate charge neutrality and are therefore acceptable. It is left to the reader to prove that the choice of anion vacancies as the compensating defect at this nonmetal activity leads to inconsistent results. As always, there is only one consistent choice. The new concentrations can be entered in Fig. 11.2 at $P(O_2) = 10^{-10}$ atm.

The Impurity-Controlled Region: Electronic Compensation

With holes as the primary compensating defects, the condition of charge neutrality can be approximated by:

$$p \approx [A_M'] \tag{11.37}$$

Introduction of this relationship into Eq. (11.31), the mass-action expression for the oxidation reaction, gives

$$[V_M'''] \approx \frac{K_p^{1/2}}{[A_M']^3} P(O_2)^{3/4} \tag{11.38}$$

This is the same dependence on $P(O_2)$ as was found for $[V_M''']$ in pure M_2O_3 in the near-stoichiometric region, so a line for this defect can be extended from its new value, parallel to its line in the pure compound. In the direction of higher $P(O_2)$, this line for $[V_M''']$ will intersect its line in the pure compound, in the highly oxidized region, at $P(O_2) = 1$ atm, exactly the value at which the horizontal lines representing p and n intersect their values in the pure compound. Further oxidation will give results that

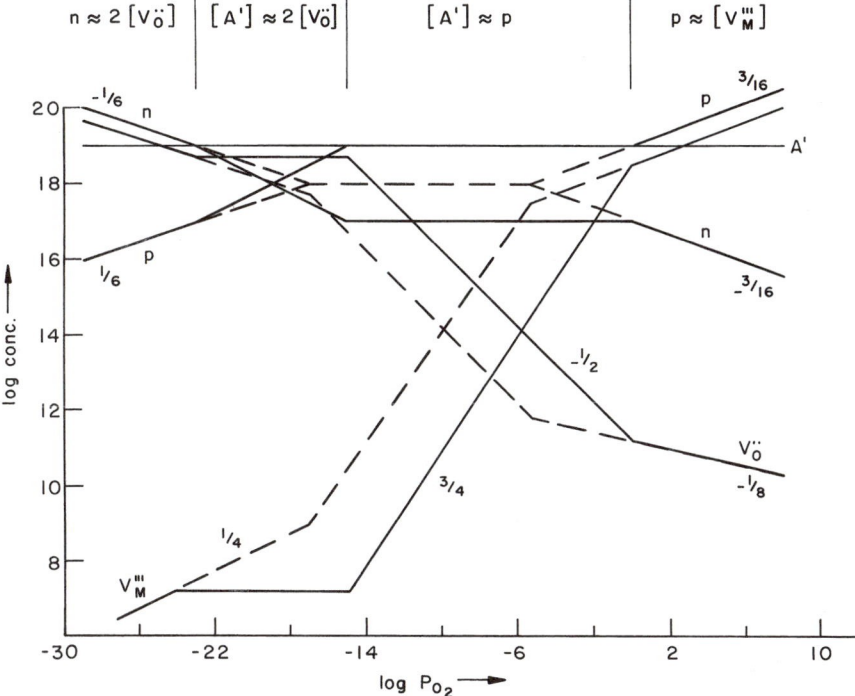

Figure 11.2 The Kröger–Vink diagram for acceptor-doped M_2O_3 with the parameters given in the text: dashed lines, diagram for pure M_2O_3.

are identical to those obtained for the pure compound, and the new approximation to charge neutrality will become

$$p \approx 3[V_M^{///}] \tag{11.39}$$

Back in the impurity-controlled region, expressions for $[V_O^{\cdot\cdot}]$ can be obtained by substituting Eq. (11.38) into the mass-action expression for Schottky disorder, Eq. (11.27):

$$[V_O^{\cdot\cdot}] \approx \left(\frac{K_S}{K_p}\right)^{1/3} [A_M']^2 P(O_2)^{-1/2} \tag{11.40}$$

or by combining Eq. (11.33), the mass-action expression for the reduction reaction, with Eqs. (11.29) and (11.37):

$$[V_O^{\cdot\cdot}] \approx K_n \left(\frac{[A_M']}{K_I}\right)^2 P(O_2)^{-1/2} \tag{11.41}$$

In either case, the dependence on $P(O_2)$ is the same as in the near-stoichiometric region in the pure compound, and a parallel line can be extended from the new value determined for $[V_O^{\cdot\cdot}]$ at $P(O_2) = 10^{-10}$ atm. This line will also intersect the line for $[V_O^{\cdot\cdot}]$ in the pure compound at $P(O_2) = 1$ atm. This completes the Kröger–Vink diagram for acceptor-doped M_2O_3 for values of $P(O_2)$ greater than 10^{-10} atm.

The Impurity-Controlled Region: Ionic Compensation

At lower values of $P(O_2)$, the lines for p and n will continue horizontally, while $[V_O^{\cdot\cdot}]$ increases with decreasing $P(O_2)$ with a slope of $-1/2$. It will increase with reduction to the value $[A_M'] \approx 2[V_O^{\cdot\cdot}]$, at which point it is temporarily constrained from further increases because there is no oppositely charged species that can increase with it to maintain charge neutrality. This leads to a region in which charge neutrality is approximated by:

$$[A_M'] \approx 2[V_O^{\cdot\cdot}] \tag{11.42}$$

In this "crisscross" region, the electron concentration rises with reduction according to:

$$n \approx \left(\frac{2K_n}{[A_M']}\right)^{1/2} P(O_2)^{-1/4} \tag{11.43}$$

while the hole concentration decreases as follows:

$$p \approx K_I \left(\frac{[A_M']}{2K_n}\right)^{1/2} P(O_2)^{1/4} \tag{11.44}$$

or, alternatively:

$$p \approx \left(\frac{K_p}{K_S}\right)^{1/6} \left(\frac{[A_M']}{2}\right)^{1/2} P(O_2)^{1/4} \tag{11.45}$$

$[V_M''']$ must remain independent of $P(O_2)$ as long as $[V_O^{\cdot\cdot}]$ is held constant at the acceptor impurity level, and is given by:

$$[V_M'''] \approx \left(\frac{2}{[A_M']}\right)^{3/2} K_S^{1/2} \tag{11.46}$$

The extensions of all these lines meet their corresponding lines in the pure compound at $P(O_2) = 10^{-23}$ atm, and further reduction proceeds as in the pure case. This leads to the highly reduced case, where

$$n \approx 2[V_O^{\cdot\cdot}] \tag{11.47}$$

This completes the Kröger–Vink diagram for this specific case of acceptor-doped M_2O_3, as shown in Fig. 11.2.

GENERAL CONSIDERATIONS

The derivation of the Kröger–Vink diagram just described was based on drawing lines of derived slopes from some initial data points until they reached concentrations that corresponded to the boundary between regions with different approximations to charge neutrality. Considerable care is necessary to obtain an accurate diagram with this approach, especially when odd slopes such as 3/16 and 3/4 are involved. Figure 11.2 was actually constructed by calculating the values of $P(O_2)$ and the defect concentrations at the boundaries between regions, and at a few other useful points, from the various mass-action constants. It is then just a matter of connecting up the points correctly. The calculation of such critical points for this specific example is left to the exercises at the end of the chapter.

As in the case of donor-doped MX, the near-stoichiometric region of the pure compound has been replaced by two impurity-controlled regions. In one of these, the impurity is compensated by an ionic defect, while in the other it is compensated by an electronic defect. In the case of an acceptor impurity, electronic compensation occurs at the higher nonmetal activities, while the reverse is true for a donor impurity. The total width of the impurity-controlled region is always wider in terms of nonmetal activity than the near-stoichiometric region of the pure compound.

On one side of the Kröger–Vink diagram, the defect concentrations in the impurity-controlled region blend directly into those for the pure compound. On the other side, there is a characteristic "crisscross" region, in which the charge compensating defect is temporarily constrained from increasing above the impurity level until its oppositely charged companion in the formation reaction has risen to its level. The side on which this occurs depends on whether the impurity is an acceptor or a donor, and whether electronic or ionic defects are the favored type of intrinsic disorder. In donor-doped M_2O_3, the "crisscross" region would appear on the high $P(O_2)$ side of the diagram.

The log–log slopes, the dependences of the defect concentrations on the impurity content, and the components of the activation enthalpies for the various defects in the impurity-controlled regions of acceptor-doped M_2O_3 are summarized in Table 11.2. It is a useful exercise to confirm that these values are consistent with the various mass-action expressions.

Once again, we see the peculiar phenomenon of one product of a reaction, the $V_O^{\cdot\cdot}$ of the reduction reaction in this case, rising above the level of its companion reaction product, the electrons, which are constrained to remain level as long as their complementary defects, the holes, are fixed by the impurity level. As in the case of the similar region in donor-doped MX, one product of the reaction, the electron, combines with and is annihilated by its complementary intrinsic defect, the hole, which is present in abundance as the primary compensating defect. This can be viewed as a two-step process, the normal reduction reaction, combined with the reverse of the formation of intrinsic electronic disorder

$$O_O \rightleftharpoons \tfrac{1}{2}O_2 + V_O^{\cdot\cdot} + 2e' \qquad \Delta H_N$$

$$-(\text{nil} \rightleftharpoons 2e' + 2h^{\cdot}) \qquad -2E_g^{\circ} \qquad (11.48)$$

$$\overline{O_O + 2h^{\cdot} \rightleftharpoons \tfrac{1}{2}O_2 + V_O^{\cdot\cdot}} \qquad \overline{\Delta H_N - 2E_g^{\circ}}$$

The net result is a new reduction reaction that proceeds by the consumption of the primary compensating defect, with a new apparent enthalpy, ΔH_{RE}, which is less

TABLE 11.2
The exponent of the oxygen activity dependences of the defect concentrations, the components of the apparent enthalpies, and the impurity dependences for the defects in acceptor-doped M_2O_3 for the two impurity-controlled regions

	$[A_M'] = 2[V_O^{\cdot\cdot}]$	$[A_M'] = p$
V_M'''	0	3/4
	$\Delta H_S/2$	$\Delta H_P/2$
	$[A_M']^{-3/2}$	$[A_M']^{-3}$
$V_O^{\cdot\cdot}$	0	$-1/2$
	0	$\Delta H_N - 2E_g^\circ$
	$[A_M']$	$[A_M']^2$
n	$-1/4$	0
	$\Delta H_N/2$	E_g°
	$[A_M']^{-1/2}$	$[A_M']^{-1}$
p	1/4	0
	$E_g^\circ - \Delta H_N/2$	0
	$[A_M']^{1/2}$	$[A_M']$

than ΔH_N, the intrinsic reduction enthalpy, by $2E_g^\circ$. The subscript RE refers to the reduction reaction in which the acceptors are compensated by an electronic defect:

$$\Delta H_{RE} = \Delta H_N - 2E_g^\circ \tag{11.49}$$

This strongly enhances the reduction reaction in this region by the reduction of the net enthalpy. In pure M_2O_3, the temperature dependence of $[V_O^{\cdot\cdot}]$ is determined by $\Delta H_N - E_g^\circ$, Eq. (10.82), while in the acceptor-doped case, it is $\Delta H_N - 2E_g^\circ$. The enthalpy for the creation of $V_O^{\cdot\cdot}$ has been reduced by an amount E_g° by the addition of the acceptor dopant. In most cases, oxidation and reduction reactions produce two types of defect, one ionic and one electronic. For this specific situation in the impurity-controlled region, the reactions produce one type of defect and consume the other. This is a much more favorable energetic situation, and the enthalpy is reduced accordingly. The mass-action expression for Eq. (11.48) can be written as follows:

$$\frac{[V_O^{\cdot\cdot}]}{p^2} \approx K_{RE} P(O_2)^{-1/2} \tag{11.50}$$

and the alternate expression for the anion vacancy concentration in this region is then:

$$[V_O^{\cdot\cdot}] \approx K_{RE}[A_M']^2 P(O_2)^{-1/2} \tag{11.51}$$

In the discussion of pure M_2O_3, the relationship between the reaction enthalpies was found to be, Eq. (10.90):

$$3\Delta H_N + \Delta H_P = \Delta H_S + 6E_g^\circ \tag{11.52}$$

and this is still valid for the doped material. Remember that the numerical coefficients result from the fact that the Schottky formation reaction and the oxidation reaction have been written such that six defect charges of each sign are created, while only two are created by the reduction reaction and one by the formation reaction for intrinsic electronic disorder. These are arbitrary choices made to creat the minimum number of defect charges without requiring fractional coefficients on the defects.

Combination of Eq. (11.48) and the oxidation reaction, Eq. (11.30), also gives us an alternative relationship:

$$\frac{3}{2}O_2 \rightleftharpoons 3O_O + 2V_M^{///} + 6h^\cdot \qquad \Delta H_P$$

$$\underline{3O_O + 6h^\cdot \rightleftharpoons \frac{3}{2}O_2 + 3V_O^{\cdot\cdot}} \qquad 3\Delta H_{RE} \tag{11.53}$$

$$nil \rightleftharpoons 2V_M^{///} + 3V_O^{\cdot\cdot} \qquad \Delta H_S$$

indicating that:

$$\Delta H_P + 3\Delta H_{RE} = \Delta H_S \tag{11.54}$$

An Arrhenius plot of any experimental parameter that is proportional to $[V_O^{\cdot\cdot}]$ will yield ΔH_{RE}. When this is combined with ΔH_P, there will be a relationship with ΔH_S, but E_g° will be missing compared with Eq. (11.52) because the concentration of holes is fixed by the acceptor content, not by a thermally activated process involving intrinsic ionization. Experimental agreement with Eq. (11.53) is proof that the hole concentration is controlled extrinsically in this region. Such relationships can be very useful in the interpretation of experimental results. It will be seen that some materials are useful for practical applications only because of the basic phenomenon just described. In some cases this results from the naturally occurring impurity content of the raw materials, a truly fortuitous situation.

As in the case of intrinsic nonstoichiometry, a few generalities can be obtained from Tables 11.1 and 11.2:

1. When a defect is compensating the charge of the impurity center, the enthalpy associated with its complementary intrinsic defect is the full intrinsic formation enthalpy per defect. Thus when $V_O^{\cdot\cdot}$ is the compensating defect in acceptor-doped M_2O_3, the entire Schottky formation enthalpy, ΔH_P, is thrown onto the two $V_M^{///}$ that result from the oxidation reaction, or $\Delta H_P/2$ for each of them.

2. When a defect is compensating the charge of the impurity center, the enthalpy associated with the other defect formed with it in a nonstoichiometric reaction is the full reaction enthalpy per defect. Thus when holes are the compensating defects in acceptor-doped M_2O_3, the full reduction enthalpy is thrown onto the two cation vacancies that result from the oxidation reaction, or $\Delta H_P/2$ for each of them.

3. The enthalpy associated with the defect that bears a charge that is opposite to that of the impurity center, but is not the compensating defect, is the full enthalpy per defect of the nonstoichiometric reaction minus the enthalpy necessary to form its complementary reaction defect. Thus when holes are the compensating defects in acceptor-doped M_2O_3, the enthalpy associated with the oxygen vacancies is the full enthalpy of the reduction reaction minus the enthalpy necessary to form the two electrons from that reaction, E_g° each, or $2E_g^\circ$ for the two of them. Likewise, when oxygen vacancies are the compensating defect, the enthalpy associated with holes is the enthalpy of the oxidation reaction per hole, $\Delta H_P/6$, minus the enthalpy necessary to form an equivalent charge of cation vacancies, $\Delta H_S/6$.

NONSTOICHIOMETRIC REACTIONS IN THE IMPURITY-CONTROLLED REGION

For the case of donor-doped MX, when the donors are compensated by cation vacancies, the reduction reaction can be viewed as consuming cation vacancies, which are present in abundance, rather than as creating anion vacancies:

$$X_X + V'_M \rightleftharpoons \tfrac{1}{2}X_2 + e' \tag{11.55}$$

It is also true that in the region where the donors are compensated by electrons, the oxidation reaction can be viewed as consuming electrons, which are present in abundance, rather than as creating holes:

$$\tfrac{1}{2}X_2 + e' \rightleftharpoons X_X + V'_M \tag{11.56}$$

But Eq. (11.56) is just the reverse of Eq. (11.55), so their enthalpies must be equal in absolute magnitude, but opposite in sign. One of these enthalpies must be negative. Similarly for the case of acceptor-doped M_2O_3, oxidation in the low $P(O_2)$ side of the impurity-controlled region can be viewed as the consumption of oxygen vacancies:

$$\tfrac{1}{2}O_2 + V_O^{\cdot\cdot} \rightleftharpoons O_O + 2h^\cdot \tag{11.57}$$

while reduction in the high $P(O_2)$ side results in the consumption of holes:

$$O_O + 2h^\cdot \rightleftharpoons \tfrac{1}{2}O_2 + V_O^{\cdot\cdot} \tag{11.58}$$

Again Eq. (11.58) is the reverse of Eq. (11.57), so their enthalpies must be equal in magnitude but opposite in sign. It is easily shown that this must be true by the principles of plane geometry. Thus if one can measure the enthalpy of one of these reactions, the enthalpy of the other is obvious.

The conceptual background has now been set for the analysis of some real systems.

☑ PROBLEMS

11.1. For the example of acceptor-doped M_2O_3 described in the text, show that oxygen vacancies cannot be the compensating defects at the $P(O_2)$ value of the stoichiometric composition of the pure compound.

11.2. For the example of acceptor-doped M_2O_3 described in the text, derive expressions and give numerical values for the oxygen activities for:

    a. The boundaries between regions having different approximations to charge neutrality.

    b. The points at which n = p and $[V_M''']= [V_O^{\cdot\cdot}]$.

11.3. Assume that an oxide M_2O_5 has anion Frenkel defects as the preferred type of intrinsic ionic disorder and that $K_{AF} = 10^{26}$ defects^2/cm^6 and $K_I = 10^{24}$ defects^2/cm^6 at 1000°C. Also assume that the pure compound is exactly stoichiometric at $P(O_2) = 1$ atm. Assuming that M_2O_5 contains a net excess of 10^{16} donor impurities/cm^3, draw a defect diagram for this compound at 1000°C.

11.4. On the defect diagram for pure AgBr derived in Problem 10.3, superimpose a diagram for AgBr containing 1 mol % $CdBr_2$. Distinguish between the two diagrams by using dashed lines or different colors.

11.5. On the defect diagram for pure M_2O derived in Problem 10.4, superimpose a diagram for M_2O containing 10^{19} cm^{-3} of a divalent cation D^{2+}. What is the effect of the impurity addition on the values of the mass-action constants?

11.6. The halide of a divalent metal, MX_2, has the fluorite structure and thus has anion Frenkel disorder as the preferred type of intrinsic ionic disorder. At 800°C the pure compound is precisely stoichiometric at $P(X_2) = 10^{-16}$ atm, where n = p = 10^{14} cm^{-3}. When doped with 10^{17} cm^{-3} of a monovalent cation, A^+, $[X_I'] = [V_X^{\cdot}] = 10^{16}$ cm^{-3} at $P(X_2) = 10^{-10}$ atm.

    a. On the same piece of graph paper, draw Kröger–Vink diagrams for both the pure and the doped MX_2 at 800°C.

    b. At what value of $P(X_2)$ is the doped MX_2 stoichiometric?

    c. Give general analytical expressions (in terms of the mass-action constants and the impurity concentration) and numerical values for the location of both the upper and lower boundaries of the impurity-controlled region, as well as the position at which the compensating defect changes from ionic to electronic.

11.7. The following figure represents the measured equilibrium electrical conductivity of a hypothetical oxide MO of unknown impurity content as a function of oxygen activity at 800°C. Assume that the conductivity is due only to electrons and/or holes and that their mobilities are equal and independent of defect concentrations.

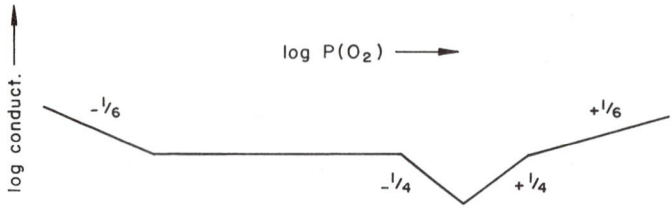

a. Complete the form of the defect diagram for MO at 800°C. Include as many defects as possible, and label each line, as well as the approximation to charge neutrality for each region. Show how the slopes for each line were determined.

b. Indicate how you would determine the value of the band gap (the enthalpy of intrinsic electronic disorder) and the enthalpy of intrinsic ionic disorder from such measurements made at several temperatures.

11.8. The following figure represents the measured equilibrium electrical conductivity of a hypothetical oxide NO of unknown impurity content as a function of $P(O_2)$ at 1000°C. Make the same assumptions as listed in Problem 11.7.

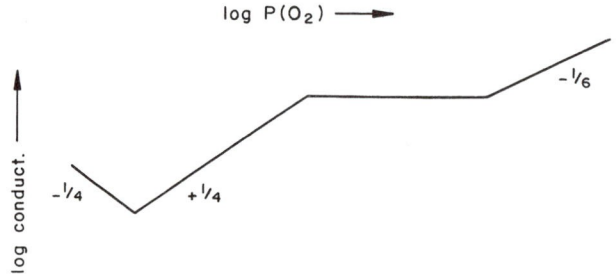

a. Complete the form of the defect diagram as described in Problem 11.7.

b. Superimpose on your diagram a diagram for ideally pure NO.

Titanium Dioxide

12

INTRODUCTION

TiO_2 is a compound of considerable usefulness. Because of its optical properties it is widely used as a pigment (e.g., as the major white pigment in exterior and interior paints, as a brightener in laundry detergents). It has also been used as the dielectric in ceramic capacitors because of its high dielectric constant, although it has been surplanted in recent years by more effective materials, the most prominent of which, $BaTiO_3$, is usually produced by the reaction of $BaCO_3$ and TiO_2. Because of the usefulness and ready availability of TiO_2, its defect chemistry has been studied extensively. In this chapter, we review the present status of our understanding of this material.

Before discussing the various experimental studies of TiO_2, let us try our predictive bag of tricks to see what might be expected for its properties. There are several crystalline forms of TiO_2, but the most stable structure, to which all the others revert at high temperatures, is the rutile structure, and this is the only form with which we need to be concerned. This is the name of the mineral form of TiO_2 found most commonly in nature. As described in Chapter 2, the rutile structure has an approximately hexagonal-close-packed oxygen sublattice, with cations occupying one half of each layer of octahedrally coordinated cation sites. With a close-packed anion sublattice, anion Frenkel disorder should be very unfavorable. This leaves cation Frenkel or Schottky disorder as possibilities for the preferred form of intrinsic ionic disorder. Since only half the octahedrally coordinated cation sites are occupied, cation Frenkel disorder is a definite possibility, although both defects have quite high formal charges:

$$Ti_{Ti} + V_I \rightleftharpoons Ti_I^{4\cdot} + V_{Ti}^{4/}$$ (12.1)

This amounts to a partial disordering of the pattern of occupation of the octahedral sites. Since oxygen vacancies are frequently found to exist in oxides, Schottky disorder is also possible, although the Ti vacancy again is highly charged:

$$nil \rightleftharpoons V_{Ti}^{4/} + 2V_O^{\cdot\cdot}$$ (12.2)

217

Without further information (i.e., experimental results), it is not possible to make a confident guess about which one of these possibilities is preferred. Because both options involve very highly charged defects, the enthalpies of formation can be expected to be quite high. There have been attempts to calculate defect energies, although there are admitted difficulties in finding a self-consistent set of interatomic potentials that accurately reproduce measured properties of the crystal (e.g., dielectric constants and elastic constants). These calculations indicate that Schottky disorder should be favored by a wide margin [2.7 eV (260 kJ/mol) per defect], while *anion* Frenkel disorder [4.4 eV (420 kJ/mol) per defect] is even more favorable than cation Frenkel disorder [7.3 eV (700 kJ/mol) per defect] (Catlow and James, 1982, Catlow et al., 1985). This ranking seems counterintuitive, and it will be seen that it is not well supported by experimental evidence.

The Ti^{4+} ion has the argon electronic structure and thus cannot be oxidized to a higher oxidation state. However, it can be reduced to Ti^{3+} and even Ti^{2+}. Aqueous solutions of Ti^{3+} are stable when they are protected from oxygen. TiO_2 belongs to the class of transition metal oxides in which the cation has no d electrons and is in its highest oxidation state. Thus the conduction band is expected to be made up of empty Ti 3d states, which are hospitable to electrons, while the valence band is composed of filled O 2p states, which are not hospitable to holes. The conduction band should lie at a lower level than it does in insulating oxides, so that the band gap should be correspondingly smaller, perhaps in the midrange of 3–4 eV (300–400 kJ/mol) that is typical of this class of oxides.

With the enthalpies for all kinds of intrinsic disorder expected to be quite high, it is likely that TiO_2 will be very sensitive to the presence of aliovalent impurities. In fact, it is probable that even for undoped material the concentrations of impurity centers and their compensating extrinsic defects will exceed those of any type of intrinsic defect under some conditions. Since the Ti^{4+} ion carries such a high charge, it is almost certain that most of the accidental metallic impurities will have smaller charges and will function as acceptor impurities (e.g., Al^{3+}, Fe^{3+}, Mg^{2+}). The net acceptor content can be charge-compensated by Ti interstitials, oxygen vacancies, or holes. Since the valence band is not receptive to holes, it is more likely that one of the ionic defects is preferred. Thus the two major possibilities are:

$$2A_2O_3 + Ti_{Ti} + V_I \xrightarrow{(3TiO_2)} 4A'_{Ti} + Ti_I^{4\cdot} + 6O_O \tag{12.3}$$

$$A_2O_3 \xrightarrow{(2TiO_2)} 2A'_{Ti} + 3O_O + V_O^{\cdot\cdot} \tag{12.4}$$

Until we have more information, it is difficult to choose between these two options.

In the case of donor impurities, which will most likely have to be deliberately added, the positively charged impurity centers can be compensated by interstitial oxygen, Ti vacancies, or electrons. The first of these is unlikely in the close-packed lattice, and the Ti vacancies carry a very high formal charge. However, the conduction band, which is derived from the empty Ti 3d states, should be quite receptive to electrons, and they are a reasonable choice for the compensating defect. The incorporation reaction would then be:

$$D_2O_5 \xrightarrow{(2TiO_2)} 2D'_{Ti} + 4O_O + {}^1\!/_2 O_2 + 2e' \tag{12.5}$$

It is possible that this electronic compensation may switch over to compensation by Ti vacancies under oxidizing conditions.

Since the valence band is inhospitable to holes, we expect the acceptor centers to be effective hole traps and to represent deep levels in the band gap. However, donor levels are expected to be quite shallow, because the conduction band is receptive to electrons. Thus we might expect acceptor-doped TiO_2 to be a light-colored insulator, while the donor-doped compound should be a dark-colored semiconductor.

For the formation of nonstoichiometric compositions, the most likely reduction reactions are:

$$O_O \rightleftharpoons {}^1\!/_2 O_2 + V_O^{\cdot\cdot} + 2e' \tag{12.6}$$

$$Ti_{Ti} + 2O_O + V_I \rightleftharpoons O_2 + Ti_I^{4\cdot} + 4e' \tag{12.7}$$

Experience has shown that oxygen vacancies are easily formed in oxides, and in this case the conduction band is receptive to electrons. Thus Eq. (12.6) should be a low enthalpy process. It remains to be seen whether Eq. (12.7) can compete with it. With an easily reducible cation and an easily formed ionic defect, we can expect to see an extensive range of oxygen-deficient nonstoichiometric compositions. The possible oxidation reactions are:

$$ {}^1\!/_2 O_2 + V_I \rightleftharpoons O_I'' + 2h^{\cdot} \tag{12.8}$$

$$O_2 \rightleftharpoons 2O_O + V_{Ti}^{4/} + 4h^{\cdot} \tag{12.9}$$

Both these possibilities suffer from the fact that the valence band is not a friendly place for holes. Equation (12.8) is particularly unfavorable because of the difficulty of forming interstitial oxygen in the rutile structure, and this reaction can be dismissed. In Eq. (12.9), the Ti vacancy is highly charged, and this property, together with the unfavorable holes, suggests that this is a high enthalpy reaction. Thus as in all oxides that do not have an oxidizable cation, we do not expect to see an extensive range of oxygen-excess nonstoichiometry. However, the oxidation reaction can be substantially enhanced in the presence of acceptor impurities. Oxidation can then proceed with the consumption of ionic defects, rather than their creation. For the two possible compensating ionic defects, the two oxidation reactions would be:

$$ {}^1\!/_2 O_2 + V_O^{\cdot\cdot} \rightleftharpoons O_O + 2h^{\cdot} \tag{12.10}$$

$$O_2 + Ti_I^{4\cdot} \rightleftharpoons 2O_O + Ti_{Ti} + V_I + 4h^{\cdot} \tag{12.11}$$

If cation Frenkel disorder is the preferred type of intrinsic ionic disorder, then the acceptor impurities should be compensated by Ti interstitials, and Eq. (12.11) would be expected to be the major oxidation reaction. It is left to the reader to show that the enthalpy for Eq. (12.11) is the enthalpy of the intrinsic oxidation reaction Eq. (12.9), reduced by the enthalpy for cation Frenkel disorder (12.1). That will result in a reduction in the oxidation enthalpy by several electron-volts. Thus TiO_2 that contains naturally occurring or deliberately added acceptor impurities may indeed show some range of oxygen-deficient nonstoichiometry.

These rather reasonable predictions should be kept in mind as we review the actual experimental results obtained with undoped and doped TiO_2. They will prove to be quite accurate, and their success demonstrates the usefulness of this predictive approach.

THE AMOUNT OF NONSTOICHIOMETRY

If the extent of nonstoichiometry is sufficiently large, the amount of oxygen that enters or leaves a sample can be observed as a change in weight. This approach is often referred to as thermogravimetry, or TGA, for thermogravimetric analysis. TGA is best done at temperature, but in some cases the samples can be quenched fast enough, and the equilibration rates are slow enough, to permit the weighing of quenched, equilibrated samples at room temperature. Obviously, this is best done for temperatures at which the equilibration times are rather long. Measurements made at temperature are plagued mainly by the agitation due to a flowing gas stream and the convection currents that result from temperature gradients in the apparatus. For that reason, the measurements are often made at reduced total gas pressures (e.g., ≤ 0.1 atm). For a change in oxygen content of one part in 10^4 (e.g., $TiO_{1.9998}$), a 10 g sample of TiO_2 would lose 0.0004 g, or 0.4 mg, of weight. This is close to the ultimate limit of sensitivity for an ordinary analytical balance. For a 1 g sample the weight loss would be 4×10^{-5} g or 40 μg, which is approaching the sensitivity of an ordinary apparatus with a microbalance. In some cases careful design of the apparatus has resulted in sensitivities of a few micrograms.

Figure 12.1 shows the TGA results of several investigators. It is gratifying to see that their results are in very good agreement. The presentation of the results in the form of x in TiO_{2-x} might imply that oxygen vacancies are the main product of reduction; if Ti interstitials are actually preferred, however, the results might be better expressed as y in $Ti_{1+y}O_2$. The oxygen activities are in a low range of values, 10^{-8}–10^{-18} atm. Such values are usually obtained by means of CO–CO_2 mixtures and depend on the equilibrium reaction:

$$CO + \tfrac{1}{2}O_2 \rightleftharpoons CO_2 \tag{12.12}$$

The mass-action expression for this can be solved for $P(O_2)$:

$$P(O_2) = \frac{1}{K(T)^2} \left[\frac{P(CO_2)}{P(CO)} \right]^2 \tag{12.13}$$

The mass-action constant $K(T)$ has been tabulated as a function of temperature, so it is only necessary to know the ratio CO_2/CO. The gas mixtures can be obtained

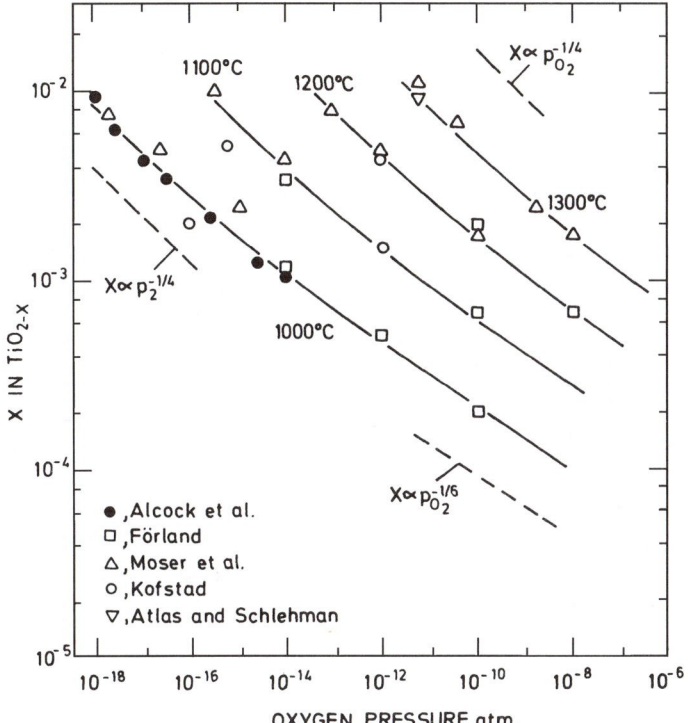

Figure 12.1 The value of x in TiO_{2-x} determined from thermogravimetric measurements by several investigators. (Reproduced from Kofstad, 1972, by permission of the editors.)

by flowing the components through precisely calibrated flow meters or by means of mechanical gas pumps that can deliver accurately known amounts. Or, as done in the author's laboratory, the oxygen activity can be measured by means of an oxygen concentration cell that uses acceptor-doped ZrO_2 as an ideal oxygen electrolyte. One merely inserts a closed-end zirconia tube, electroded inside and out at the tip with fired-on Pt paint, and measures the emf generated by the difference in oxygen activities near the sample in the furnace and inside the zirconia tube, either air or pure oxygen. In the case of metered gas flow rates, the mixtures can be measured accurately down to 1 part in 10^4. So for mixtures ranging from dilute CO in CO_2 to dilute CO_2 in CO, and given the square relationship shown in Eq. (12.13), one can in principle cover 16 orders of magnitude in oxygen activities with these mixtures. However, there is a restriction at the low activity end due to the disproportionation reaction of CO-rich mixtures:

$$2CO \rightleftharpoons CO_2 + C \qquad (12.14)$$

This can be detected by observing the deposition of soot on the furnace tube. The results shown in Fig. 12.1 cover most of the range that is accessible with these gas mixtures. Lower values of $P(O_2)$ can be reached by the use of H_2–H_2O mixtures, which establish an equilibrium mixture through the reaction:

$$H_2 + \frac{1}{2}O_2 \rightleftharpoons H_2O \qquad (12.15)$$

An oxide will lose weight with decreasing $P(O_2)$ whether it is on the oxygen-deficient or oxygen-excess side of stoichiometry. If the TiO_2 represented in Fig. 12.1 were on the oxygen-excess side [i.e., if it were moving from TiO_{2+x} toward TiO_2 as the $P(O_2)$ is decreased], then the amount of weight loss for a given increment of $\log P(O_2)$ would decrease as $P(O_2)$ decreases. Since, however, experimental results clearly show that the reverse is true, the samples must be moving away from the stoichiometric composition as $P(O_2)$ decreases. The samples are thus on the oxygen-deficient side of stoichiometry, as we predicted.

The measured weight change indicates how much oxygen entered or left the oxide after a change in the conditions of equilibration, but it gives no information on what those compositions are. One needs a reference composition that is achieved under known equilibration conditions. In the case of TiO_2, the data show that the amount of nonstoichiometry is approaching the limit of detectability at the lowest temperature and highest value of $P(O_2)$. Thus no weight change can be detected at 1000°C as the $P(O_2)$ approaches 1 atm. A sample equilibrated at that temperature in pure oxygen then is indistinguishable from a stoichiometric sample by this technique, and that condition can be used as the reference state from which the weight changes are compared. For some oxides, it is possible to determine the exact composition of a reference state by chemical analysis. This involves measurement of the chemical oxidizing or reducing power of the nonstoichiometric material that results from the gain or loss of oxygen from the stoichiometric composition. Measurement of the weight loss that results from the reduction of an oxide to the pure metal has also been used (e.g., by high temperature reduction in hydrogen).

The experimental results show a maximum deviation from stoichiometry of $x = 10^{-2}$, or $TiO_{1.98}$. It is not certain that the ionic defects are still randomly distributed at that composition, since it is known that even relatively small deviations from stoichiometry in TiO_2 are accommodated by a structural modification, the so-called Magnéli shear structures, described in Chapter 15.

It is seen that the amount of nonstoichiometry increases with increasing temperature. Dashed lines representing various postulated dependences on the oxygen activity are also shown in Fig. 12.1. Detailed interpretation is deferred until additional data have been examined.

THE EQUILIBRIUM ELECTRICAL CONDUCTIVITY OF UNDOPED TiO_2

The equilibrium electrical conductivity of undoped and doped TiO_2 has been measured by many investigators. We will focus on the very precise results shown in Fig. 12.2, which were obtained by Baumard et al. (1975) at a French national research laboratory (CNRS) in Orléans. It is easy to measure even small levels of conductivity, so the

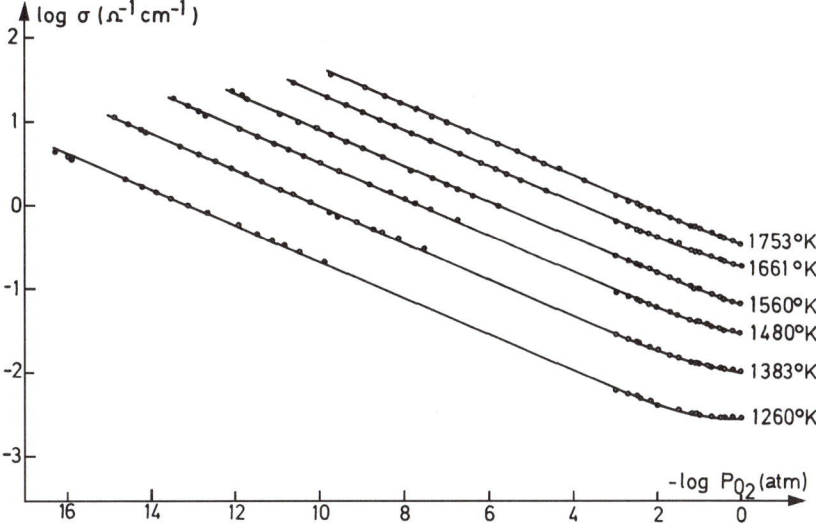

Figure 12.2 The equilibrium electrical conductivity of an undoped TiO_2 single crystal as a function of oxygen activity. (Reproduced from Baumard et al., 1975, by permission of Masson Editeur.)

results extend all the way to $P(O_2) = 1$ atm, where the TGA technique was too insensitive to be useful. The triangular gap in the data, which widens with decreasing temperature, is commonly seen. In the range of high $P(O_2)$, the gas mixtures consist of oxygen diluted with argon, while the low values of $P(O_2)$ were obtained with CO–CO_2 mixtures. The two ranges do not overlap because of the limitation on measuring very low flow rates. The gap is bounded on the high activity side by the most dilute mixtures of oxygen in argon, while on the low activity side it is bounded by the most dilute addition of CO to CO_2. The latter mixtures give a temperature-dependent oxygen activity. More recently, this gap is often bridged with the use of an electrochemical oxygen sensor, that is, an oxygen concentration cell with an electrolyte of acceptor-doped ZrO_2. Ironically, the region of the data gap often contains important information, as is the case for TiO_2.

The results clearly show a conductivity that is increasing with decreasing $P(O_2)$, which is characteristic of n-type conductivity resulting from oxygen-deficient non-stoichiometry. Thermoforce measurements (Seebeck coefficients) confirm that the conductivity is n type. It has been found that at these high temperatures the carrier mobilities are not sensitive to the amount of nonstoichiometry, so the conductivities are tracking changes in the carrier concentration. As the TGA results indicated, the deviation from stoichiometry is increasing with increasing temperature.

Baumard et al. concluded that their data in the region of low $P(O_2)$ are best fit by a log–log slope of $-1/5$. This suggests that the ionic defects created by reduction are predominantly fully ionized Ti interstitials, according to Eq. (12.7). The mass-action expression for this can be written as follows:

$$[Ti_I^{4\cdot}]n^4 = K_n P(O_2)^{-1} \qquad (12.16)$$

If Eq. (12.7) is the major source of defects, then charge neutrality can be approximated by

$$n \approx 4[Ti_I^{4\cdot}] \qquad (12.17)$$

and these two equations together give

$$n \approx (4K_n)^{1/5} P(O_2)^{-1/5} \qquad (12.18)$$

This agreement with the experimental results implies that there is no interaction between the electrons and the Ti interstitials, that is, that the interstitials are fully ionized.

These results imply that Ti interstitials are preferred over oxygen vacancies as the product of reduction, since a log–log slope of $-1/6$ would have been expected if fully ionized oxygen vacancies had been the major product of reduction. If Ti interstitials are the preferred product of reduction, then the predominant form of ionic disorder must be cation Frenkel, Eq. (12.1), with the mass-action expression:

$$[Ti_I^{4\cdot}] [V_{Ti}^{4/}] = K_{CF}(T) \qquad (12.19)$$

The concentration of Ti interstitials for any equilibration condition can be obtained from the TGA data, which allow us to find the electron concentration through Eq. (12.17). This can then be combined with the conductivity to calculate the electron mobility. It was found that the mobility was 0.16 to $0.18 cm^2/V \cdot s$, independent of temperature and $P(O_2)$. If Eq. (12.17) were valid even in the higher range of $P(O_2)$, so that the electron concentration always followed Eq. (12.18), then it could easily be shown that the aliovalent impurity content must be well below 1 ppm to avoid being a significant part of charge neutrality. This is an unreasonably low impurity content for a refractory oxide, however, and in fact earlier measurements on TiO_2 to which 200 ppm of iron had been added gave conductivity curves very similar to those in the work we are discussing (Rudolph, 1959). Thus in the high $P(O_2)$ range, it is reasonable that aliovalent impurities have become important, and as we predicted, it was found that the results are best explained by the presence of acceptor-type impurities such as Al^{3+} or Fe^{3+}, incorporated according to Eq. (12.3). Charge neutrality in this region can then be approximated by:

$$[A_{Ti}'] \approx 4[Ti_I^{4\cdot}] \qquad (12.20)$$

where A_{Ti}' is a generalized acceptor impurity (e.g., a trivalent impurity on a Ti site). When this is combined with the mass-action expression, Eq. (12.16), the electron concentration becomes:

$$n \approx \left(\frac{4K_n}{[A_{Ti}']} \right)^{1/4} P(O_2)^{-1/4} \qquad (12.21)$$

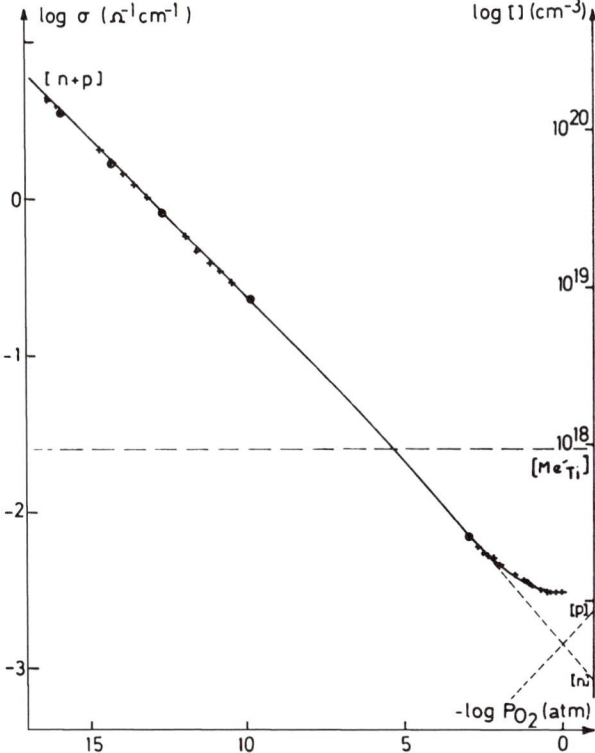

Figure 12.3 The fit of the Ti interstitial model to the equilibrium electrical conductivity of TiO_2. (Reproduced from Baumard et al., 1975, by permission of Masson Editeur.)

Since the electron concentration is decreasing with increasing $P(O_2)$, the hole concentration must be increasing with the complementary slope. The conductivity slopes are flattening as they approach $P(O_2) = 1$ atm, increasingly so with decreasing temperature. This is indicative of the formation of a conductivity minimum, as the hole concentration becomes comparable to the electron concentration, and the conductivity subsequently increases with increasing $P(O_2)$. The hole concentration can be obtained by combination of the mass-action expression for intrinsic electronic disorder:

$$np = K_I \tag{12.22}$$

and Eq. (12.21):

$$p \approx K_I \left(\frac{[A'_{Ti}]}{4K_n} \right)^{1/4} P(O_2)^{1/4} \tag{12.23}$$

When all this is put together, the fit to the experimental results is shown in Fig. 12.3. It is quite satisfactory. A Kröger–Vink diagram for this model is shown in Fig. 12.4. If the acceptor centers are all singly charged, then the net acceptor excess is equal to

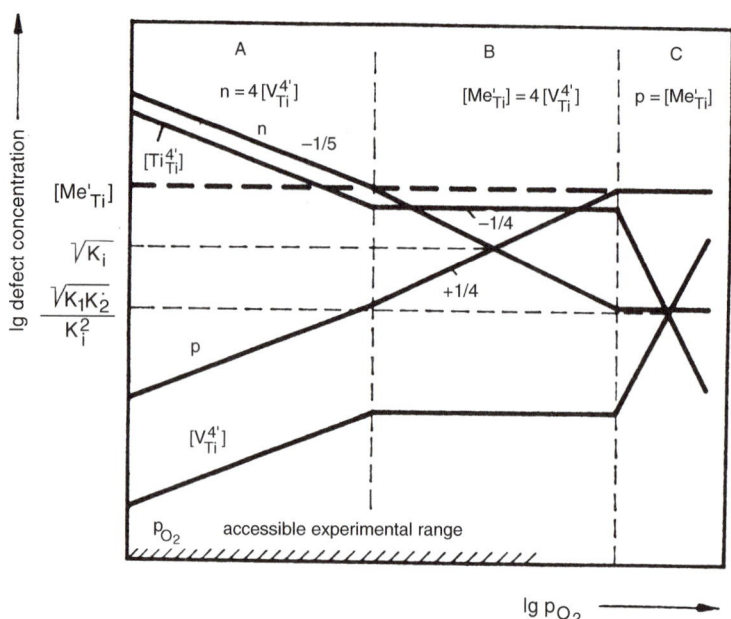

Figure 12.4 A Kröger–Vink diagram of the defect concentrations of TiO_2 as a function of oxygen activity for the Ti interstitial model. (Reproduced from Baumard and Tani, 1977, by permission of Wiley VCH, STM.)

the electron concentration at the point where the slope of the conductivity changes from $-1/5$ to $-1/4$. This was mathematically determined to be about 30 ppm, and is shown on Fig. 12.3.

Logothetis and Hetrick (1979) of the Ford Motor Company, measured the electrical resistivity of TiO_2 as a function of $P(O_2)$ at lower temperatures, and their results are shown in Fig. 12.5. The conductivity minima (resistance maxima) move clearly into view as the temperature decreases, and the p-type conduction persists to lower $P(O_2)$. The results down to 500°C probably represent equilibrium conditions, but the authors question whether equilibrium was achieved at lower temperatures.

The enthalpy of the reduction reaction, Eq. (12.7), can be obtained from an Arrhenius plot of the conductivity, since the electron mobility has no detectable temperature dependence. Two such plots are shown in Fig. 12.6. Note that the slopes should be $\Delta H_N/5k$; the results give an enthalpy of 10.6 eV (5.3 eV per oxygen atom) [1020 kJ/mol (510 kJ/mol per oxygen atom)].

Previous studies of the self-diffusion of oxygen, using ^{18}O, indicated that the diffusion constant is independent of $P(O_2)$ between 1 and 10^{-6} atm (Haul and Dumbgen, 1965). This is the region in which we expect the concentration of Ti interstitials to be determined by the impurity content, Eq. (12.20). Then the oxygen vacancy concentration must also be independent of $P(O_2)$ according to the mass-action expression for Schottky disorder, Eq. (12.2). This expression must be obeyed even though Schottky disorder may not be the preferred form of intrinsic ionic

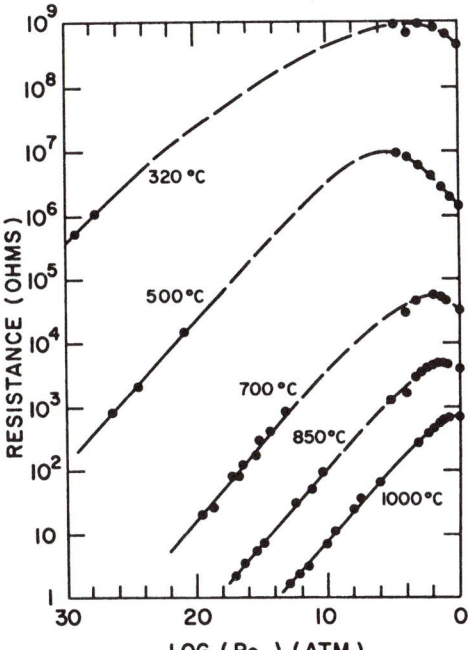

Figure 12.5 The electrical resistance of undoped TiO_2 ceramic as a function of oxygen activity at several temperatures showing the appearance of a resistance maximum (conductivity minimum) at the lower temperatures. (Reproduced from Logothedis and Hetrick, 1979, by permission of Elsevier Science.)

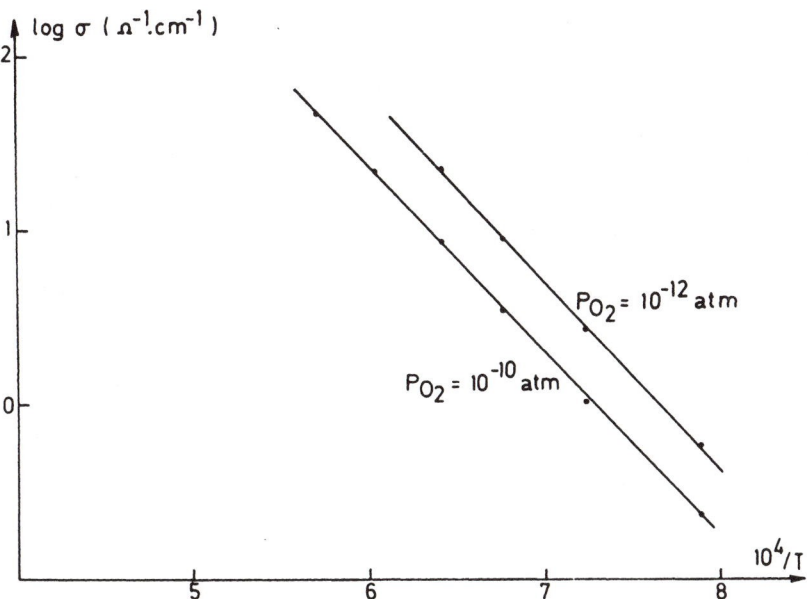

Figure 12.6 Arrhenius plots of the conductivity of TiO_2 at two oxygen activities. (Reproduced from Baumard et al., 1975, by permission of Masson Editeur.)

disorder. If the oxygen diffuses by way of vacancies, then the experimental observation is explained.

There has been a very clever confirmation that Ti interstitials are an important product of reduction in TiO_2. Yagi et al. (1977) fired 4 meV protons at both stoichiometric and slightly reduced single crystals of TiO_2. The pattern of cation occupation of the octahedral sites leaves linear channels of unoccupied cation sites in the [001] direction. The investigators found a peak in the backscattering of the protons in just that direction *only* in the reduced sample. This was interpreted as an indication that the channels of normally empty interstitial sites had become partially blocked by Ti ions that had moved into them as a result of reduction. However, note that this indicates only that reduction results in some titanium interstitials; it does not preclude the presence of some oxygen vacancies at the same time.

THE SEEBECK COEFFICIENT OF UNDOPED TiO$_2$

The Seebeck coefficient, sometimes called the thermoforce, is obtained from a measurement of the open-circuit emf that results from a temperature gradient across the sample. In effect, the sample is turned into a thermocouple. The rigorous expression for the Seebeck coefficient, Q, which has the units of volts per degree kelvin, is rather complex but can be greatly simplified by assuming that the electron and hole mobilities are equal, that their transport numbers are equal, that the effective density of states at the bottom of the conduction band and at the top of the valence band are equal, and that $n \gg p$. With these assumptions, the expression becomes

$$Q = \frac{-k}{e}\left(\ln\frac{N_c}{n} + A_n\right) \tag{12.24}$$

where N_c is the effective density of states in the conduction band, and A_n is the transport number. The latter depends on the conduction mechanism and typically varies between 0 and 4. It is seen that the Seebeck coefficient is closely related to the electron concentration. A comparable expression for holes can be obtained for the case of $p \gg n$. Note that the relationship is inverse in that a large Q corresponds to a small carrier concentration. A negative Q indicates n-type conduction, while a positive value indicates p-type conduction. Baumard and Tani (1977), measured the Seebeck coefficient for undoped TiO_2, and their results, shown in Fig. 12.7, correspond very closely to the conductivity results shown in Fig. 12.2. There is n-type conduction over most of the range of $P(O_2)$, with p-type conductivity beginning to appear near $P(O_2) = 1$ atm at the lowest temperatures. The slope in the n-type region is $-1/5$, the same as in the conductivity results.

IONIC CONDUCTION IN TiO$_2$

Singheiser and Auer (1977) used a variation of the Tubandt experiment to find evidence for a small amount of ionic conduction in TiO_2 at high temperatures.

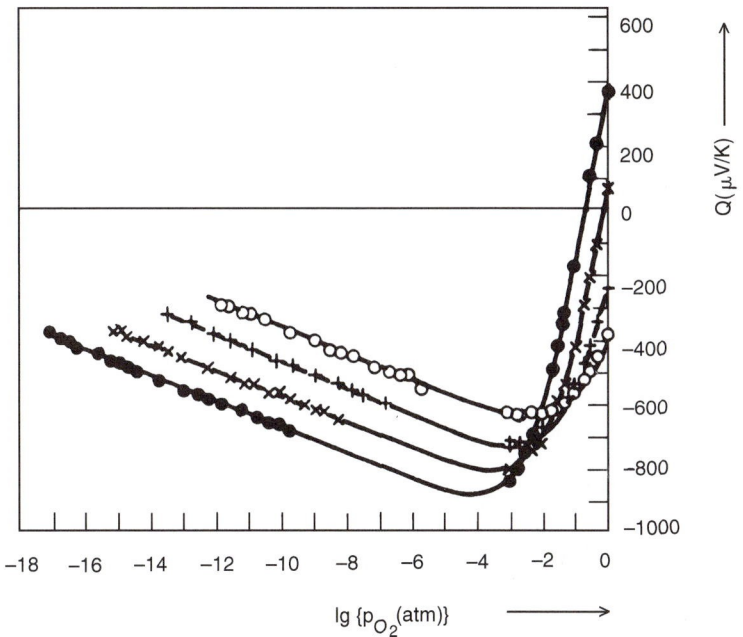

Figure 12.7 Seebeck coefficient of undoped TiO_2 as a function of oxygen activity. (Reproduced from Baumard and Tani, 1977, by permission of Wiley VCH, STM.)

They placed together three sintered pellets of TiO_2 in an oxygen partial pressure of 3.4×10^{-13} atm at temperatures between 842 and 982°C. Under these conditions the TiO_2 is quite conducting. A constant dc current was applied for several hours, after which the weight changes of the anode and cathode pellets were determined. The cathode pellets gained weight, while the anode pellets lost a corresponding amount of weight, indicating that a transport of mass had occurred. The weight change was proportional to the amount of charge passed through the pellets. The ionic transport number is given by the ratio of the chemical equivalents of ionic charge measured as the weight change to the total chemical equivalents passed. If it is assumed that the transported species is Ti^{4+}, the transport numbers increased with increasing temperature from 1.24×10^{-4} to 2.83×10^{-4} at an oxygen partial pressure of 3.4×10^{-13} atm; of course the transport number will be highly dependent on the oxygen activity because the total conductivity is changing. The actual partial ionic conductivities, obtained from the transport numbers and the total conductivities, were found to have an activation energy of 2.7 eV (260 kJ/mol). This value is substantially higher than what is usually observed for conduction by oxygen vacancies, suggesting that the major ionic charge carrier is the Ti interstitial. In any case, this very interesting experiment shows quite clearly that some ionic species makes a small contribution, about 0.1%, to the total conductivity.

THE EFFECT OF DOPANTS ON TiO$_2$

Acceptor Dopants

The minimum in the equilibrium electrical conductivity occurs where the electron and hole conductivities are equal. The oxygen activity at the minimum, $P(O_2)^\circ$ can be obtained by setting Eqs. (12.21) and (12.23) equal (each multiplied by the appropriate carrier mobility)

$$P(O_2)_0 = 4 \left(\frac{\mu_n}{\mu_p} \right)^2 \frac{K_n}{K_I^2 [A_{Ti}']} \tag{12.25}$$

It is apparent that the conductivity minimum should move to lower values of $P(O_2)$ as the acceptor content is increased. The equilibrium conductivities of TiO$_2$ with 0–5 mol % Cr^{3+} as an acceptor impurity are shown in Fig. 12.8 (Carpentier et al., 1986). For an acceptor concentration of about 2 mol %, the conductivity minimum has moved to a value of $P(O_2)$ that is lower by approximately 2.5 orders of magnitude, relative to the undoped material. From Eq. (12.25), it would appear that the added acceptor must have also increased the net acceptor excess by 2.5 orders of magnitude. This suggests that the net acceptor content of the undoped TiO$_2$ used in this study must have been about 0.007 atom %, or 70 ppm. This is similar to the 30 ppm level found by Baumard et al. for their undoped material. Thus the effect of the acceptor dopant is very close to what we expect. The authors found that the $P(O_2)$ at the conductivity minimum was inversely proportional to the acceptor content, as predicted by Eq. (12.25). It is left to the reader to show that this dependence is inconsistent with the alternative model of compensation of the acceptors by oxygen vacancies.

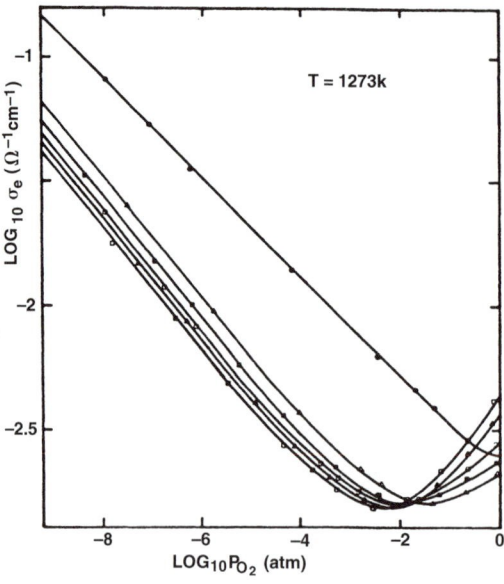

Figure 12.8 The equilibrium electrical conductivity of TiO$_2$ as a function of oxygen activities for various chromium concentrations. Values of Cr/Cr + Ti from top to bottom at low oxygen activities are: 0, 0.01, 0.02, 0.03, 0.04, and 0.05. (Reproduced from Carpentier et al., 1986, by permission of Editions de Physique.)

Donor Dopants

Again, we will use excellent data obtained by Baumard and his colleagues (Baumard and Tani, 1977). The equilibrium conductivity of TiO_2 doped with various levels of Nb^{5+} is shown in Fig. 12.9. The results can be described by three regions:

1. A region at the lowest values of $P(O_2)$, where the conductivity increases with decreasing $P(O_2)$. The conductivities in this region are exactly the same as in undoped TiO_2. This region becomes more pronounced with decreasing dopant concentration and increasing temperature.

2. A central flat plateau that becomes higher and wider with increasing Nb content. The conductivity in this region was found to be independent of temperature.

3. A region at higher values of $P(O_2)$, where the conductivity decreases with increasing $P(O_2)$. This region is most pronounced at high dopant concentrations and lower temperatures.

Obviously, in region 1 the products of the reduction reaction have become the major defects, and the dopant can be neglected. As a result, the behavior is identical to that of the undoped material and can be described by Eqs. (12.16), (12.17), and (12.18).

As shown in Fig. 12.10, the conductivity in region 2 is proportional to the Nb concentration, and it also was found to be independent of temperature. Therefore, this region corresponds to the compensation of the donors by electrons:

$$n \approx \left[Nb_{Ti}^{\cdot} \right] \tag{12.26}$$

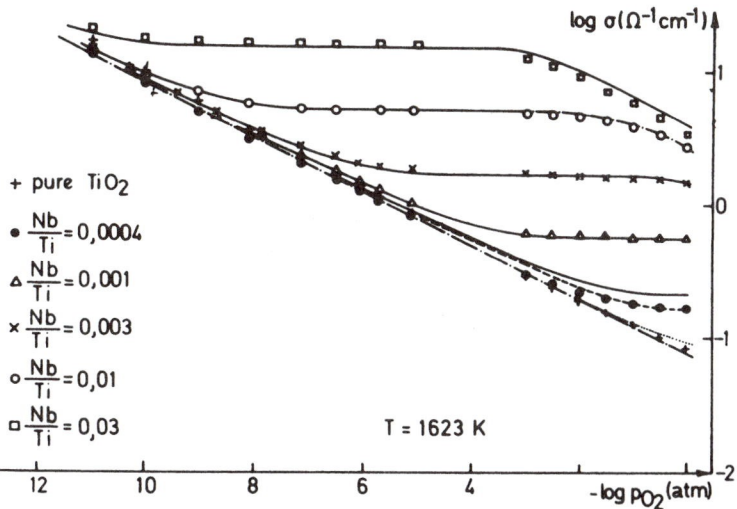

Figure 12.9 The equilibrium electrical conductivity of TiO_2 doped with niobium as a function of oxygen activity. (Reproduced from Baumard and Tani, 1977, by permission of the American Institute of Physics.)

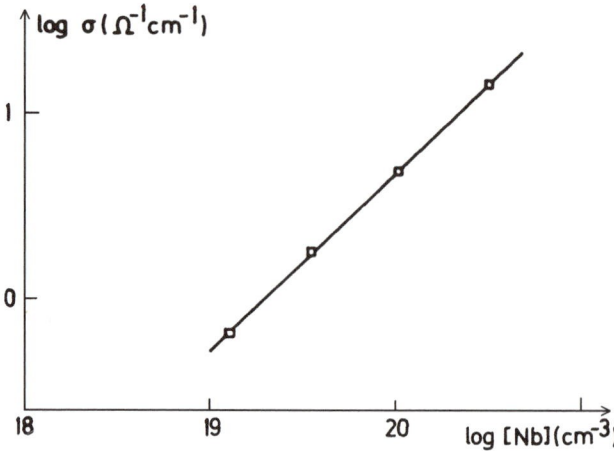

Figure 12.10 The equilibrium conductivity of TiO_2 as a function of niobium content at 1273 K and an oxygen activity of 10^{-10} atm (Reproduced from Baumard and Tani, 1977, by permission of the American Institute of Physics.)

and corresponds to the incorporation of the Nb_2O_5 according to Eq. (12.5). The temperature independence indicates that the electron mobility is also independent of temperature; and from the conductivity values and Eq. (12.26), the magnitude of the mobility can be calculated to be 0.1 $cm^2/V \cdot s$. This value, which is typical for transition metal oxides, is similar to that quoted earlier based on the TGA and conductivity results for undoped TiO_2. From the mass-action expression, Eq. (12.16), and the approximate condition of charge neutrality, Eq. (12.26), the Ti interstitial concentration in this region is given by:

$$\left[Ti_I^{4\cdot}\right] \approx \frac{K_n}{\left[Nb_{Ti}^{\cdot}\right]^4}P(O_2)^{-1} \tag{12.27}$$

and is thus dropping off rapidly with increasing $P(O_2)$. It is also suppressed by the donor content. On the other hand, assuming that cation Frenkel disorder is the preferred form of ionic disorder, the Ti vacancy concentration is rising rapidly with the complementary slope. This eventually leads to region 3.

In region 3, the Ti vacancies have risen from the depths to take over the role of compensating defect, and their concentration has become frozen by the impurity level. The approximation to charge neutrality has then become:

$$4\left[V_{Ti}^{4/}\right] \approx \left[Nb_{Ti}^{\cdot}\right] \tag{12.28}$$

The reduction reaction in this region can be obtained by subtracting the equation for cation Frenkel disorder, Eq. (12.1), from the intrinsic reduction reaction, Eq. (12.7), to give:

$$2O_O + V_{Ti}^{4/} \rightleftharpoons O_2 + 4e^{/} \tag{12.29}$$

which indicates that reduction in this region proceeds by the consumption of some of the large reservoir of Ti vacancies. The enthalpy of reduction in this region is less than that of the intrinsic reduction reaction, Eq. (12.7), by the enthalpy of formation for cation Frenkel disorder. Again, this will result in a reduction in enthalpy by several electron volts. The mass-action expression for Eq. (12.29) can be combined with the local expression of charge neutrality, Eq. (12.28), to give:

$$n \approx \left(\frac{K_n \left[Nb_{Ti}^{\cdot} \right]}{4 K_{CF}} \right)^{1/4} P(O_2)^{-1/4} \tag{12.30}$$

and the n-type conductivity drops off with increasing $P(O_2)$ with a log–log slope of $-1/4$ as observed.

Although the data in this region are somewhat sparse, there seems to be little dependence of the conductivity on the Nb content, in agreement with the fourth-root dependence predicted by Eq. (12.30). The hole concentration is rising with increasing $P(O_2)$, with a log–log slope of $+1/4$, but the conductivity minimum has been pushed out of sight above $P(O_2) = 1$ atm.

A Kröger–Vink diagram for donor-doped TiO_2 is shown in Fig. 12.11. It is apparent that lines for the Ti vacancy and interstitial concentrations can be constructed with knowledge of only the concentrations of electrons. This has been done for the region of electronic compensation for the data in Fig. 12.10, as shown in Fig. 12.12. (Strictly speaking, these lines should start at the point at which the slopes change at magnitudes that are a factor of 4 below the donor concentrations at those points, since the ionic defects are quadruply charged. That detail is not necessary for the following treatment.) It is possible to do this construction for only the three highest dopant levels,

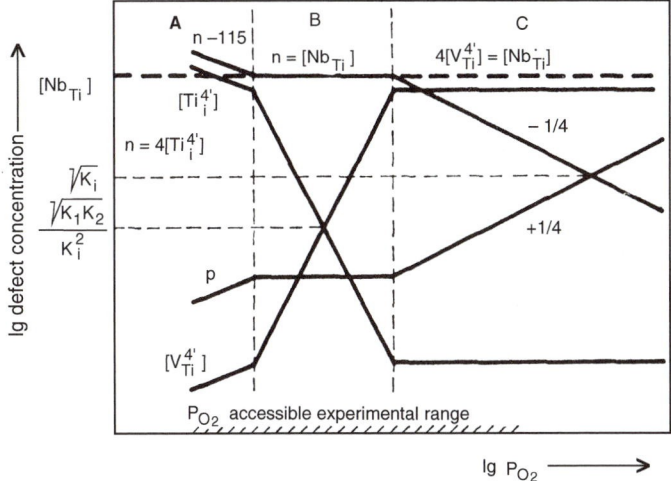

Figure 12.11 Kröger–Vink diagram for Nb-doped TiO_2. (Reproduced from Baumard and Tani, 1977, by permission of Wiley VCH, SEM.)

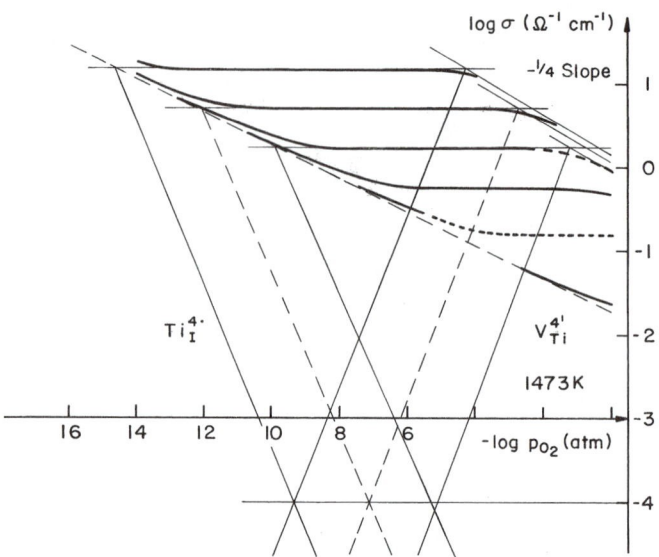

Figure 12.12 The equilibrium electrical conductivity of Nb-doped TiO$_2$ showing the derived relative concentrations of the cation Frenkel defects.

but it is gratifying that the points where $[V_{Ti}^{4/}] = [Ti_i^{4\cdot}]$ lie at about the same level for all dopant concentrations, as they should. Baumard and Tani show data for two temperatures, and when this construction is done for both sets, and an Arrhenius plot is made for the conductivities where $[V_{Ti}^{4/}] = [Ti_i^{4\cdot}]$, the slope should be $\Delta H_{CF}/2$. This gives an enthalpy for cation Frenkel disorder of 4.7 eV (450 kJ/mol). Of course the accuracy of this value is diminished by the availability of data at only two temperatures (but that does guarantee a perfectly linear Arrhenius plot!). From their experimental data, Baumard and Tani obtained numerical values for K_n and for K_p/K_I^4 at both experimental temperatures, where K_n and K_p are the mass-action constants for the intrinsic reactions for reduction and oxidation, Eqs. (12.7) and (12.9), and K_I is the mass-action constant for intrinsic electronic disorder, Eq. (12.22). Ikeda and Chiang (1993) noted that the product $K_n \times K_p/K_I^4$ gives K_{CF}, the mass-action constant for cation Frenkel disorder. Then at each temperature they obtained the free energy of formation, ΔG_{CF}, by solving the relationship:

$$K_{CF} = \frac{K_n K_p}{K_I^4} = e^{-\Delta G_{CF}/kT} \tag{12.31}$$

They obtained values for ΔG_{CF} of 4.5 and 4.4 eV (430 and 420 kJ/mol) at 1200 and 1350°C, respectively. An Arrhenius plot of the mass-action constants for cation Frenkel disorder (Fig. 12.12) gives an enthalpy of 4.7 eV (450 kJ/mol). Again with data available at only two temperatures, this cannot be considered to be a very accurate value. However, the similarity of the values obtained from two entirely different approaches suggests that about 5 eV (480 kJ/mol) is a reasonable value for the enthalpy of cation Frenkel disorder in TiO_2, assuming that the entropy term is relatively small.

GENERAL COMMENTS ON THE DEFECT CHEMISTRY OF TiO$_2$

Based on the following pieces of evidence, it is concluded that Ti interstitials are preferred over oxygen vacancies as the lattice defect created by reduction, and as the compensating defect for acceptor impurities:

- The log–log slope of $-1/5$ for the equilibrium conductivity in the most highly reduced region is consistent with the formation of Ti interstitials. A slope of $-1/6$ would be expected if oxygen vacancies were the major product of reduction.

- The finding that the normally vacant octahedrally coordinated sites in the rutile structure are partially blocked in reduced TiO_2, is consistent with the presence of Ti interstitials in these sites.

- The observation that the oxygen activity at the conductivity minimum is inversely proportional to the concentration of added acceptor impurities is predicted for the case of Ti interstitials but is not consistent with the oxygen vacancy model.

Thus the suggested equations for ionic disorder, intrinsic reduction, and acceptor-impurity incorporation are Eqs. (12.1), (12.7), and (12.3), respectively. The enthalpy of intrinsic reduction has been determined to be 10.6 eV (5.3 eV per oxygen lost) [1020 kJ/mol (510 kJ/mol per oxygen lost)], while the equilibrium conductivities of donor-doped material suggest an enthalpy for cation Frenkel disorder of about 5 eV (480 kJ/mol). Baumard and Tani report a value for the band gap of

$$E_g = 3.2 - 6.6 \times 10^{-4}T \quad \text{eV} \tag{12.32}$$

Thus the enthalpy of intrinsic electronic ionization (or the band gap at 0 K) is 3.2 eV (310 kJ/mol).

Because we have information on three enthalpies of reaction, we can calculate the enthalpies of other possible reactions. Thus the addition of the intrinsic reduction reaction to the extrinsic oxidation reaction in the region of electronic compensation of donors, which proceeds by the consumption of electrons, gives the reaction for cation Frenkel disorder:

$$2O_O + Ti_{Ti} + V_I \rightleftharpoons O_2 + Ti_I^{4\cdot} + 4e' \qquad \Delta H_N$$

$$\underline{O_2 + 4e' \rightleftharpoons 2O_O + V_{Ti}^{4/}} \qquad \underline{\qquad \Delta H_{ode} \qquad} \tag{12.33}$$

$$Ti_{Ti} + V_I \rightleftharpoons Ti_I^{4\cdot} + V_{Ti}^{4/} \qquad \Delta H_{CF}$$

where the subscript ode denotes the oxidation reaction in the region where the donors are compensated by an electronic defect. Equation (12.33) indicates that

$$\Delta H_{CF} - \Delta H_N = \Delta H_{ode}$$

$$5 - 10.6 = -5.6 \tag{12.34}$$

This gives a value of -5.6 eV for ΔH_{ode} (-2.8 eV per oxygen added) [-540 kJ/mol (-270 kJ/mol per oxygen added)]. This is obviously a very favorable reaction, since it is consuming electrons rather than creating holes, as in the case of intrinsic oxidation.

Since the enthalpy of extrinsic reduction in the region of ionic compensation of donors, ΔH_{rdi}, must be equal in magnitude and opposite in sign to the enthalpy of extrinsic oxidation in the region of electronic compensation, ΔH_{rdi} must be $+5.6$ eV ($+540$ kJ/mol). This is confirmed by subtracting the reaction for cation Frenkel disorder from the intrinsic reduction reaction:

$$2O_O + Ti_{Ti} + V_I \rightleftharpoons O_2 + Ti_I^{4\cdot} + 4e' \qquad \Delta H_N$$

$$-\left(Ti_{Ti} + V_I \rightleftharpoons Ti_I^{4\cdot} + V_{Ti}^{4/}\right) \qquad -\Delta H_{CF} \tag{12.35}$$

$$\overline{2O_O + V_{Ti}^{4/} \rightleftharpoons O_2 + 4e'} \qquad \Delta H_{rdi}$$

which gives:

$$\Delta H_N - \Delta H_{CF} = \Delta H_{rdi}$$

$$10.6 - 5 = 5.6 \tag{12.36}$$

[2.8 eV (270 kJ/mol)per oxygen lost]. This is also lower than the enthalpy of intrinsic reduction because it consumes Ti vacancies rather than producing Ti interstitials.

Similarly, the addition of the intrinsic reduction reaction, Eq. (12.7), to the extrinsic oxidation reaction in the region of ionic compensation of acceptors, Eq. (12.11), indicates that:

$$4E_g^{\circ} - \Delta H_N = \Delta H_{oai}$$

$$4(3.2) - 10.6 = 2.2 \tag{12.37}$$

and thus that $\Delta H_{oai} = 2.2$ eV (1.1 eV per added oxygen) [210 kJ/mol (105 kJ/mol per added oxygen)]. This is in the middle of the range of experimentally determined enthalpies of extrinsic oxidation of acceptor-doped titanates such as $BaTiO_3$ and $SrTiO_3$.

The usual tests for self-consistency can be applied. For example, the mass-action expression for intrinsic reduction, Eq. (12.16), requires that the activation energy for Ti interstitials plus four times the activation energy for holes always add up to 10.6 eV (1020 kJ/mol). This is seen to be true. In the region of electronic compensation, the high activation energy for Ti vacancies, combined with the low activation energy for the point at which $[V_{Ti}^{4/}] = [Ti_I^{4\cdot}]$ requires that $[Ti_I^{4\cdot}]$ decrease with increasing temperature in accord with the assigned negative activation energy. This requires that the horizontal conductivity plateau move toward higher oxygen activities with increasing temperature, and this is confirmed by the data.

The enthalpy for intrinsic oxidation, Eq. (12.9), has not been discussed in terms of these experimental results. This can be obtained by combination of the enthalpies of extrinsic oxidation in the region of electronic compensation, ΔH_{ode} and the band gap, E_g°, to be $\Delta H_P = 7.2$ eV (3.6 eV per oxygen) [690 kJ/mol (340 kJ/mol per oxygen)]. Thus the enthalpy of oxidation in the region of electronic compensation, ΔH_{ode} is reduced by $4E_g^\circ = 12.8$ eV (6.4 eV per oxygen) [1230 kJ/mol (620 kJ/mol per oxygen)] from that of intrinsic oxidation, ΔH_P.

This example of the defect chemistry of TiO_2 shows the very strong advantage of having data for both acceptor- and donor-doped samples, as well as for the undoped material.

As will be discussed in Chapter 15, some metal oxides, including TiO_2, accommodate large deviations from stoichiometry and large concentrations of aliovalent impurities by subtle structural modifications, rather than by randomly distributed point defects. However, it appears that the experimental results discussed in this chapter, with the possible exception of data below 800°C, are safely in the single-phase, point defect regime (Blanchin et al., 1980).

REFERENCES

Baumard, J.-F., D. Panis, and D. Ruffier. Conductivité électrique du rutile monocristallin à haute température. *Rev. Int. Hautes Temp. Refract.* 12:321–327, 1975.

Baumard, J.-F., and E. Tani. Electrical conductivity and charge compensation in Nb doped rutile. *J. Chem. Phys.* 67:857–860, 1977.

Baumard, J.-F., and E. Tani. Thermoelectric power in reduced pure and Nb-doped TiO_2 rutile at high temperatures. *Phys. Stat. Sol.* 39:373–382, 1977.

Blanchin, M. G., P. Faisant, C. Picard, M. Ezzo, and G. Fontaine. Transmission electron microscope observations of slightly reduced rutile. *Phys. Stat. Sol.* A60:357–362, 1980.

Carpentier, J.-L., A. Lebrun, and F. Perdu. Electronic conduction in pure and chromium-doped rutile at 1273 K. *J. Phys, Colloq. C1*, (suppl. to no. 2) 47: C1-819–C1-823, 1986.

Catlow, C. R. A., and R. James. Disorder in TiO_{2-x}. *Proc. Ro. Soc. London* A384:157–173, 1982.

Catlow, C. R. A., C. M. Freeman, and R. L. Royle. Recent studies using static simulation techniques. *Physica* 131B:1–12, 1985.

Haul, R., and G. Dumbgen. Sauerstoff-selbstdiffusion in Rutilkristallen. *J. Phys. Chem. Solids* 26:1–10, 1965.

Ikeda, J. A., and Y.-M Chiang. Space charge segregation at grain boundaries in titanium dioxide. I. Relationship between lattice defect chemistry and space charge potential. *J. Am. Ceram. Soc.* 76:2437–2446, 1993.

Kofstad, P. *Nonstoichiometry, Diffusion, and Electrical Conductivity in Binary Metal Oxides.* New York: Wiley-Interscience, 1972, p. 141.

Logothetis, E. M., and R. E. Hetrick. Oscillations in the electrical resistivity of TiO_2 induced by solid/gas interactions. *Solid State Commun.* 31:167–171, 1979.

Rudolph, J. Über den leitungsmechanismus oxidischer Halbleiter bei höhen Temperaturen. *Z. Naturforsch.* 14a:727–737, 1959.

Singheiser, L., and W. Auer. Untersuchung der Fehlordnung von TiO_2 (Rutil) mit Hilfe von Leitfähigkeits- und Überführungsmessungen. *Ber. Bunsen-Ges. Phys. Chem.* 81:1167–1171, 1977.

Yagi, E., A. Koyama, H. Sakairi, and R. R. Hasiguti. Investigation of Ti interstitials in slightly reduced rutile (TiO$_2$) by means of channeling method. *J. Phys. Soc. Jpn.* 42:939–946, 1977.

☑ PROBLEMS

12.1. Show that the proportionality of the oxygen activities at the conductivity minima to the acceptor concentration in TiO$_2$ is inconsistent with the model whereby the acceptors are compensated by oxygen vacancies.

12.2. Show that the enthalpy of oxidation of acceptor-doped TiO$_2$ in the region of compensation by Ti interstitials is related to the enthalpies of intrinsic oxidation and cation Frenkel disorder.

Cobalt Oxide and Nickel Oxide

13

INTRODUCTION

As we move through the sequence of the 3d transition metals, Sc, Ti, V, Cr, Mn, Fe, Co, Ni, Cu, and Zn, it becomes increasingly difficult to ionize the atoms back to the electronic configuration of argon, the preceding rare gas. This is easy for Sc, for which only the $+3$ state is stable, and relatively easy for Ti which is $+4$ when processed in air but can be reduced to $+3$ or $+2$. For V, full ionization begins to fade, and either $+5$ or $+4$ may be favored, depending on the environment. The stable simple oxide for Cr is Cr_2O_3, and the $+6$ state is present in the chromates, but the latter are reasonably strong oxidizing agents, reflecting their desire to get some electrons back into the 3d states. Mn_2O_7 is a powerful, unstable oxidizing liquid, and the ultimately stable oxide after ignition in air is MnO. This starts a series of divalent metal oxides, MnO, FeO, CoO, NiO, CuO, and ZnO, although $+3$ and even $+4$ states can be achieved for some of these. With the exception of Cu, these represent the lowest available oxidation states, and the cations cannot be reduced to the monovalent state. Therefore, the first four of the series represent cations that can be oxidized, but not reduced, and they contain 5, 6, 7, and 8 electrons in the 3d states, respectively. The valence band is then made up primarily of the filled metal 3d states, which are reasonably receptive to oxidation (i.e., to the presence of holes), and it thus lies at a relatively high level. As a result, acceptor states are expected to be quite shallow. The conduction band consists primarily of empty cation 3d and possibly 4s states, which are resistant to reduction (i.e., do not favor the presence of electrons). These bands are at a relatively high level, and donor states are expected to be rather deep. This band structure is the reverse of that of TiO_2. In TiO_2 the band gap is reduced from the values typical of insulators because the conduction band is at a lower level, while in MnO, FeO, CoO, and NiO it is reduced because the valence bands are at higher levels. Thus the band gap for NiO is generally accepted to be about 3.5 eV (340 kJ/mol), only slightly higher than the 3 eV (290 kJ/mol) band gap of TiO_2.

For these cations that can be oxidized but not reduced, we expect to see predominantly oxygen-excess, p-type nonstoichiometry. This is indeed the case, and, when equilibrated in air, NiO has an oxygen excess of the order of 0.01%, while in CoO it is about 1%, in MnO about 5%, and in FeO about 15%. The latter figure is extraordinarily high, and FeO even decomposes with the separation of metallic Fe for compositions below 5% excess oxygen. Thus stoichiometric FeO does not exist at ordinary pressures. This material, the mineral wustite, has been the subject of many studies because of its extraordinarily high degree of nonstoichiometry. We shall take CoO and NiO as our model materials; their modest deviations from stoichiometry give them the best chance for ideal behavior. We start with a discussion of CoO based on the very thorough study by Dieckmann (1977), followed by a comparative discussion of NiO.

COBALTOUS OXIDE, CoO

The defect chemistry of CoO has been studied by many groups. There is general agreement on the basic behavior, but there are some differing views on specific details. Dieckmann (1977) has made a very thorough study of this material and has correlated his results with those of other researchers. We base this summary on his work; those interested in more details can refer to the publications referenced in his paper.

The stable phase field of CoO is bracketed by metallic Co at low oxygen activities and Co_3O_4 at high oxygen activities. Based on the results of several investigators, Dieckmann summarizes the phase field in Fig. 13.1. The boundaries can be expressed as follows:

$$\log P(O_2)\frac{Co}{CoO} = 7.2 - \frac{24,100}{T}$$

$$\log P(O_2)\frac{CoO}{Co_3O_4} = 16.5 - \frac{20,300}{T}$$

(13.1)

with T expressed in degrees kelvin.

The deviations from stoichiometry, x in $Co_{1-x}O$ (x is represented by δ in Dieckmann's paper), as a function of $P(O_2)$ at several temperatures, is shown in Fig. 13.2.

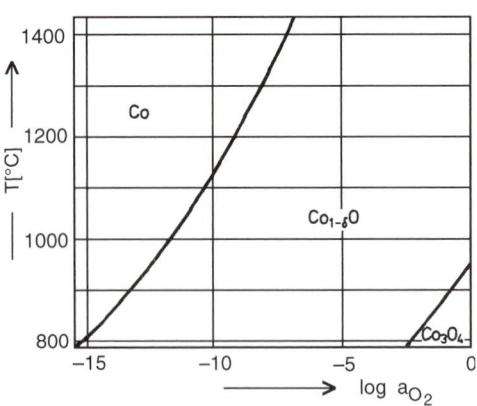

Figure 13.1 The stability range of $Co_{1-x}O$ as a function of temperature and oxygen activity. (Reproduced from Dieckmann, 1977, by permission of R. Oldenbourg Verlag.)

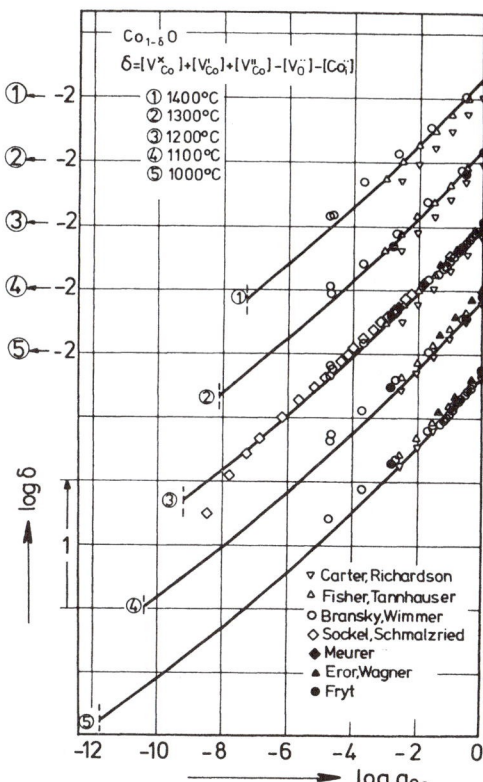

Figure 13.2 The deviation from stoichiometry, x, in $Co_{1-x}O$ as a function of temperature and oxygen activity as determined by several groups. (Reproduced from Dieckmann, 1977, by permission of R. Oldenbourg Verlag.)

These results by several investigators were mostly obtained by thermogravimetric analysis, and there is very gratifying agreement among them. The short dashed vertical lines at the low pressure end of each data set indicate the location of the Co/CoO phase boundary. Only in the case of the 1200°C data do the results extend nearly to this boundary. The solid lines are calculated from a model presented by Dieckmann to be described shortly. Figure 13.2 is a bit tricky. The left-hand side indicates that the values of log x have been displaced upward by 0.5 for each 100°C increase in temperature. Without this displacement, the data at all temperatures would fall almost directly on one another, creating a hopelessly messy plot. In other words, there is almost no temperature dependence to x. Note that at all temperatures, the maximum deviation from stoichiometry is very close to 10^{-2}, 1%, and it occurs at $P(O_2) = 1$ atm. One can certainly expect intrinsic nonstoichiometry to prevail at such a high defect concentration. The lowest measured value of x is about 10^{-4}, 100 ppm, where it is possible that impurity effects might start to come into play; at the lowest extrapolated nonstoichiometry, where x is about 10^{-5}, 10 ppm, at the Co/CoO boundary at 1000°C, impurity effects would certainly have to be considered.

The equilibrium electrical conductivity of CoO as measured by several groups is shown in Fig. 13.3 as a function of $P(O_2)$ at several temperatures. Once again, there

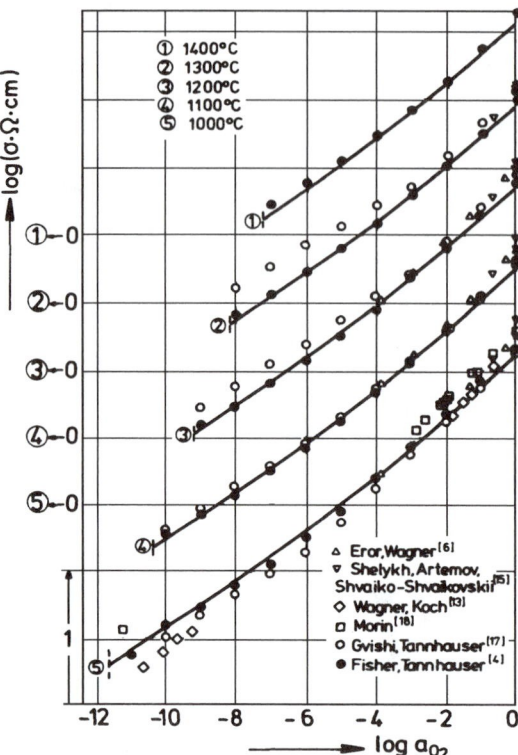

Figure 13.3 The equilibrium electrical conductivity of $Co_{1-x}O$ as a function of temperature and oxygen activity as determined by several groups. (Reproduced from Dieckmann, 1977, by permission of R. Oldenbourg Verlag.)

is quite good agreement. The lines for each temperature are again offset to avoid overlap, and as in the case of the deviations from stoichiometry, there is very little temperature dependence.

Cation tracer diffusion has been measured at relatively high oxygen activities by several groups using the radioactive isotope ^{60}Co, and these measurements have been extended to cover the full width of the CoO phase by Dieckmann, whose results are shown in Fig. 13.4. These results are in good agreement with the earlier work. In this case there is a significant temperature dependence, as would be expected, since the mobility of the cation vacancies should be thermally activated.

All investigators agree that CoO is a cation deficient, p-type oxide and that the major defects are thus cation vacancies and holes. The cation deficiency varies from 1% in 1 atm of oxygen to values at least three to four orders of magnitude less at the Co/CoO boundary. Earlier investigators differed in their opinions about the charge state of the cation vacancies. Some proposed that they are singly charged, while others proposed mixtures of doubly and singly charged vacancies. Dieckmann proposes not only that there is such a mixture, but also includes neutral vacancies in his model. Thus we can start with an oxidation reaction for the formation of doubly charged cation vacancies and free holes that initially ignores interactions between these oppositely charged defects:

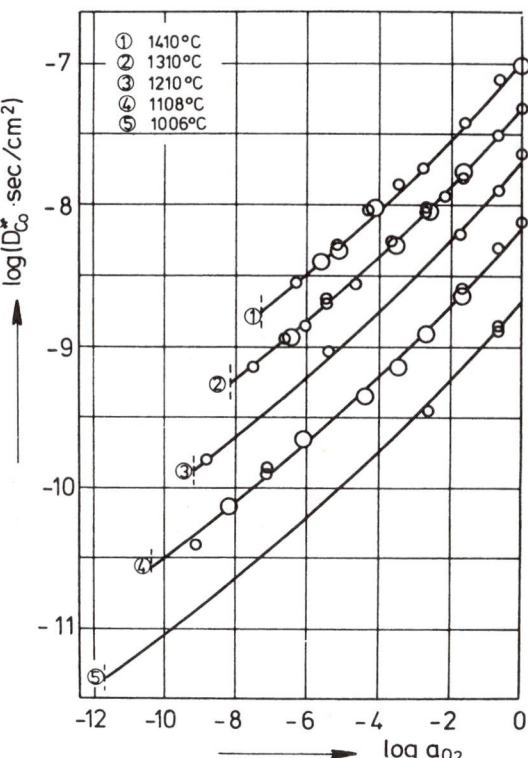

Figure 13.4 Cobalt tracer diffusion coefficients in $Co_{1-x}O$ as a function of temperature and oxygen activity. (Reproduced from Dieckmann, 1977, by permission of R. Oldenbourg Verlag.)

$$\tfrac{1}{2}O_2 \rightleftharpoons O_O + V_{Co}^{//} + 2h^\cdot \tag{13.2}$$

and the mass-action expression:

$$[V_{Co}^{//}]p^2 = K_P P(O_2)^{1/2} \tag{13.3}$$

Combined with the approximation to charge neutrality:

$$p \approx 2[V_{Co}^{//}] \tag{13.4}$$

this gives for the value of x in $Co_{1-x}O$, and the cation vacancy concentration:

$$x = [V_{Co}^{//}] \approx \left(\frac{K_P}{4}\right)^{1/3} P(O_2)^{1/6} \tag{13.5}$$

and for the hole concentration (hence the oxygen activity dependence of the p-type conductivity):

$$p \approx (2K_P)^{1/3} P(O_2)^{1/6} \tag{13.6}$$

The deviation from stoichiometry, the equilibrium conductivity, and the tracer diffusion coefficient should then all vary as $P(O_2)^{1/6}$. However, observation of the $P(O_2)$

dependences seen in Fig. 13.2, 13.3, and 13.4 show that in all cases, the logarithmic slopes change gradually from 1/4 at high values of $P(O_2)$ to approximately 1/5 at low oxygen activities. Thus Dieckmann adds two additional reactions for the interactions between cation vacancies and holes:

$$V_{Co}'' + h^{\cdot} \rightleftharpoons V_{Co}' \tag{13.7}$$

$$V_{Co}' + h^{\cdot} \rightleftharpoons V_{Co}^{x} \tag{13.8}$$

and their mass-action expressions:

$$\frac{[V_{Co}']}{[V_{Co}'']p} = K_1 \tag{13.9}$$

$$\frac{[V_{Co}^{x}]}{[V_{Co}']p} = K_2 \tag{13.10}$$

Additional relationships include the full expression for charge neutrality:

$$p = 2[V_{Co}''] + [V_{Co}'] \tag{13.11}$$

and the expression for the total cation vacancy content:

$$[V_{Co}]_{total} = [V_{Co}''] + [V_{Co}'] + [V_{Co}^{x}] \tag{13.12}$$

Dieckmann was then able to fit the data in Figs. 13.2, 13.3, and 13.4 with the values for the mass-action constants:

$$
\begin{aligned}
K_P &= 6.5 \times 10^{-3} \, e^{(-1.55/kT)} \\
K_1 &= 5.9 \, e^{(0.75/kT)} \\
K_2 &= 0.42 \, e^{(0.53/kT)}
\end{aligned}
$$

[Dieckmann uses a different formalism from which the preceding interpretation has been derived. He starts with an oxidation reaction for the formation of *neutral* cation vacancies and then adds two successive ionization reactions, i.e., the reverse of Eqs. (13.7) and (13.8). We have usually started with a reaction that forms a major defect, V_{Co}'' in this case, then proceeding to consider the interactions between the defects. Dieckmann's approach has been adjusted to conform to this preference. There is no right or wrong approach; it is merely a matter of which is more comfortable.]

As part of this analysis, comparison of the temperature dependences of the concentration of holes and the equilibrium conductivity indicates that the hole mobility is slightly thermally activated, with an activation enthalpy of 0.09 eV (9 kJ/mol). Also, comparison of the temperature dependences of the cation vacancy concentration and the tracer diffusion constant indicates an activation enthalpy of 1.4 eV (130 kJ/mol) for the vacancy mobility.

There has been a report, based on high temperature x-ray diffraction studies, that 2–5% of the cations in NiO–CoO mixed crystals reside in interstitial sites (Stiglich et al.,

1973). This would imply a comparably high contribution from cation Frenkel disorder. Moreover, studies of oxygen diffusion in Li_2O- and Al_2O_3-doped CoO by Chen and Jackson (1969) indicated that the diffusion was by way of oxygen vacancies. This implies that intrinsic Schottky disorder might play a role. Since cation interstitials and oxygen vacancies are both positively charged defects, they should have similar effects on the properties. If they are present in signicant concentrations, they should have observable effects on the measured properties at low oxygen activities (i.e., where the amount of nonstoichiometry is least). Dieckmann found no such effects on the amount of nonstoichiometry, the equilibrium conductivity, or the cation tracer diffusion, and he concluded that these defects could only be minority defects that do not contribute significantly to the condition of charge neutrality, nor to the cation diffusivity.

The enthalpies in the mass-action expressions K_1 and K_2 should be the trap depths for the second and first ionization steps from the neutral cation vacancies [i.e., 0.53 and 0.75 eV (51 and 72 kJ/mol) for the successive ionization reactions]. These are surprisingly deep acceptor levels for an oxide with an oxidizable cation. In fact, they are so deep that p-type CoO should be insulating at room temperature, whereas it is actually quite a good conductor. This discrepancy is not presently resolved.

The model for nonstoichiometry in CoO is summarized in a Kröger–Vink diagram (Fig. 13.5), that is consistent with the three mass-action expressions, Eqs. (13.3), (13.9), and (13.10), and the expressions for charge neutrality, Eq. (13.11), and total vacancy content, Eq. (13.12). This diagram shows that doubly ionized cation vacancies predominate at low oxygen activities, but singly ionized vacancies, with their faster rise with increasing $P(O_2)$, eventually prevail at high oxygen activities. Neutral vacancies are approaching significant levels only at the highest oxygen activities. Although there are quite a large number of adjustable parameters in the model, it is quite impressive that a single set of mass-action expressions can accurately fit the temperature and oxygen activity dependences of the amount of nonstoichiometry, the equilibrium conductivity, and the cation tracer diffusion. This

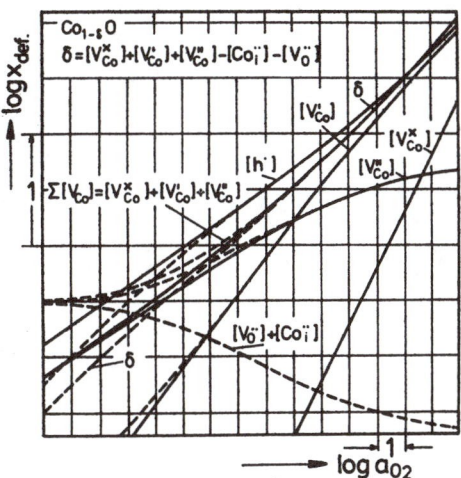

Figure 13.5 A Kröger–Vink diagram for the defect concentrations as a function of oxygen activity in $Co_{1-x}O$. (Reproduced from Dieckmann, 1977, by permission of R. Oldenbourg Verlag.)

is shown by the agreement of the calculated lines with the data points in Figs. 13.2, 13.3, and 13.4.

The Hall mobilities of the charge carriers in undoped and Ti-doped CoO are shown in Fig. 13.6 as a function of $q = P(CO_2)/P(CO)$ at 1140°C [$P(O_2)$ varies as q^2] (Gvishi and Tannhauser, 1972). The results correspond to positive charge carriers, holes, at high oxygen activities but a switch to negative charge carriers, electrons, at low oxygen activities. Thus there is a change from p-type to n-type conduction with decreasing $P(O_2)$. Support for this change comes from earlier measurements of the thermoelectric power (Fisher and Tannhauser, 1966; Henri le Brusq et al., 1968). This contribution by electrons was not included in the Dieckmann model described above. However, the switch does not require that the electron concentration exceed the hole concentration, since the electrons are much more mobile: 0.36–0.6 cm²/V · s for electrons versus 0.06 cm²/V · s for holes. The switch in major charge carrier moves to higher values of $q \propto [P(O_2)^{1/2}]$ in the Ti-doped samples. This is to be expected, since the donor impurities, $Ti_{Co}^{\cdot\cdot}$, should increase the electron concentration and decrease the hole concentration. The two doped samples, 0.3 and 1% Ti, behave similarly, and this is in accord with the estimated solubility limit of 0.5%.

Evidence for defect interaction is found in the dielectric loss spectrum for Li-doped CoO shown in Fig. 13.7 (Bosman and Crevecoeur, 1968). This measures the energy absorption (tan δ) of ac signals at three different frequencies as a function of temperature. The positions of the absorption peaks do not change with Li content, but their magnitudes increase with increasing Li concentration, indicating that this is a specific bulk process involving the Li. The frequency dependence of the temperatures

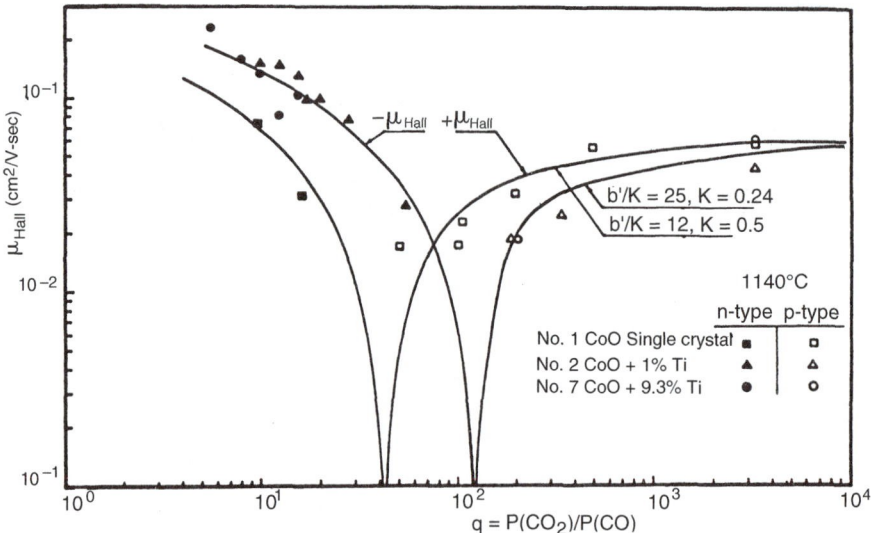

Figure 13.6 Hall mobility versus $q = P(CO_2)/P(CO)$ for undoped and Ti-doped $Co_{1-x}O$ at 1140°C. (Reproduced from Dieckmann, 1977, by permission of R. Oldenbourg Verlag.)

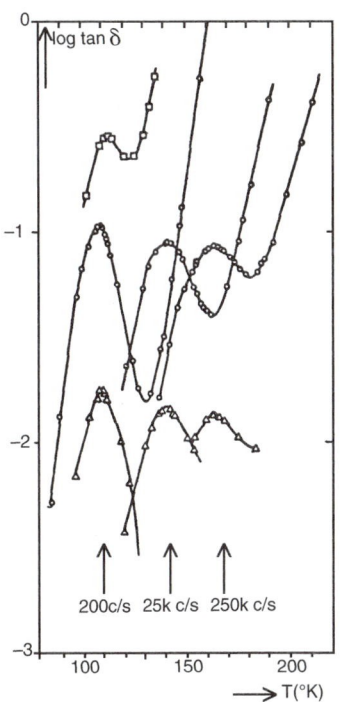

Figure 13.7 Dipole relaxation losses of Li-doped $Co_{1-x}O$ as a function of temperature at various frequencies; atom % Li from bottom to top: 0.015, 0.08, and 0.45. (Reproduced from Bosman and Crevecoeur, 1968, by permission of Elsevier Science.)

at the peaks indicates an activation energy of 0.2 eV (19 kJ/mol). This is attributed to the reorientation of holes among the Co^{2+} ions that surround a substitutional Li^+, Li'_{Co}. This corresponds to the dipolar defect complex ($Co^{\cdot}_{Co}Li'_{Co}$).

NICKELOUS OXIDE, NiO

The behavior of NiO is very similar to that of CoO, except that the deviation from stoichiometry is about two orders of magnitude less in the former, reflecting the greater difficulty in oxidizing Ni^{2+}. Tretyakov and Rapp (1969) used a technique called "Coulometric titration" to measure the deviation from stoichiometry (i.e., the value of x in $Ni_{1-x}O$). Their experimental cell is shown in Fig. 13.8. The sample is contained in a gas-tight chamber constructed from a closed-end alumina tube. The open end of the tube is covered by a CaO-doped zirconia slab that is electroded inside and out with platinum electrodes fired on from a commercial platinum paste. The zirconia slab is separated from the alumina tube by a thin ring of Pyrex glass that softens at the experimental temperatures and effects a hermetic seal (sometimes). A platinum wire connected to the inside electrode is brought out through the glass seal. The electroded zirconia serves as an oxygen concentration cell whereby the emf generated by the difference in oxygen activities inside and outside the cell can be measured. Since the activity outside the cell, usually air, is known, the

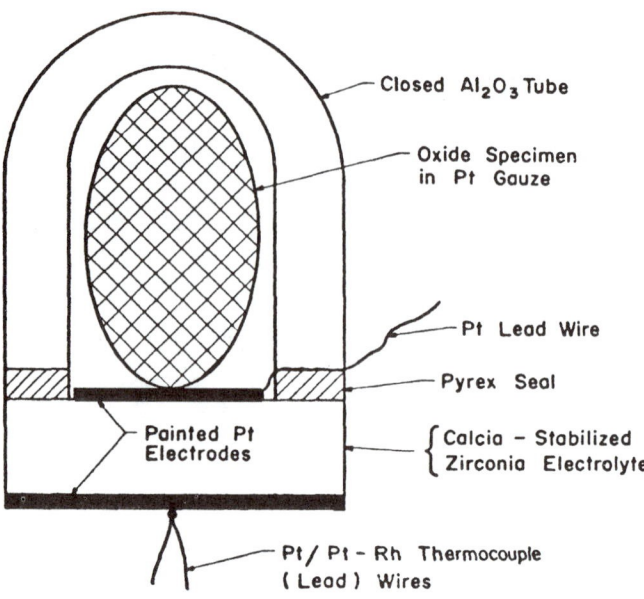

Figure 13.8 The sealed cell used for the determination of x in $Ni_{1-x}O$ by coulometric titration. (Reproduced from Tretyakov and Rapp, 1969, by permission of The Materials Society.)

activity inside the cell can be calculated. In addition, by application of an electrical potential across the zirconia, oxygen can be pumped into or out of the cell. Since the zirconia is very nearly an ideal electrolyte, the amount of charge passed gives an accurate measure of the amount of oxygen transported. Thus one can measure the change in equilibrium oxygen activity due to a known change in the oxygen content of the sample. In the case that the dominant ionic defects are doubly charged cation vacancies and holes, the deviation from stoichiometry, x, can be expressed as follows:

$$x = [V_{Ni}]_{total} = x_0 P(O_2)^{1/n} \tag{13.13}$$

where $[V_{Ni}]_{total}$ represents the concentration of vacancies in all charge states. The change in x that results from a change in oxygen activity is then:

$$\Delta x = x_0 \Delta P(O_2)^{1/n} \tag{13.14}$$

and a plot of Δx versus $\Delta P(O_2)^{1/6}$ should give a staight line whose slope is x_0, the deviation from stoichiometry at $P(O_2) = 1$ atm. Values at other values of $P(O_2)$ can be obtained by reference to that value. The titration process measures only the number of moles of oxygen that are moved into or out of the cell: most is incorporated into the sample, but some remains in the gas phase to establish the new equilibrium activity. It is necessary to know the latter amount so that only the part that enters the sample is considered in the change in x. For that it is necessary to know the volume of the gas phase. The volume of the cell can be obtained by measuring the change in oxygen activity as a function of the amount of oxygen added to the empty cell:

$$V = RT \frac{\Delta n}{\Delta P} \qquad (13.15)$$

The volume of the sample can be obtained from its weight and density. Subtraction of this value from the total cell volume gives the volume of the gas phase around the sample, and the correction is then easily made.

Having plotted their data according to Eq. (13.14) for values of $n = 4$, 5, and 6, Tretyakov and Rapp concluded that the best linearity was obtained for $n = 6$ and thus assumed that doubly charged cation vacancies are the major ionic defects. They concluded that $x = [V_{Ni}^{//}] = 2.7 \times 10^{-4}$ at 1000°C in 1 atm of oxygen. That value will be seen to be larger by about a factor of 4 than values for x subsequently obtained by others. The discrepancy may result from the assumption that $n = 6$ and that the cation vacancies are all doubly ionized.

Meier and Rapp (1971) subsequently measured the equilibrium conductivity of NiO at temperatures up to 1200°C across the entire stability range. They concluded that $n = 6$ for very pure NiO, but decreased with decreasing purity and reached a value of 4 for Cr-doped NiO. This would be expected if the $Cr_{Ni}^{\cdot}$ donor centers are compensated by doubly charged cation vacancies.

At about the same time, Osburn and Vest (1971) studied NiO by both thermogravimetry and measurements of the equilibrium conductivity. Their conductivity results are shown in Fig. 13.9. Figure 13.9a shows results for a sample of "high" purity (40 ppm Fe), while Fig. 13.9b represents a sample of "intermediate" purity (200 ppm Fe). They concluded that n was approaching 6 with increasing purity and decided that doubly charged cation vacancies are the dominant ionic defect.

Stroud et al. (1973) carried out very detailed measurements of the equilibrium conductivity of "high" purity NiO from 1 to 10^{-4} atm of oxygen at temperatures from 950 to 1350°C and concluded that $n = 5.33$. They attributed this noninteger value to the presence of a mixture of singly and doubly charged cation vacancies. Their expression for $[V_{Ni}^{/}]/[V_{Ni}^{//}]$ gives values of 1.8 and 1.3 at 1100 and 1200°C, respectively at $P(O_2) = 1$ atm. This implies clear domination by singly charged cation vacancies at high oxygen activities, much as was found for CoO.

Finally, Farhi and Petot-Ervas (1978) performed careful measurements of the equilibrium conductivity of NiO single crystals made from "high" purity powder. Their experimental results are shown in Fig. 13.10. They find that n increases from 4.17 to 4.66 as the temperature is raised from 1000 to 1400°C in 1 atm of oxygen, and from 4.78 to 5.46 over the same temperature range in an oxygen activity of 1.89×10^{-4}. They treat this as the result of a mixture of doubly and singly charged cation vacancies and derive an expression for the ratio of doubly charged to singly charged vacancies:

$$k = \frac{[V_{Ni}^{//}]}{[V_{Ni}^{/}]} = \frac{1}{2}\left(\frac{n-4}{6-n}\right) \qquad (13.16)$$

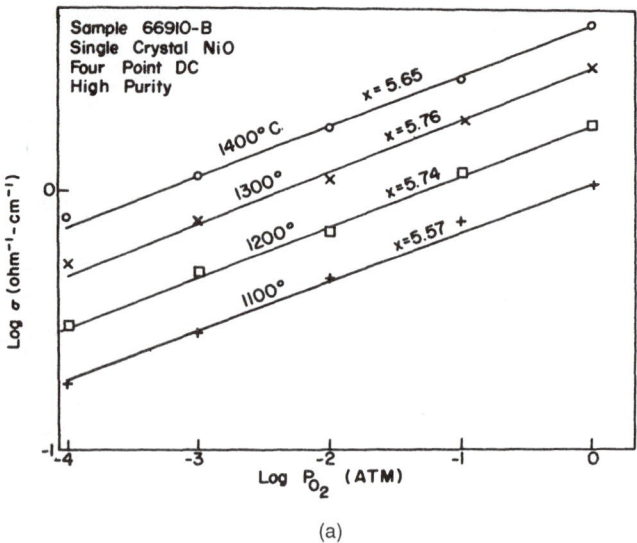

(a)

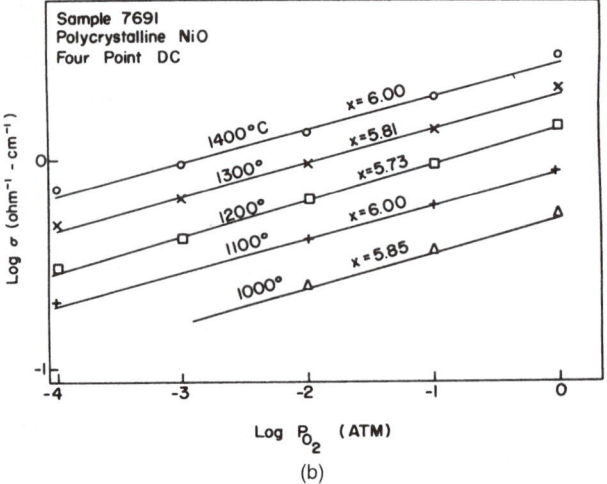

(b)

Figure 13.9 The equilibrium electrical conductivity of $Ni_{1-x}O$ as a function of oxygen activity for (a) Single crystal designated as "high purity" and (b) material designated as "polycrystalline." (Reproduced from Osburn and Vest, 1971, by permission of Elsevier Science.)

They then show a plot of iso-k lines (% of doubly charged vacancies) as a function of oxygen activity and temperature (Fig. 13.11). Only in a small corner of their experimental range does the concentration of doubly charged vacancies exceed that of singly charged vacancies. This plot predicts that doubly charged vacancies make

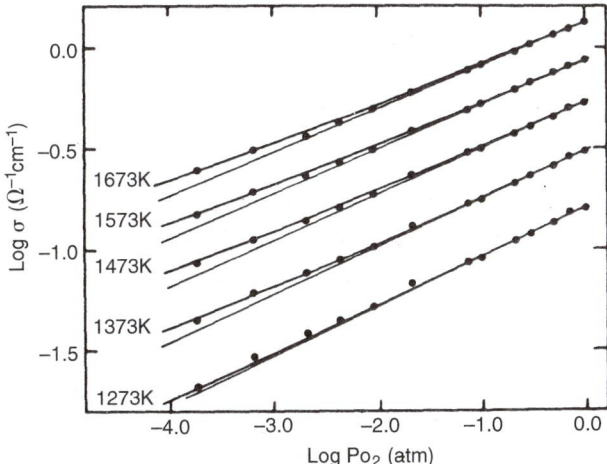

Figure 13.10 The equilibrium electrical conductivity of an undoped $Ni_{1-x}O$ single crystal as a function of oxygen activity. The lines through the data points were calculated from the authors' model; the thinner lines are extrapolations of the slopes measured near $P(O_2) = 1$ atm. (Reproduced from Farhi and Petot-Ervas, 1978, by permission of Elsevier Science.)

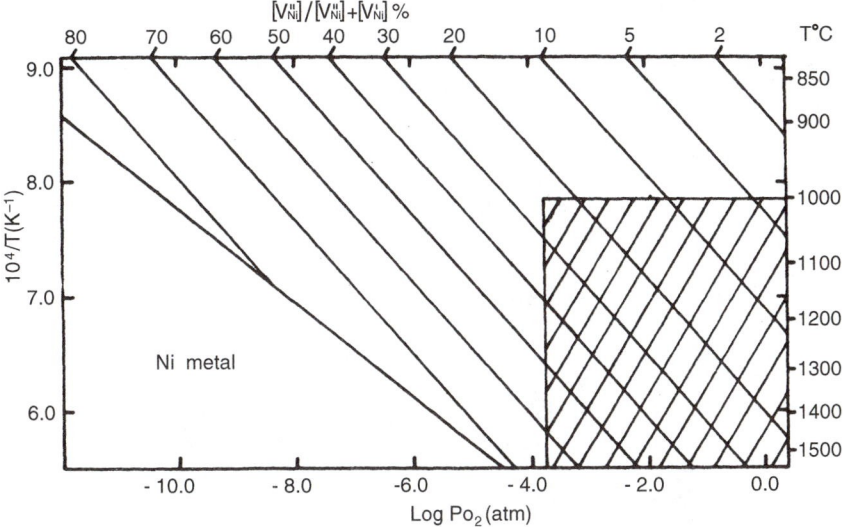

Figure 13.11 Calculated iso-k lines ($k = \%$ of cation vacancies in the doubly charged state) as a function of temperature and oxygen activity. (Reproduced from Farhi and Petot-Ervas, 1978, by permission of Elsevier Science.)

up only about 10% of the total at 1200°C and 1 atm, whereas Stroud et al. found about 45%. The two groups agree that the percentage of doubly charged vacancies increases with increasing temperature. The lines through the data points in Fig. 13.10

were calculated from their model, while the thinner lines are a linear extrapolation of the slope near $P(O_2) = 1$ atm, demonstrating the nonlinearity of the data.

SUMMARY

Both CoO and NiO exhibit cation-deficient nonstoichiometry and p-type conduction, although there is some evidence for a contribution to the equilibrium conductivity from electrons in CoO near the Co/CoO boundary due to their higher mobility. The major defects are cation vacancies and holes, with the former being a mixture of doubly and singly ionized vacancies, with possibly some small contribution from neutral vacancies in CoO. The fraction of doubly charged vacancies increases with increasing temperature and decreasing oxygen activity.

REFERENCES

Bosman, A. J., and C. Crevecoeur. Dipole relaxation losses in CoO doped with Li or Na. *J. Phys. Chem. Solids* 29:109–113, 1968.

Chen, W. K., and R. A. Jackson. Oxygen self-diffusion in undoped and doped cobaltous oxide. *J. Phys. Chem. Solids* 30:1309–1314, 1969.

Dieckmann, R. Cobaltous oxide point defect structure and nonstoichiometry, electrical conductivity, cobalt tracer diffusion. *Z. Phys. Chem. N.F.* 107:189–210, 1977.

Farhi, R., and G. Petot-Ervas. Electrical conductivity and chemical diffusion coefficient measurements in single crystalline nickel oxide at high temperatures. *J. Phys. Chem. Solids* 39:1169–1173, 1978.

Farhi, R., and G. Petot-Ervas. Thermodynamic study of point defects in single crystalline nickel oxide: Analysis of experimental results. *J. Phys. Chem. Solids* 39:1175–1179, 1978.

Fisher, B. and D. S. Tannhauser. Electrical properties of cobalt monoxide. *J. Chem. Phys.* 44:1663–1672, 1966.

Gvishi, M., and D. S. Tannhauser. Hall mobility and defect structure in undoped and Cr or Ti-doped CoO at high temperature. *J. Phys. Chem. Solids* 33:893–911, 1972.

Le Brusq, H., J. J. OEhlig, and F. Marion. Sur l'évolution de la nature des défauts des oxydes MnO et CoO en fonction de la pression partielle d'oxygène à haute température. *C. R. Acad. Sci. Paris, Serie C* 266:965–968, 1968.

Meier, G. H. and R. A. Rapp. Electrical conductivities and defect structures of pure NiO and chromium-doped NiO. *Z. Physikalische Chemie, Neue Folge* 74:168–189, 1971.

Osburn, C. M., and R. W. Vest. Defect structure and electrical properties of NiO. I. High temperature. *J. Phys. Chem. Solids* 32:1331–1342, 1971.

Stiglich, J. J., Jr., J. B. Cohen, and D. H. Whitmore. Interdiffusion in CoO–NiO solid solutions. *J. Am. Ceram. Soc.* 56:119–126, 1973.

Stiglich, J. J., Jr., J. B. Cohen, and D. H. Whitmore. Defect structure of NiO–CoO solid solutions. *J. Am. Ceram. Soc.* 56:211–213, 1973.

Stroud, J. E., I. Bransky, and N. M. Tallan. On the pressure dependence of the electrical conductivity of NiO at high temperatures. *J. Chem. Phys.* 58:1263–1264, 1973.

Tretyakov, Y. D., and R. A. Rapp. Nonstoichiometries and defect structures in pure nickel oxide and lithium ferrite. *Trans. Metall. Soc. AIME* 245:1235–1241, 1969.

Barium Titanate

14

INTRODUCTION

The ferroelectric oxide barium titanate, $BaTiO_3$, is the basic dielectric material in ceramic capacitors, which are manufactured at the rate of over 100 billion devices per year. Thus it is one of the most important materials used in electronic applications. The most popular configuration for these capacitors is the multilayer ceramic capacitor (MLC) in which an interdigitated electrode pattern separates multiple dielectric layers, as shown in Fig. 14.1. The total capacitance is the sum of the capacitances of each layer, giving a very high volume efficiency. The number of dielectric layers may be as low as 10 or as high as several hundred. The dielectric constant of the oxide varies from 1000 to nearly 10,000, depending on the exact formulation; this is much higher

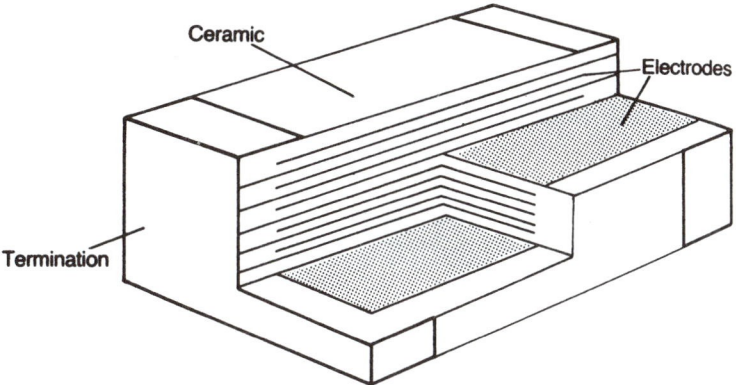

Figure 14.1 Schematic cross-sectional view of a multilayer ceramic capacitor. The $BaTiO_3$-based dielectric layers are separated by an interdigitated array of thin metallic electrodes. The capacitor has been sectioned both parallel and perpendicular to the plane of the layers. The total capacitance is the sum of all the electroded layers. (Reproduced from Moulsen and Herbert, 1990, by permission of Kluwer Academic Publishers.)

than the dielectric constants of widely used nonferroelectric oxides such as Al_2O_3 ($k = 10$) and Ta_2O_5 ($k = 26$). The dielectric layers are usually formed by a tape-casting process using a slurry of the dielectric powder, while the metal electrodes are applied by screen-printing an ink that contains a dispersion of the electrode metal. As manufacturing processes have steadily improved, the minimum dielectric thickness has decreased from about 20 μm to 5 μm, further enhancing the capacitance density. These capacitors are highly efficient, inexpensive, and stable. They are amenable to automated mounting techniques and are extremely compatible with silicon-based microcircuits, where they serve as signal filters and as local reservoirs of charge (to increase device speed, since it is no longer feasible to wait for charge to arrive from a more distant source). This combination of favorable properties has led to the incredible volume of production cited above. An understanding of the defect chemistry of $BaTiO_3$ has made it possible to dope the material so that it can be sintered in an atmosphere that is reducing enough for nickel electrodes to be used instead of the traditional, and much more expensive, palladium–silver alloys. For the larger capacitors this can represent a substantial reduction in material costs. It has also been clearly shown that the gradual degradation of insulation resistance that occurs after prolonged and severe voltage–temperature stress is related to the migration of point defects (i.e., oxygen vacancies). In addition to these practical aspects of defect chemistry, $BaTiO_3$ has proved to be an excellent model material for studies of nonstoichiometery and doping effects. It follows ideal behavior quite closely and demonstrates the basic principles of defect chemistry very clearly. For these reasons, and because the author has spent much of his career studying this material, as well as other closely related compounds, this chapter covers the defect chemistry of $BaTiO_3$ in considerable detail.

Above its Curie temperature of about 130°C, $BaTiO_3$ has the cubic perovskite structure and, because the unit cell is charge-symmetric, is paraelectric (i.e., non-ferroelectric). Below that temperature there is a slight tetragonal distortion and the material becomes ferroelectric, which means that the unit cell is not charge-symmetric and has a dipole moment. There is a pronounced maximum in the dielectric constant at the transition temperature. There are two further distortions at lower temperatures, but they are of less significance. Obviously, any study of the defect chemistry of $BaTiO_3$ under equilibrium conditions will involve only the high temperature, perovskite structure.

$BaTiO_3$ is a ternary compound; it contains three different elements, and the phase rule then requires that four parameters be defined to fully characterize equilibrium. These parameters are typically temperature, pressure, and the activity of one component, as in a binary oxide, plus the activity of one of the other components. Thus for $BaTiO_3$ we need to know not only the oxygen activity, but also the activity of either Ba^{2+} or Ti^{4+}, or its binary oxide. The conditions of full thermodynamic equilibrium for ternary compounds have been discussed in the literature (Schmalzried, 1965). Since, however, accurate determination of a cationic activity is not easily achieved, and is usually not attempted, some compromise of thermodynamic rigor is required in most cases. In the few cases in which one of the component binary oxides is volatile (e.g., PbO in $PbTiO_3$ or Li_2O in $LiNbO_3$), one can consider the possibility of equilibrating the compound with an external source of fixed cationic activity, but even in those cases, most studies have been carried out at relatively low temperatures, where volatilization is minimized. In a ternary oxide such as $BaTiO_3$, where neither binary oxide component has significant volatility, it is assumed that the Ba/Ti ratio is fixed in a given sample. This ratio may or may not be known accurately to the level of the defect concentrations. If the ratio is not exactly unity in the single-phase material, there must be appropriate lattice defects to accommodate this nonideality. In this chapter on $BaTiO_3$, we assume that the Ba/Ti ratio is invariant; it is the best that we can do. However, it should be noted that the defect chemistry would be slightly different if the $BaTiO_3$ were in thermodynamic equilibrium with a fixed external activity of one of the cationic components (Smyth, 1976, 1977). Similar arguments prevail for doped ternary compounds. The addition of a dopant, another component, requires knowledge of yet another activity to define equilibrium rigorously. As in the case of doped binary oxides, the usual assumption is that the dopant concentration is fixed in a given sample of the compound.

GENERAL EXPECTATIONS

The perovskite structure, as described in Chapter 2, can be viewed as a cubic-close-packed array of Ba^{2+} and O^{2-} ions of comparable size, with Ti^{4+} ions occupying all the octahedral sites that have only O^{2-} as nearest neighbors. The other octahedral sites are much too small for Ba^{2+} or O^{2-} ions and are electrostatically inhospitable to Ti^{4+} ions because of the presence of Ba^{2+} in some of the nearest-neighbor sites. Thus it is to be expected that all three forms of Frenkel disorder will require rather large enthalpies. Indeed theoretical calculations of the enthalpies of intrinsic ionic disorder give values of 4.49, 5.94, and 7.56 eV (432, 572, and 728 kJ/mol) per defect for oxygen, barium, and titanium Frenkel disorder, respectively, whereas Schottky disorder requires only 2.29 eV (221 kJ/mol) per defect [the enthalpy of formation for a full formula unit of Schottky disorder has been calculated to be $5 \times 2.29 = 11.45$ eV (1103 kJ/mol)] (Lewis and Catlow, 1983). So we can expect Schottky disorder to be the preferred form of intrinsic ionic disorder:

$$\text{nil} \rightleftharpoons V_{Ba}'' + V_{Ti}^{4/} + 3V_O^{\cdot\cdot} \tag{14.1}$$

with the mass-action expression:

$$[V_{Ba}''] \, [V_{Ti}^{4/}] \, [V_O^{\cdot\cdot}]^3 = K_S e^{-\Delta H_S / kT} \tag{14.2}$$

with $\Delta H_S = 11.45$ eV (1103 kJ/mol). The exponential term per defect then gives values of 8.6×10^{-10} and 1.2×10^{-7} at 1000 and 1400°C, respectively. The former is about 1 ppb and is far below any reasonable impurity content, while the latter is about 0.1 ppm at a temperature commonly used for sintering polycrystalline $BaTiO_3$, where equilibrium might be possible. Considering that the entropy term may increase these numbers by one to two orders of magnitude, the concentration of intrinsic ionic defects may approach the aliovalent impurity level of highly pure material at the sintering temperature. There have been suggestions that there might be Schottky disorder involving only the ions of one of the binary constituents (i.e., only Ba and O vacancies, or only Ti and O vacancies). Of course these proposals violate the conservation of lattice site ratios and should be rejected.

Redox equilibria in $BaTiO_3$ should be related primarily to the reducibility of the Ti^{4+} ion. Neither Ba^{2+} nor O^{2-} is expected to be redox-active. Thus $BaTiO_3$ can be expected to be quite similar to TiO_2 in terms of nonstoichiometry, with the BaO component being essentially an inert diluent. This suggests that we should see primarily oxygen-deficient nonstoichiometry, and n-type semiconductivity. With the valence band made up of filled O 2p states and the conduction band of empty Ti 3d states, the band gap should be similar to that of TiO_2, in the 3–4 eV (300–400 kJ/mol) range.

Acceptor impurities (e.g., large monovalent cations such as K^+ substituted for Ba^{2+}, or small trivalent cations such as Al^{3+} or Fe^{3+} substituted for Ti^{4+}), could be charge-compensated by interstitial cations, oxygen vacancies, or holes. Cation interstitials are not favorable in the perovskite structure. Since the valence band lies at a rather deep level because of the nonoxidizability of O^{2-}, holes are also not particularly favorable. Oxygen vacancies are always possible and seem to be the best choice for $BaTiO_3$. Donor impurities—for example, large trivalent cations such as La^{3+} substituted for Ba^{2+}, or small pentavalent cations such as Nb^{5+} substituted for Ti^{4+} (or large monovalent anions such as F^- substituted for O^{2-})—could be charge-compensated by cation vacancies, anion interstitials, or electrons. Oxygen interstitials can be dismissed, and it would appear that cation vacancies should not be very favorable, although they cannot be completely ignored in view of the preference for Schottky disorder. Since the conduction band, made up of reducible Ti 3d states, lies at a fairly low level, compensation by electrons is a distinct possibility. Because of the low-lying conduction band and deep valence band, donor levels should be shallow and acceptor levels should be deep. Acceptor dopants, if compensated by oxygen vacancies, should reduce the enthalpy of oxidation by providing convenient sites for excess oxygen. On the other hand, donor dopants should enhance reduction through loss of the excess oxygen brought in by the donor oxide.

Now let us check on the accuracy of our predictions.

THE EQUILIBRIUM CONDUCTIVITY OF UNDOPED BaTiO$_3$

The equilibrium conductivity of undoped, polycrystalline $BaTiO_3$ between 750 and 1000°C and 10^{-21}–1 atm of oxygen activity is shown in Fig. 14.2. This sample was slightly Ti-rich, which has no discernible effect on the results. These are the data of Dr. Ning-Huat Chan, a former postdoctoral research associate in the author's laboratory, and his first academic collaborator (N.-H. Chan et al., 1981). The apparatus was constructed by Dr. Chan, and the elegance of the data is a tribute to his skill and perseverance. Over the middle range of oxygen activities, the flow rate of the minor constituent (i.e., O_2 in Ar–O_2 or CO in CO–CO_2), is too small to measure accurately, and the oxygen activity was measured by means of an oxygen concentration cell using acceptor-doped ZrO_2 as the solid oxide electrolyte. This made it possible to avoid the

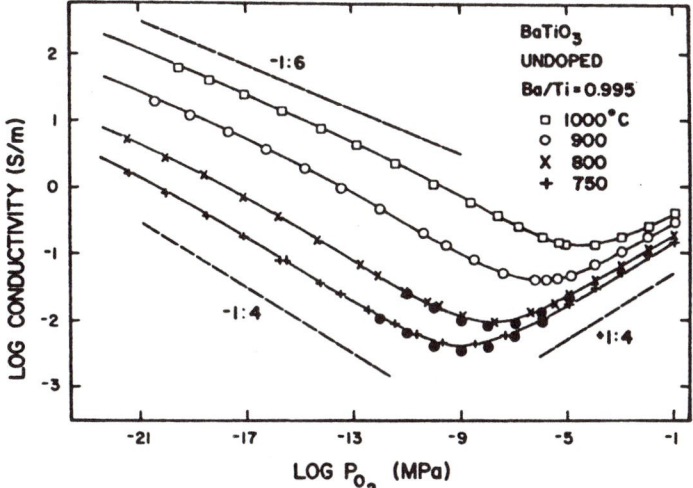

Figure 14.2 The equilibrium electrical conductivity of undoped, polycrystalline BaTiO₃ from 750 to 1000°C over the $P(O_2)$ range 10^{-21}–1 atm. Calculated conductivities from the defect model described in the text are shown as solid lines; solid circles represent subtraction of the ionic contribution from the measured total conductivity. (Reproduced from N.-H. Chan et al., 1981, by permission of the American Ceramic Society.)

commonly seen data gap in this region (e.g., Fig. 12.2). The temperature range was chosen for practical reasons. The tube furnace used in this work had a maximum rated temperature of 1100°C, and a furnace should never be used at its rated temperature if you want it to last very long. At 1000°C, the samples equilibrated with the gas phase faster than either the temperature or the oxygen activity could be changed, but at 600°C the equilibration took several hours for each data point, and this was the maximum that could be tolerated even by the very patient Dr. Chan. The measurements were made with a standard 4-point dc technique, using both polarities to ensure that there was no significant polarization due to ionic contributions to the conductivity.

The data are very similar to those of acceptor-doped TiO₂, as shown earlier (Fig. 12.8). In fact they correspond to the data of undoped TiO₂ pushed down to lower oxygen activities by a few orders of magnitude. This reflects the dilution of the reducibility of Ti⁴⁺, due to expansion of the structure by the BaO component. As expected, there is an extensive region of oxygen-deficient, n-type conductivity, and there is also a significant region of oxygen-excess, p-type conductivity. The log–log slopes appear to be +1/4 and −1/4 above and below the conductivity minima, with a transition to −1/6 at the lowest oxygen activities and highest temperatures. The latter is consistent with a reduction reaction that produces oxygen vacancies, as expected for a compound that has Schottky disorder as its preferred intrinsic ionic disorder, and free electrons:

$$O_O \rightleftharpoons {}^1\!/_2 O_2 + V_O^{\cdot\cdot} + 2e' \tag{14.3}$$

with the mass-action expression:

$$[V_O^{\cdot\cdot}]n^2 = K_n P(O_2)^{-1/2} \tag{14.4}$$

If the reduction reaction is the major source of defects in the most highly reducing conditions, then

$$n \approx 2[V_O^{\cdot\cdot}] \tag{14.5}$$

and

$$n \approx (2K_n)^{1/3} P(O_2)^{-1/6} \tag{14.6}$$

in agreement with the observed slope. For many years the intermediate slope of $-1/4$ was attributed to the formation of singly ionized oxygen vacancies

$$O_O \rightleftharpoons \tfrac{1}{2}O_2 + V_O^{\cdot} + e' \tag{14.7}$$

which would require a change in the state of ionization of the vacancies with changing $P(O_2)$ to account for the two different slopes in the n-type region. However, as will be seen, this explanation is not necessary, and there is no evidence for singly ionized oxygen vacancies under equilibrium conditions.

What about the oxidation reaction that gives rise to the p-type behavior at high oxygen activities? Either type of intrinsic oxidation reaction would be expected to require very high enthalpies:

$$\tfrac{1}{2}O_2 \rightleftharpoons O_I'' + 2h^{\cdot} \tag{14.8}$$

$$\tfrac{3}{2}O_2 \rightleftharpoons 3O_O + V_{Ba}'' + V_{Ti}^{4/} + 6h^{\cdot} \tag{14.9}$$

However, note that the temperature dependence of the conductivity in the p-type region is much less than that of the n-type region. This indicates that the enthalpy of oxidation is less than that of reduction and that the material is more easily oxidized than reduced. Reduction with the formation of oxygen vacancies causes no conceptual difficulties, but oxidation by either Eq. (14.8) or (14.9) should be difficult. This implies that oxidation occurs by an extrinsic reaction that involves the filling of extrinsic oxygen vacancies

$$\tfrac{1}{2}O_2 + V_O^{\cdot\cdot} \rightleftharpoons O_O + 2h^{\cdot} \tag{14.10}$$

This reaction consumes lattice defects rather than creating them and should have a correspondingly lower enthalpy. As in the case of cation interstitials in TiO_2, it has been suggested that the extrinsic oxygen vacancies result from an excess of naturally occurring acceptor impurities. If one looks at the abundances of the stable elements, it is seen that the elements that would act as acceptors in $BaTiO_3$, plus those that would be inactive, account for 99.7% by weight of the earth's crust. Thus it is reasonable to expect that the impurity content would be highly biased on the side of acceptors. The author is aware of only a single exception to the domination by acceptor impurities. Many years ago, a Japanese manufacturer of ceramic capacitors based on $BaTiO_3$ had some difficulty with high leakage currents. This was traced to an unusually high Nb content in the TiO_2 raw material. Nb^{5+} substitutes for Ti^{4+} as a donor impurity, and in small concentrations is compensated by electrons that lead to some conductivity. When acceptor dopants were added to the dielectric formulation to compensate the donor impurities, highly insulating behavior was restored. Apparently the specific ore deposit from which the TiO_2 was obtained had an unusually high level of Nb. It has also been suggested that the extrinsic oxygen vacancies could result from frozen-in Schottky disorder that was created at the processing temperature (i.e., crystal growth temperature or sintering temperature). This possibility cannot be excluded, but the usual impurity levels in these materials are easily adequate to provide the necessary oxygen vacancies.

The incorporation of a generic acceptor oxide, A_2O_3, where A^{3+} substitutes for Ti^{4+}, can be written as follows:

$$A_2O_3 \xrightarrow{(2TiO_2)} 2A'_{Ti} + 3O_O + V_O^{\cdot\cdot} \tag{14.11}$$

Near the conductivity minima, the condition of charge neutrality will be dominated by the acceptor centers and their compensating oxygen vacancies:

$$[A'_{Ti}] \approx 2[V_O^{\cdot\cdot}] \tag{14.12}$$

When this is combined with the mass-action expression for the oxidation reaction:

$$\frac{p^2}{[V_O^{\cdot\cdot}]} = K_p P(O_2)^{1/2} \tag{14.13}$$

one obtains:

$$p \approx \left(\frac{K_p [A'_{Ti}]}{2} \right)^{1/2} P(O_2)^{1/4} \tag{14.14}$$

and from Eqs. (14.4) and (14.12) one obtains:

$$n \approx \left(\frac{2K_n}{[A'_{Ti}]} \right)^{1/2} P(O_2)^{-1/4} \tag{14.15}$$

Both are in accord with the observed dependences of the oxygen activity on either side of the minima.

To fit the equations for the electron and hole concentrations to the experimental conductivity data, one needs values for the mobilities. The electron mobility has been obtained from measurements of the Hall coefficient for both single- crystalline and polycrystalline $BaTiO_3$ (Seuter, 1974). The data in Fig. 14.2 for large-grained poly-crystalline $BaTiO_3$ have been found to be in very good agreement with measurements made on a single crystal (Eror and Smyth, 1978), so the single-crystal mobilities have been used in the present evaluation. These were subsequently fit by the expression (Ihrig, 1976):

$$\mu_n = 8080 T^{-3/2} e^{-0.021 eV/kT} cm^2/V \cdot s \tag{14.16}$$

This is primarily a $T^{-3/2}$ dependence, since the exponential term varies only from 0.76 to 0.83 over the experimental temperature range. The electron mobility decreases from 0.24 to 0.15 $cm^2/V \cdot s$ as the temperature increases from 600 to 1000°C. There is no comparable information on the hole mobility in $BaTiO_3$. As is the usual case, it was assumed that the holes are less mobile than the electrons. However, if they were more than three times less mobile, the hole concentration would have to be so high that a significant fraction of the extrinsic oxygen vacancies would have been filled by the oxidation reaction. In that case the hole concentration, and the p-type conductivity, should begin to flatten out toward high oxygen activities from its log–log slope of 1/4, contrary to what is observed. It was thus assumed that the hole mobility is half the electron mobility at all temperatures. Ironically, more recent work on the very similar $SrTiO_3$ has indicated that this is almost exactly the case (Choi and Tuller, 1988). Using these mobilities, the conductivity data were fit to the mass-action expressions for the reduction and oxidation reaction, Eqs. (14.4) and (14.13). The results of the fit, shown by the calculated lines in Fig. 14.2, are seen to be quite satisfactory (calculated points are also shown near the minima for the case that a small ionic contribution to the conductivity has been subtracted from the total measured value). Arrhenius plots of the mass-action constants for reduction and oxidation are shown in Figs. 14.3 and 14.4. Data from four samples are included: an undoped sample and samples with three different levels of added acceptor, in this case Al^{3+}. The Arrhenius plots are linear and are independent of the dopant concentration, as they should be. The enthalpy of reduction is 5.90 eV (568 kJ/mol) and that of oxidation is 0.92 eV (89 kJ/mol), the

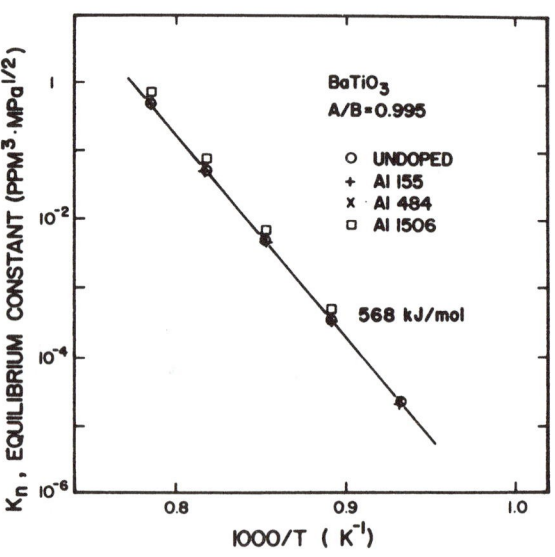

Figure 14.3 Arrhenius plot of the mass-action constant for the reduction reaction derived from the equilibrium conductivities. Results are shown for undoped $BaTiO_3$ and for three samples with various levels of Al^{3+} substituted for Ti^{4+} (in ppm). The enthalpy of reduction is 5.90 eV (568 kJ/mol). (Reproduced from N.-H. Chan et al., 1982, by permission of the American Ceramic Society.)

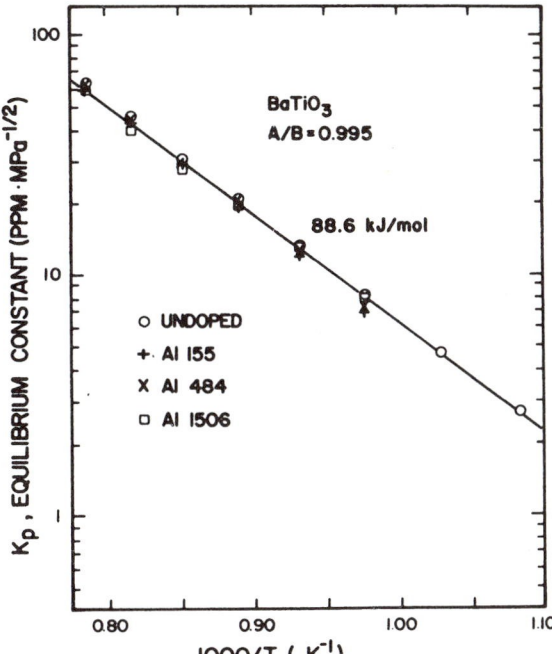

Figure 14.4 Arrhenius plot of the mass-action constant for the oxidation reaction derived from the equilibrium conductivities. The compositions are the same as in Fig. 14.4. The enthalpy of oxidation is 0.92 eV (89 kJ/mol). (Reproduced from N.-H. Chan et al., 1982, by permission of the American Ceramic Society.)

latter indicating the ease of oxidation in the presence of extrinsic oxygen vacancies. The sum of the oxidation and reduction reactions, Eqs. (14.3) and (14.10), gives twice the intrinsic ionization reaction, so the enthalpies are linked by the following relationship:

$$\Delta H_{\mathrm{N}} + \Delta H_{\mathrm{P}} = 2E_{\mathrm{g}}^{\circ} \tag{14.17}$$

This gives a band gap (at 0 K) of 3.41 eV (328 kJ/mol), and this is in excellent agreement with the value obtained from an Arrhenius plot of the electronic conductivities at the minima (corrected for the small ionic contribution). This agreement confirms that there is no temperature dependence for the concentrations of the defects that dominate the condition of charge neutrality near the minima, as is expected for an acceptor-doped material. Thus it is incorrect to argue that some form of thermally activated intrinsic ionic disorder, in thermodynamic equilibrium, plays a role in the defect chemistry of $BaTiO_3$ (this does not preclude a frozen-in concentration of Schottky defects that were created at the processing temperature).

The model thus far can be summarized by a schematic Kröger–Vink diagram as shown in Fig. 14.5, which shows the concentrations of electrons, holes, oxygen vacancies, and acceptor centers. Note that intrinsic nonstoichiometry is seen only under the most reducing conditions. The acceptor centers and their compensating oxygen vacancies dominate the defect chemistry over most of the experimental range. While the hole concentration is rising with increasing oxygen activity, it never reaches a level high enough to replace the oxygen vacancies in the approximation to

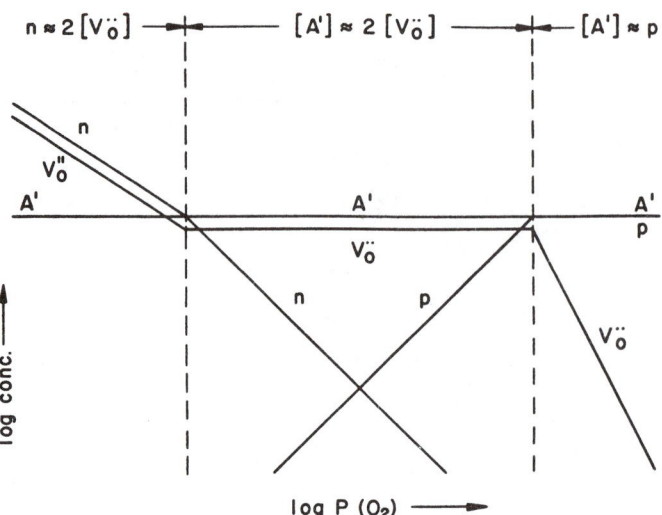

Figure 14.5 Kröger–Vink diagram of the defect concentrations in undoped or acceptor-doped BaTiO₃ as a function of oxygen activity under equilibrium conditions. (Reproduced from Smyth, 1993, by permission of Marcel Dekker.)

charge neutrality, and this indicates that only a minor fraction of the extrinsic oxygen vacancies are ever filled.

INSULATING PROPERTIES OF BaTiO₃

To be useful as a capacitor dielectric, $BaTiO_3$ must be an insulator. In fact, undoped and acceptor-doped $BaTiO_3$ is a light-colored insulator when processed or equilibrated under oxidizing conditions (e.g., in air). However, after equilibration under highly reducing conditions, it is a black semiconductor with resistivities of the order of 0.1 ohm · cm. Thus there is a strong asymmetry between the conduction properties of n-type and p-type $BaTiO_3$. The relatively high conductivity of highly reduced $BaTiO_3$ indicates that the electrons that are the basis for the high temperature equilibrium conductivity in the n-type region are not trapped when the material is cooled to room temperature or below. Thus the donor levels associated with the oxygen vacancies must be very close to the conduction band, which is typical of a compound with a reducible cation. While the equilibrium conductivity in the p-type region is similar to that in the n-type region—as it must be, since both originate from the conductivity minimum with complementary slopes—the p-type conductivity is reduced to very small values as the material is cooled to the vicinity of room temperature. Thus the holes must be trapped during the cooling process, and this must happen because the acceptor levels are relatively deep (high above the valence band), as is usual for an oxide that does not have an oxidizable cation.

The interaction of acceptor centers and holes can be characterized by the ionization reaction for trapped holes

$$A^x \rightleftharpoons A' + h^\cdot \tag{14.18}$$

and its mass-action expression

$$\frac{[A']p}{[A^x]} = K_A(T) = K_A' e^{-E_A/kT} \tag{14.19}$$

where E_A is the ionization energy for holes, that is, the height of the acceptor levels above the valence band edge. The linearity of the p-type conduction in Fig. 14.2 indicates that the concentration of charged acceptor centers is not significantly affected by the oxidation reaction. In the low temperature, insulating state, almost all holes are trapped and the concentration of trapped holes is essentially independent of temperature. Therefore, the hole concentration is simply proportional to the exponential term in Eq. (14.19). One can then easily calculate the acceptor depth necessary to reduce the conductivity from the observed equilibrium value at high temperatures to the level of an insulator at room temperature. The latter must be no more than 10^{-8} $(\Omega \cdot cm)^{-1}$ and is probably lower. Table 14.1 indicates the acceptor levels that are required to reduce the observed equilibrium conductivity of 10^{-3} $(\Omega \cdot cm)^{-1}$ at 750°C to various assumed levels at 25°C.

It is apparent that quite substantial acceptor depths are required to account for the observed insulating properties. These values are typical of those obtained for $BaTiO_3$ by a variety of experimental approaches and theoretical calculations. The third column in Table 14.1 gives the value of the exponential term at 1000°C for the three calculated acceptor depths; this is a measure of the fractional amount of ionization to be expected at that temperature. These indicate that the amount of ionization even at that high temperature is very small, from 0.27 to 0.025%. Thus we have left something out of the defect model depicted in Fig. 14.5. Most of the holes that are created by the oxidation reaction are trapped at the equilibration temperatures, and the p-type conductivity is due to the small fraction of mobile, ionized holes. The hierarchy of concentrations then becomes $[A'] \gg [A^x] \gg p$.

Since the major electronic product of the oxidation reaction is trapped holes, the reaction might better be written as follows:

$$\frac{1}{2}O_2 + V_O^{\cdot\cdot} + 2A' \rightleftharpoons O_O + 2A^x \tag{14.20}$$

TABLE 14.1
Acceptor levels in BaTiO$_3$

Conductivity at 25°C $(\Omega \cdot cm)^{-1}$	Required acceptor depth [eV (kJ/mol)]	$e^{(-E_A/kT)}$ at 1000°C
10^{-8}	0.65 (63)	2.7×10^{-3}
10^{-9}	0.78 (75)	8.3×10^{-4}
10^{-10}	0.91 (88)	2.5×10^{-4}

with its mass-action expression:

$$\frac{[A^x]^2}{[A']^2[V_O^{\cdot\cdot}]} = K_{ox}(T) = K'_{ox}e^{-\Delta H_{ox}/kT} \tag{14.21}$$

where ΔH_{ox} is the enthalpy of oxidation to form trapped holes. The previously determined $\Delta H_P = 0.92$ eV (89 kJ/mol) is the enthalpy of oxidation to form free holes, since it was determined by measurement of the conductivity. Equation (14.21), together with the approximate condition of charge neutrality, Eq. (14.12), gives for the equilibrium concentration of trapped holes:

$$[A^x] \approx \left(\frac{K'_{ox}}{2}\right)^{1/2} [A']^{3/2}e^{-\Delta H_{ox}/2kT}P(O_2)^{1/4} \tag{14.22}$$

Thus the trapped hole concentration has the same dependence on oxygen activity as the free hole concentration, Eq. (14.14), but varies as $[A']^{3/2}$, rather than the $[A']^{1/2}$ dependence of free holes. A line for trapped holes can now be added to Fig. 14.5 to give a more complete Kröger–Vink diagram as shown in Fig. 14.6.

The oxidation reactions to form free holes, Eq. (14.8), and trapped holes, Eq. (14.20), are linked by the acceptor ionization reaction, Eq. (14.18). If we know the enthalpies for any two of these reactions, the enthalpy of the third is readily obtained. We know that $\Delta H_P = 0.92$ eV (89 kJ/mol), and if we combine that with the middle value of the ionization energy from Table 14.1, 0.78 eV (75 kJ/mol), we obtain $\Delta H_{ox} = -0.64$ eV (-62 kJ/mol). The enthalpy of oxidation to form trapped holes, which are the major product of the excess oxygen, is a negative number! This means that the concentration of trapped holes, and of excess oxygen, is decreasing with increasing temperature of equilibration. The conductivity is increasing with increasing temperature because of an increasing amount of ionization of a decreasing population of trapped holes.

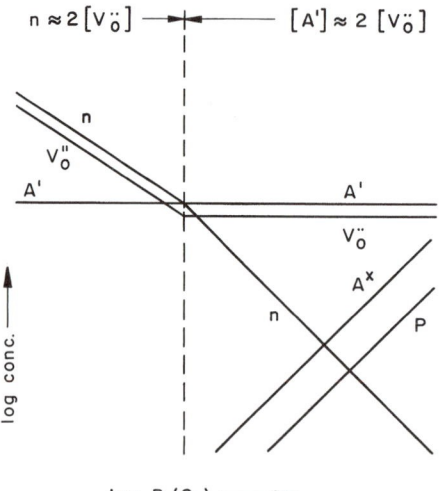

Figure 14.6 Kröger–Vink diagram of the defect concentrations in undoped or acceptor-doped $BaTiO_3$ as a function of oxygen activity. A line for acceptor centers with a trapped hole, A^x, is included.

It should be possible to obtain an experimental value for ΔH_{ox} from a thermogravimetric measurement, since the concentration of trapped holes should be approximately twice the amount of the stoichiometric amount of excess oxygen. To the author's knowledge, this has not been done; indeed, such a measurement might be difficult because of the rather small deviation from stoichiometry. A different approach has been used in the author's laboratory (Raymond and Smyth, 1994) and is based on a rearrangement of Eq. (14.22) to:

$$P(O_2) \approx \left(\frac{2}{K'_{ox}}\right)^2 \frac{[A^x]^4}{[A']^6} e^{-2\Delta H_{ox}/kT} \qquad (14.23)$$

If the equilibrium oxygen activity can be measured as a function of temperature while keeping the concentration of charged and neutral acceptor centers constant, then an Arrhenius plot should give a straight line with a slope of $-2\Delta H_{ox}$. In other words, the amount of oxygen entering or leaving the sample over the experimental temperature range must be too small to affect the defect concentrations. We have called this a "constant composition oxygen activity" experiment. This was accomplished in a sealed cell as shown in Fig. 14.7. A large sample was placed in an alumina tube that had hermetically sealed lids on top and bottom. The hermetic seal was achieved by use of glass gaskets that soften at the experimental temperature so that the application of a small mechanical pressure causes the viscous gasket to form a seal (if the operator is having a lucky day). The bottom lid was merely an inert disk of alumina, but the top lid was acceptor-doped zirconia with porous platinum electrodes on both sides. An electrical lead (Pt) was brought out from the inner electrode through the softened glass seal. The oxygen activity inside the cell could then be obtained from the emf of the oxygen concentration cell formed by the oxygen activities inside and outside (air) the cell. The hermeticity of the seal could be assured if the results proved to be reversible with temperature. With a large sample and very little residual gas volume,

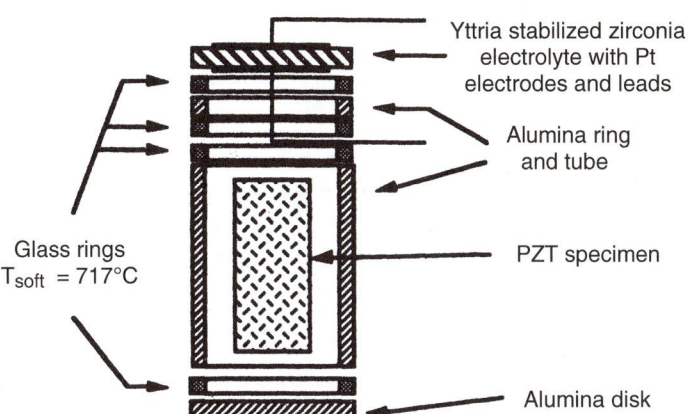

Figure 14.7   The sealed cell arrangement used to measure the equilibrium oxygen activity as a function of temperature for acceptor-doped BaTiO₃ at constant defect concentrations. (Reproduced from Raymond and Smyth, 1994, by permission of Gordon & Breach.)

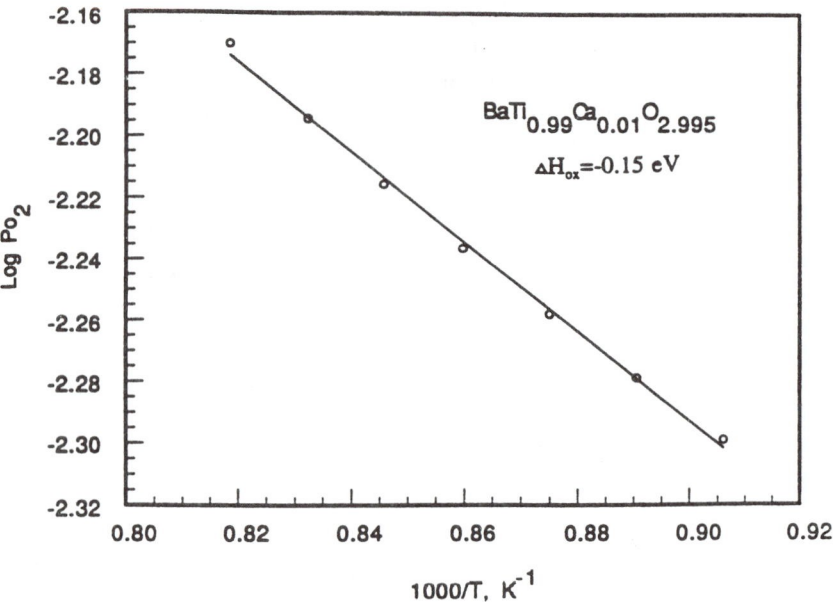

Figure 14.8 Arrhenius plot of the equilibrium oxygen activity for acceptor-doped $BaTiO_3$ at constant defect concentrations. (Reproduced from Ma. 1995.)

it was estimated that the amount of oxygen that had to enter or leave the sample to establish the new equilibrium activity after a change in temperature would affect the defect concentrations by less than 1%. A highly acceptor-doped sample was used to maximize the defect concentrations, and the experimental result is shown in Fig. 14.8 (Ma, 1995). The Arrhenius plot is linear and indicates a value of ΔH_{ox} for $BaTiO_3$ of -0.15 eV (-14 kJ/mol). Together with ΔH_P, this gives an acceptor depth of 0.54 eV (52 kJ/mol), somewhat smaller than the calculated values in Table 14.1. The depth of the acceptor levels may differ according to the source of the sample and the types of acceptor impurity present. Similar experiments done on $SrTiO_3$ and $CaTiO_3$ gave acceptor depths of 0.69 and 1.22 eV (66 and 117 kJ/mol), respectively.

Inclusion of the concentration of trapped holes in the defect model has an important practical consequence. If there were no trapped holes, the model could be represented by Fig. 14.5, and the change in conductivity level in quenched samples should occur when the equilibrium oxygen activity crosses that of the conductivity minima. Samples quenched from the n-type side should be conducting, while those quenched from the p-type side should be insulating. When most of the holes are trapped, as shown in Fig. 14.6, this transition will occur for the equilibration conditions under which the concentration of all electron species, almost all free, equals that of all hole species, mostly trapped. This occurs for equilibrium oxygen activities considerably lower than the conductivity minima. This is shown in Fig. 14.9, which includes lines for the electron and hole concentrations for both high temperature equilibrium conditions (HT), and for quenched samples at low temperatures (LT) (Waser, 1991).

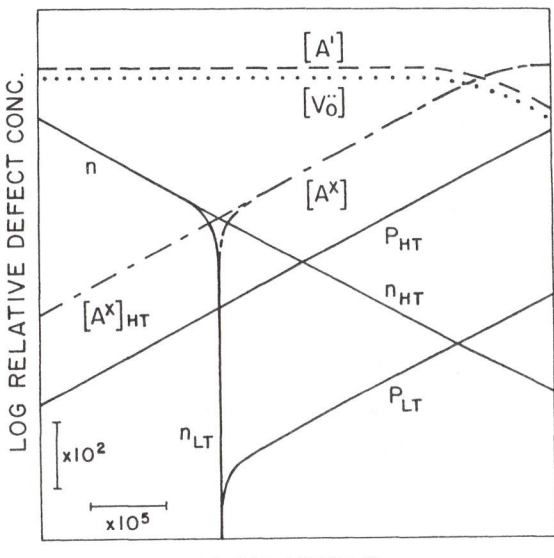

Figure 14.9 Kröger–Vink diagram for undoped or acceptor-doped BaTiO₃, including trapped holes, for high temperature, equilibrium conditions (HT), and low temperature, quenched samples (LT). (Reproduced from Waser, 1991, by permission of the American Ceramic Society.)

This has been confirmed by measurement of the conductivity of samples quenched from various equilibrium oxygen activities (Dong et al., 1994). The low temperature conductivities were measured by an ac impedance analysis technique that allows the bulk conductivity to be separated from electrode contact and grain boundary effects. The results are shown in Fig. 14.10. For equilibration at high $P(O_2)$, the low temperature conductivity varies as $P(O_2)^{1/4}$ and has an activation energy of the order of 0.6 eV (58 kJ/mol). Both values are to be expected for hole conduction, since the activation energy should correspond to the acceptor ionization energy, previously determined to be 0.54 eV (52 kJ/mol). The conductivity is independent of $P(O_2)$ in the midrange of oxygen activity and has an activation energy of about 1 eV (100 kJ/mol), characteristic of ionic conduction by oxygen vacancies (to be discussed later in this chapter). Finally, at low $P(O_2)$, the conductivity jumps abruptly to high values, and this corresponds to the transition to the highly conducting region. The equilibrium conductivities at the equilibration temperatures are shown for comparison, and it is seen that the low temperature conductivity transition occurs for equilibration values of $P(O_2)$ that are substantially lower than those at the equilibrium conductivity minima.

The location of the conductivity transition at such low values of $P(O_2)$ has made it possible to replace the usual Pd–Ag alloy electrodes with the much cheaper Ni in multilayer ceramic capacitors based on BaTiO₃. When the conductivity transition is moved by acceptor doping to the lowest $P(O_2)$ possible, the capacitors can be fired in atmospheres that are sufficiently reducing to maintain metallic Ni, while still retaining an insulating dielectric. If the conductivity transition corresponded to the conductivity minima, this would not be possible, since the conditions necessary to maintain metallic Ni electrodes would result in conducting BaTiO₃ layers. For the larger capacitors, the use of Ni electrodes can represent a considerable cost saving

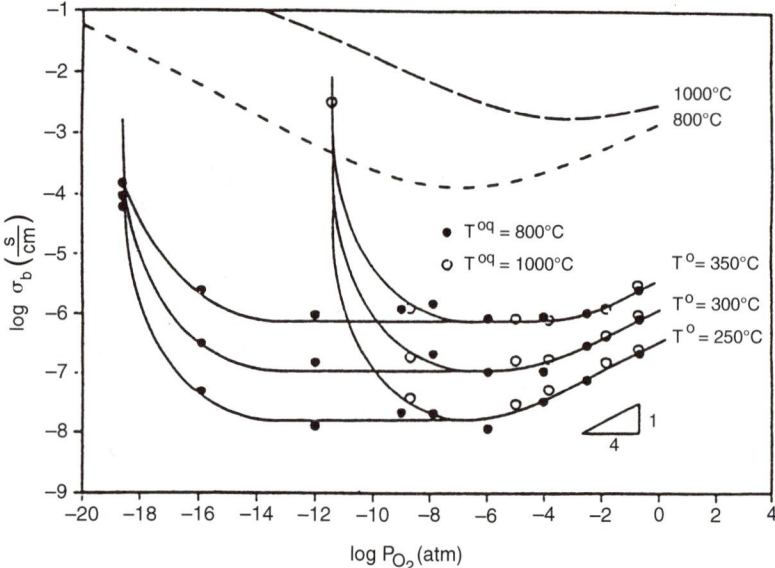

Figure 14.10 The conductivities of undoped $BaTiO_3$ measured for both equilibrium conditions at 800 and 1000°C, and for samples quenched from those temperatures and measured at 250, 300, and 350°C. Note that the rapid increase of conductivity in the quenched samples occurs at oxygen activities far below the equilibrium conductivity minima. (Reprinted from Dong et al., 1994, with permission of the editor.)

that more than compensates for the added cost of controlling the oxygen activity in the firing process. The rather unusual type of acceptor doping used to move the conductivity transition to lower values of $P(O_2)$ is described in the next section.

ACCEPTOR-DOPED BaTiO$_3$

Since the behavior of undoped $BaTiO_3$ indicates that it already has an excess of acceptor impurities, acceptor-doped $BaTiO_3$ merely represents an extension of that behavior. Examination of the equations for the electron and hole concentrations in the near-stoichiometric region, Eqs. (14.14) and (14.15), indicates that an increase in the acceptor concentration should increase the hole concentration and decrease the electron concentration, proportional to the square root of the net acceptor excess. This will result in a movement of the conductivity minima to lower oxygen activities. This movement can be quantified by equating the expressions for the hole and electron concentrations and solving for the oxygen activity for that condition, $P(O_2)^\circ$, to give

$$P(O_2)^\circ \approx \frac{4K_n}{K_p[A']^2} \tag{14.24}$$

The conductivity minima will differ from this expression for n = p by the mobility ratio, which appears to be a relatively minor correction. This expression indicates that the conductivity minima will move to lower values of $P(O_2)$ with the square

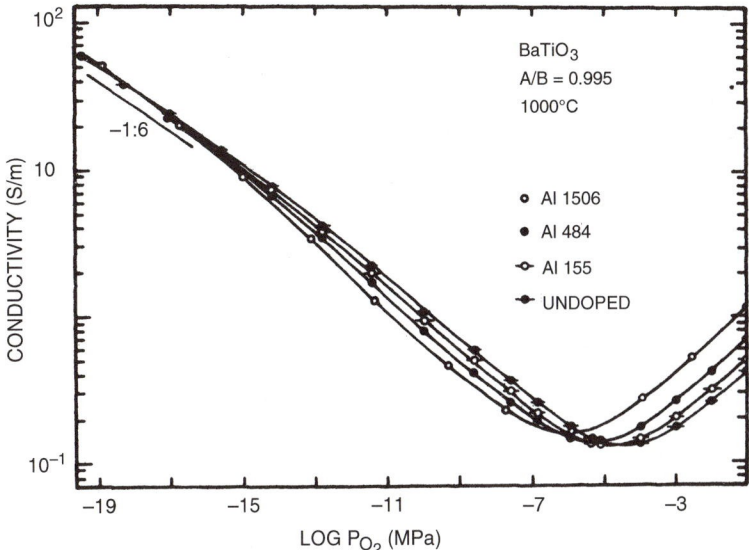

Figure 14.11 The equilibrium conductivities of undoped and Al^{3+}-doped BaTiO$_3$ as a function of oxygen activity at 1000°C. The acceptor (Al^{3+}) concentrations are in ppm replacement for Ti^{4+}. (Reproduced from N.-H. Chan et al., 1982, by permission of the American Ceramic Society.)

of the net acceptor excess. This is shown in Fig. 14.11 for BaTiO$_3$ doped with Al^{3+} replacing Ti^{4+} up to a concentration level of 0.15 atom % (N.-H. Chan et al., 1982). The behavior is as expected, and all the equilibrium conductivities of all samples converge to the same values at very low $P(O_2)$, where the oxygen deficiency exceeds the acceptor content and the approximation to charge neutrality reverts to Eq. (14.5). Note that the Arrhenius plots of the mass-action constants for the reduction and oxidation reactions (Figs. 14.3 and 14.4), show that the mass-action constants are independent of the impurity level, indicating ideal thermodynamic behavior.

It was mentioned in the preceding section that acceptor-doped BaTiO$_3$ formulations can be used in multilayer ceramic capacitors that have cofired Ni electrodes rather than the more conventional, and much more expensive, Pd–Ag alloys. This requires the movement of the insulator–conductor transition to lower values of $P(O_2)$ so that the sintering process can take place in a reducing atmosphere that will not oxidize the Ni electrodes, while ensuring the retention of the insulating property of the dielectric. As stated earlier, the conductivity transition is determined by the value of $P(O_2)$ at which the concentration of all electron species, mostly free, equals the concentration of all hole species, mostly trapped. This value of $P(O_2)$ can be obtained by equating Eqs. (14.15) and (14.22):

$$P(O_2)^t \approx \frac{4K_n}{K_{ox}[A']^4} \qquad (14.25)$$

It is seen that the position of the conductivity transition is affected much more strongly by the acceptor content than is the conductivity minimum. It was shown in Fig. 14.10

that the conductivity transition takes place at equilibrium $P(O_2)$ values that are much lower than those at the conductivity minima.

The first successful development of MLCs with BMEs (multilayer ceramic capacitors with base metal electrodes, i.e., Ni) was achieved by the Murata Manufacturing Company in Kyoto, Japan (Sakabe et al., 1987). Murata is the world's largest manufacturer of MLCs. For the capacitors with BMEs, the company uses a dielectric formulation containing Ca^{2+} that can be represented in simplified form as $[(Ba_{1-x}Ca_x)O]_m TiO_2$. Insulating material is obtained when this composition is fired in a reducing atmosphere only if the Ca^{2+} content is substantial and $m > 1$. At that time it was well known that in solid solutions in the $BaTiO_3$–$CaTiO_3$ system, the Ca^{2+} substitutes for the Ba^{2+} on the A sites of the perovskite structure and has essentially no effect on the Curie temperature (the temperature of the ferroelectric–paraelectric transition). We were able to show in our laboratory that when $(Ba + Ca)$ exceeds Ti, some Ca^{2+} can be induced to substitute for Ti^{4+} on the B sites, where it acts as an acceptor, Ca_{Ti}'' (Han et al., 1987):

$$CaO \xrightarrow{(TiO_2)} Ca_{Ti}'' + O_O + V_O^{\cdot\cdot} \qquad (14.26)$$

This also causes a substantial reduction in the Curie temperature, which confirms that not all the Ca^{2+} is on the A sites. Ca^{2+} will certainly prefer to occupy A sites, if the composition allows space there; but if the A sites are overloaded, Ca^{2+} can be forced to occupy up to about 2% of the B sites. It acts as an extremely effective acceptor that moves the conductivity transition to very low values of the equilibrium $P(O_2)$. In Fig. 14.12, which shows the effect on the conductivity minimum, we see the equilibrium conductivity at 1000°C of two samples in which some of the Ba^{2+} has been replaced by Ca^{2+}, and two samples in which CaO has been added to $BaTiO_3$. In the latter samples it is expected that the Ca^{2+} will divide itself equally between the A and B sites to complete a proper perovskite structure. The Ca^{2+} that occupies B sites acts as an acceptor. A discussion of the conductivity at the minima and the shape of the minima is deferred to the next section. Figure 14.13 shows the resistivity at room temperature of various samples fired in air or in a reducing atmosphere of 2×10^{-12} MPa (2×10^{-11} atm) of oxygen as a function of m, the ratio of large cations to small cations (Sakabe et al., 1987). The samples fired in air have high resistivities for all values of m, but when fired in the reducing atmosphere, high resistivity is obtained only for the sample that contains Ca^{2+} and for which $m > 1$. Such compositions can be fired in reducing atmospheres that are substantially below the Ni–NiO equilibrium, while still being on the insulating side of the conductivity transition for the dielectric, thus permitting the use of Ni electrodes in the MLCs. This rather clever application of defect chemistry results in a considerable saving in material cost for the larger capacitors. The solubility of Ca^{2+} on the B sites is surprisingly high, nearly 2%. Its formal charge is 2 less than that of Ti^{4+}, and it is much larger than the latter. On the other hand, Al^{3+} is only one charge below Ti^{4+} and is much closer in size; yet its solubility is about 10 times less. There is no obvious explanation for this difference. We have found no evidence of solubility of the larger Sr^{2+} on the B sites of $BaTiO_3$. Ca^{2+} thus proves to be a transitional cation in the alkaline earth titanates. Whereas

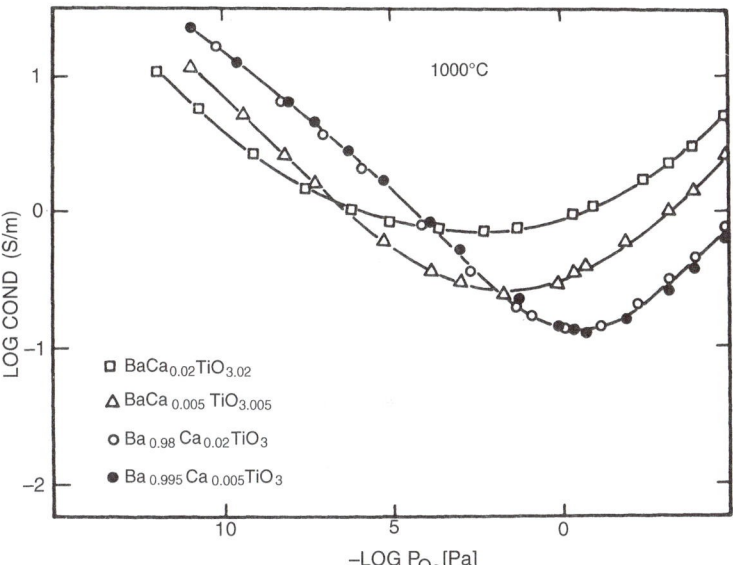

Figure 14.12 The equilibrium electrical conductivity of Ca-doped BaTiO$_3$ at 1000°C. The results compare samples with 0.5 and 2 mol % Ca^{2+}, in one case substituted for Ba^{2+} (open and solid circles) and in the other case added to stoichiometric BaTiO$_3$ (triangles and squares). In the latter samples, it appears that the Ca^{2+} divides itself equally between the Ba^{2+} and Ti^{4+} sites. (Reproduced from Han et al., 1987, by permission of the American Ceramic Society.)

SrTiO$_3$ and BaTiO$_3$ both have ideal perovskite structures, the smaller Ca^{2+} is less comfortable with the 12-coordinate A site, and CaTiO$_3$ (the mineral perovskite!) is slightly distorted from the ideal cubic structure. In the case of the even smaller Mg^{2+}, MgTiO$_3$ has the ilmenite structure, where both cations are six-coordinate.

IONIC CONDUCTION IN BaTiO$_3$

A number of studies over the years have indicated the presence of measurable ionic conduction in BaTiO$_3$, but only in the past 10 years has the phenomenon been studied in detail. The newer work started with the observation that attempts to fit the equilibrium conductivity of BaTiO$_3$ (Fig. 14.2), to the appropriate mass-action expressions were quite successful except in the region of the conductivity minima. The observed minima are higher and less sharply curved than the ideal electronic minima. An "excess" conductivity was derived from this discrepancy that gave linear Arrhenius plots with activation energies of 1.1 eV (106 kJ/mol) and increased with the concentration of acceptor dopants (Chan et al., 1982). For example, Fig. 14.11 showed that the conductivity at the minima increase with increasing acceptor concentration. It was proposed that this behavior corresponded to ionic conduction due to extrinsic oxygen vacancies. The observed activation energy is typical for the mobility of

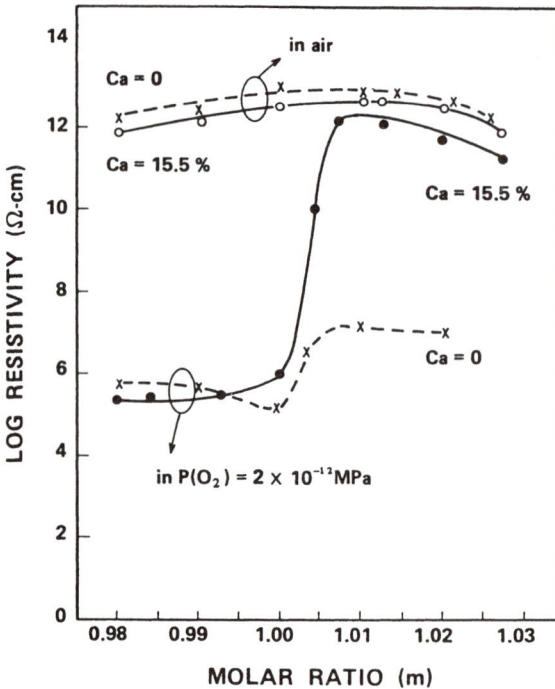

Figure 14.13 The room temperature resistivity as a function of the molar ratio m in $[(Ba_{1-x}Ca_x)O]_m TiO_2$ ceramics with $x = 0$, sintered in air (crosses) and in an oxygen activity of 2×10^{-12} MPa (2×10^{-11} atm) (crosses), and with $x = 0.155$, sintered in the same two atmospheres (open and solid circles). The resistivity of the samples sintered in the reducing atmosphere is high only for the Ca-doped sample and when $m > 1$ (i.e., when there is a crystallographic excess of alkaline earth cations). (Reproduced from Sakabe et al., 1987, by permission of the American Ceramic Society.)

oxygen vacancies in the perovskite structure. This approach was supplemented by the following argument (Chang et al., 1988). The total conductivity is the sum of the contributions due to electrons, holes, and mobile ions:

$$\sigma_T = \sigma_n + \sigma_p + \sigma_i \tag{14.27}$$

On the high $P(O_2)$ side of the minima, the contribution due to electrons can be neglected, and the total conductivity can be approximated as follows:

$$\sigma_T \approx \sigma_i + \sigma_p^{\circ} P(O_2)^{1/4} \tag{14.28}$$

where σ_p° is the hole conductivity in 1 atm of oxygen. Similarly, on the low $P(O_2)$ side of the minima, the hole contribution can be neglected and the total conductivity can be approximated as follows:

$$\sigma_T \approx \sigma_i + \sigma_n^{\circ} P(O_2)^{-1/4} \tag{14.29}$$

Thus a plot of the total conductivity on the high $P(O_2)$ side of the minimum versus $P(O_2)^{1/4}$ should give a straight line with slope σ_p° and an intercept of σ_i. Likewise, a plot of the total conductivity on the low $P(O_2)$ side of the minima should give a straight line with slope σ_n° and an intercept again of σ_i. Plots from either side of the conductivity minima should have the same intercept: that is, the $P(O_2)$-independent ionic conductivity. Such plots are shown in Fig. 14.14 for undoped and acceptor-

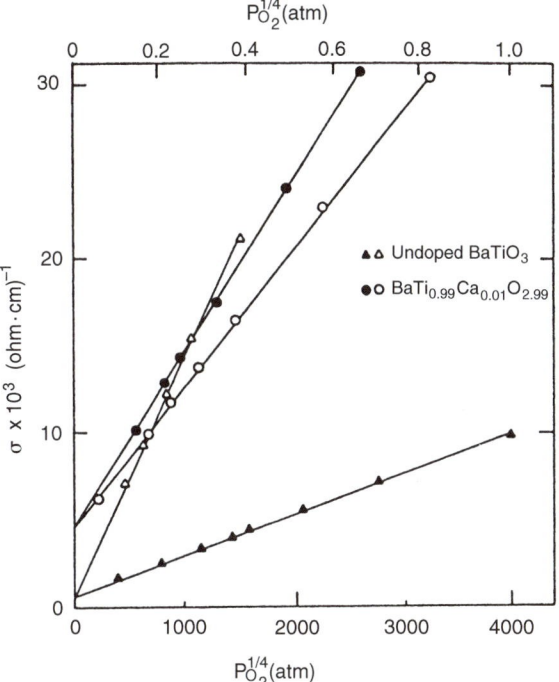

Figure 14.14 The total equilibrium conductivity for undoped BaTiO₃ (triangles) and BaTi₀.₉₉Ca₀.₀₁O₂.₉₉ (circles) above the conductivity minima (solid symbols and upper abscissa) and below the conductivity minima (open symbols and lower abscissa). The intercepts represent the ionic component of the conductivity. (Reproduced from Chang et al., 1988, by permission of the American Ceramic Society.)

doped $BaTiO_3$, and the expected behavior is observed. Similar results were obtained for $SrTiO_3$ and $CaTiO_3$. The ionic conductivity obtained in this way can be divided by the total conductivity at any value of $P(O_2)$ in the near-stoichiometric region to obtain the transport number for that condition. Plots of the transport number in undoped $BaTiO_3$ as a function of $P(O_2)$ at several temperatures are shown in Fig. 14.15. Transport numbers in excess of 0.6 are observed, while values of up to 0.9 were seen in acceptor-doped samples. The validity of the interpretation was checked by measurement of the emf of an oxygen concentration cell with the sample sandwiched between oxygen at 1 atm on one side, and varying values of $P(O_2)$ on the other. Figure 14.16 shows that there is excellent agreement between the measured values and the values calculated from the conductivities obtained by deconvolution of the curvature near the conductivity minimum. Note that the latter technique gives both transport numbers and the absolute values of the ionic conductivity, while the more complex oxygen concentration cell technique gives only transport numbers. In some cases (e.g., analysis of the degradation of leakage resistance in ceramic capacitors), the absolute value of the ionic conductivity is important. This would require measurement of the total equilibrium conductivity regardless of which approach is used. Similar agreement between the two techniques has been observed for measurements made on $LiNbO_3$.

The preceding section described the substitution of Ca^{2+} for Ti^{4+} as a very effective acceptor dopant. Figure 14.17 shows the equilibrium conductivity of $BaTiO_3$ doped

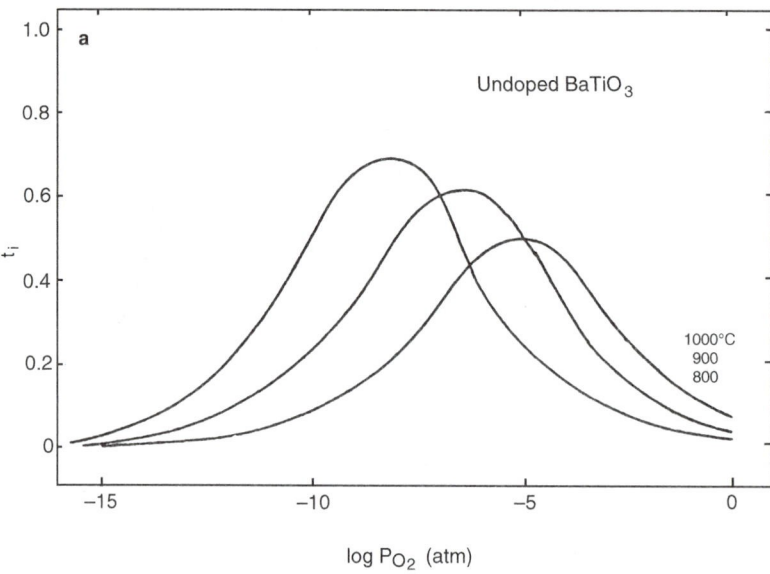

Figure 14.15 The ionic transport number for $BaTi_{0.99}Ca_{0.01}O_{2.99}$ as a function of oxygen activity at 800, 900, and 1000°C. (Reproduced from Chang et al., 1988, by permission of The Electrochemical Society.)

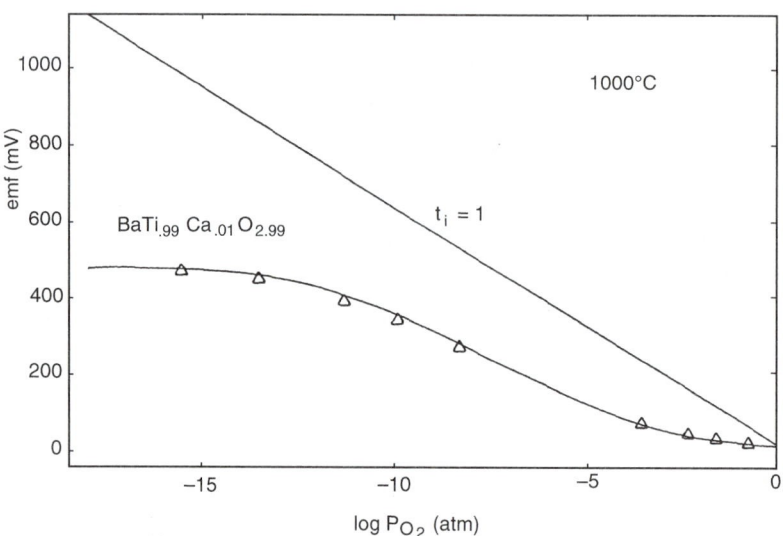

Figure 14.16 Comparison of the emf of an oxygen concentration cell for $BaTi_{0.99}Ca_{0.01}O_{2.99}$ (the symbols) with the values predicted from the ionic transport numbers derived from the total equilibrium conductivity (the line). One side of the concentration cell was held at an oxygen activity of 1 atm, while the activity on the other side was varied as shown. (Reproduced from Chang et al., 1988, by permission of The Electrochemical Society.)

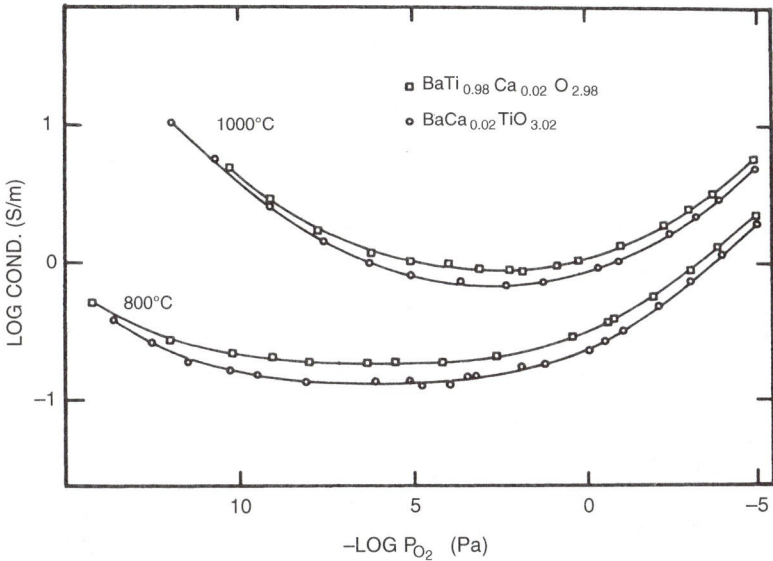

Figure 14.17 The equilibrium electrical conductivity of highly Ca-doped BaTiO₃ as a function of oxygen activity. The very broad and flat minima result from the large fraction of ionic conduction. (Reproduced from Han et al., 1987, by permission of the American Ceramic Society.)

by either the substitution of 2% CaO for TiO_2 or the addition of 2% CaO to $BaTiO_3$. The minima are very high and flat, indicating a very large ionic contribution to the conductivity, especially at the lower temperature. The ionic transport number increases with decreasing temperature because the ionic conductivity is decreasing with an activation energy of about 1 eV (100 kJ/mol), the activation energy for the mobility of oxygen vacancies, while the electronic conductivity at the minima is decreasing as $E_g^\circ/2 = 1.7$ eV (164 kJ/mol). Note also in Fig. 14.15 that the temperature dependence of the transport number has different signs on the two sides of the minima. This is because the ionic conduction is dropping faster with temperature than the p-type conduction but slower than the n-type conduction, which has a higher effective activation energy.

DONOR-DOPED BaTiO₃

The effect of donor dopants in $BaTiO_3$ is one of the least understood aspects of the defect chemistry of that compound. Large trivalent cations that substitute for Ba^{2+}, such as La^{3+}, or small pentavalent or hexavalent cations that substitute for Ti^{4+}, such as Nb^{5+}, or W^{6+}, clearly act as donors, but the mode of compensation is peculiar. For donor substitution levels up to a few tenths of an atom %, the excess oxygen contained in the donor oxide is lost, even when the material is equilibrated in an oxidizing atmosphere, and compensation is by electrons:

$$La_2O_3 \xrightarrow{(2BaO)} 2La_{Ba}^{\cdot} + 2O_O + \tfrac{1}{2}O_2 + 2e' \tag{14.30}$$

As expected, the donor levels are shallow, and the result is a dark-colored, conducting material. However, at higher concentrations, the excess oxygen is retained, and compensation appears to be primarily by Ti vacancies:

$$2Nb_2O_5 + BaO \xrightarrow{(4TiO_2)} 4Nb_{Ti}^{\cdot} + Ba_{Ba} + V_{Ti}^{4/} + 11O_O \tag{14.31}$$

This gives a light-colored insulating material. Thus as the donor concentration is increased, there is a fairly abrupt switch from conducting to insulating material, as shown in Fig. 14.18. This transition is accompanied by a drastic change in microstructure; at small concentrations the grain size is large, tens of micrometers, similar to that of undoped $BaTiO_3$, while at larger concentrations the grain size is much smaller, around 1 μm. The connection between the transitions in conductivity and the grain size is not clear. Whereas the conductivities associated with small donor concentrations are incompatible with application as capacitor dielectrics, larger concentrations of donors such as Nb^{5+} are often used in dielectric formulations. They

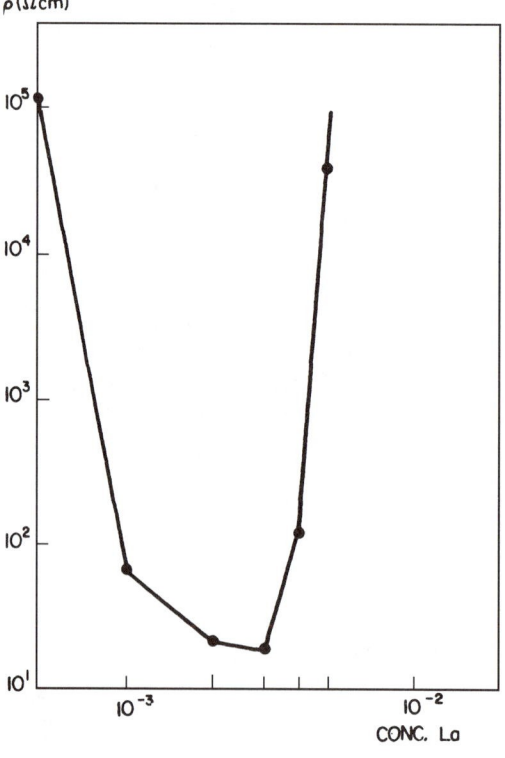

Figure 14.18 The room temperature resistivity of $BaTiO_3$ as a function of donor concentration. (Reproduced from Jonker, 1964, by permission of Elsevier Science.)

serve the double role of shifting the Curie temperature down toward room temperature, and of improving load-life stability. The latter occurs because the positively charged donor centers suppress the concentration of positively charged oxygen vacancies, the mobile species that leads to degradation of the leakage resistance.

When $BaTiO_3$ with large donor concentrations is equilibrated with a highly reducing atmosphere, the excess oxygen is lost and electrons become the charge-compensating species. Thus for small concentrations of a donor such as Nb^{5+}, the single-phase composition is $BaTi_{1-x}Nb_xe_xO_3$ under both oxidizing and reducing conditions, while for larger concentrations the single-phase compositions are $BaTi_{1-x}Nb_xe_xO_3$ for reducing conditions but $BaTi_{1-5x/4}Nb_xO_3$ for oxidizing conditions (H. M. Chan et al., 1986). Compensation by titanium vacancies under oxidizing conditions for the larger donor concentrations was confirmed by determining which compositions give single-phase material. Unless the composition corresponded to the proper concentration of compensating titanium vacancies, the material self-adjusted by splitting out the necessary amount of second phase. Thermogravimetric measurements have confirmed the expected change in oxygen content between the latter two equilibration conditions (Eror and Smyth, 1970).

The equilibrium conductivity of $BaTiO_3$ containing a small concentration of Nb^{5+} is shown in Fig. 14.19 (N.-H. Chan and Smyth, 1984). The plateau at higher values of $P(O_2)$ corresponds to electronic compensation. The relative temperature independence indicates that the electron concentration is fixed by the donor concentration and

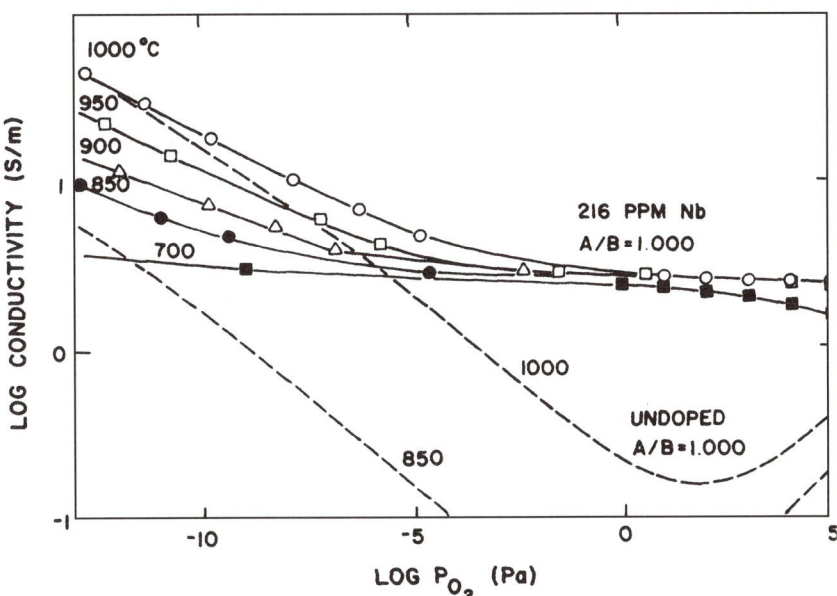

Figure 14.19 Equilibrium conductivity of $BaTiO_3$ as a function of oxygen activity with 216 ppm Nb^{5+} replacing Ti^{4+}. The dashed lines show the conductivity of undoped $BaTiO_3$ for comparison. (Reproduced from N.-H. Chan and Smyth, 1984, by permission of the American Ceramic Society.)

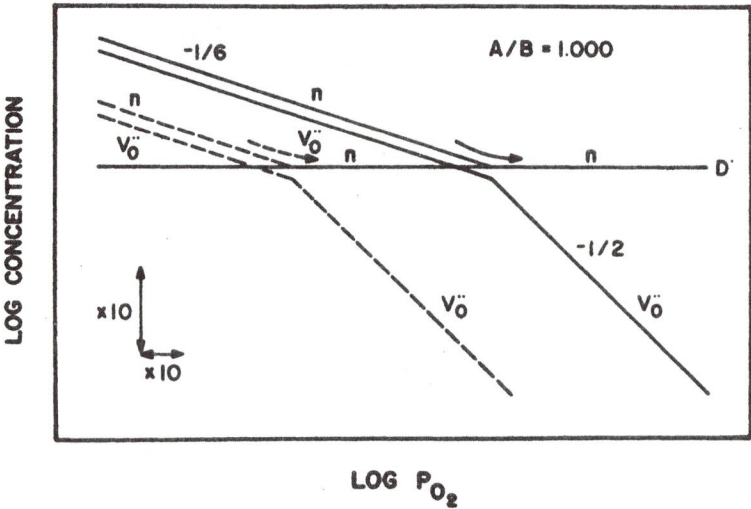

Figure 14.20 Simplified Kröger–Vink diagram for BaTiO$_3$ containing a small amount of donor dopant. Concentrations are shown for two different equilibration temperatures. (Reproduced from N.-H. Chan and Smyth, 1984, by permission of the American Ceramic Society.)

that there is little temperature dependence of the electron mobility. The slight turn-down near $P(O_2) = 1$ atm at the lower temperatures indicates that ionic compensation is becoming significant with a corresponding reduction in the electron concentration. In the low $P(O_2)$ range, oxygen loss becomes the dominant factor and the behavior converges with that of undoped BaTiO$_3$. The defect model can be summarized by a Kröger–Vink diagram as shown in Fig. 14.20. The equilibrium conductivities at 1000°C for samples with various Nb^{5+} concentrations are shown in Fig. 14.21. The smallest substitution, 61 ppm, only partially compensates for the naturally occurring acceptor impurity content and is insufficient to give donor-doped behavior.

The equilibration times in the region of electronic compensation are much longer than those for undoped or acceptor-doped material, and the experiments are correspondingly more tedious. This is consistent with the suppression of the oxygen vacancy concentration by the donor dopant, since the oxygen vacancies are the mobile species that support the diffusion of oxygen into and out of the lattice.

TRIVALENT DOPANTS IN BaTiO$_3$

We have discussed Al^{3+} substituted for Ti^{4+} as a typical acceptor dopant, and La^{3+} substituted for Ba^{2+} as a typical donor dopant. On either site the formal defect charge is unity, Al$'_{Ti}$ or La$^{\cdot}_{Ba}$. These substitutions can be expected, owing to the similarities in ionic radii between the dopant and the replaced cation. However, there is a wide range of sizes for trivalent cations, and these two examples can be considered to

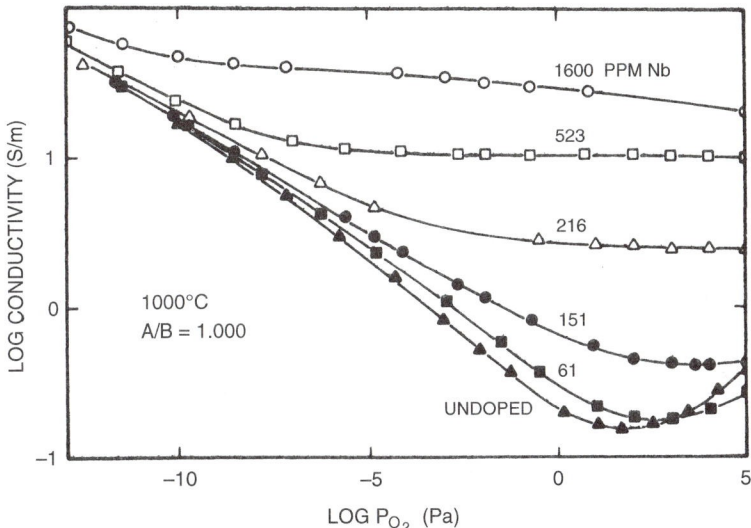

Figure 14.21 The equilibrium conductivity of donor-doped BaTiO₃ as a function of oxygen activity with several different donor levels (in ppm replacement of Nb⁵⁺ for Ti⁴⁺). (Reproduced from N.-H. Chan and Smyth, 1984, by permission of the American Ceramic Society.)

be end members. As we move down group III in the periodic table, the radii (six-coordinate) increase from Al³⁺ (0.0535 nm) to Sc³⁺ (0.0745 nm) to Y³⁺ (0.0900 nm) to La³⁺ (0.1032 nm). The rare earth cations give a finely tuned series of trivalent radii from La³⁺ (0.1032 nm) to Lu³⁺ (0.0861 nm). (The ionic radii for the trivalent rare earths decrease with increasing atomic number as the 4f shell is gradually filled. The f electrons do not shield the outer electrons effectively from the increasing charge of the nuclei, so that the electronic cloud is pulled inward. This is sometimes referred to as the lanthanide contraction.) The equilibrium electrical conductivity can be used to determine how a given dopant behaves. We know that the conductivity of acceptor-doped BaTiO₃ should show a distinct p-type region and a conductivity minimum, while that of donor-doped material should exhibit a plateau, when the donor is compensated by electrons. Fig. 14.22 shows the equilibrium conductivity at 1000°C of BaTiO₃ doped with 0.12 atom % of various trivalent impurities; this is within the range of electronic compensation for donor dopants (Takada et al., 1987). The samples shown in Fig. 14.22a had 1% excess TiO₂ while those in Fig. 14.22b had 1% excess BaO. It is seen that there is a steady shift from acceptor-doped behavior for the smallest cation, Yb³⁺ (0.0868 nm) to donor-doped behavior for the largest, Nd³⁺ (0.0983 nm). The cations of intermediate size seem to be occupying both A and B sites to varying degrees, depending on their radii. Er³⁺ (0.0900 nm) is particularly interesting as a transition size. In the presence of excess TiO₂ it behaves as a weak donor, while in the presence of excess BaO it acts as an acceptor. Apparently it can be pushed from one site to the other depending on the A/B ratio. Y³⁺ (0.0890 nm) behaves very similarly, as would be expected from the similarity in radii. The radii

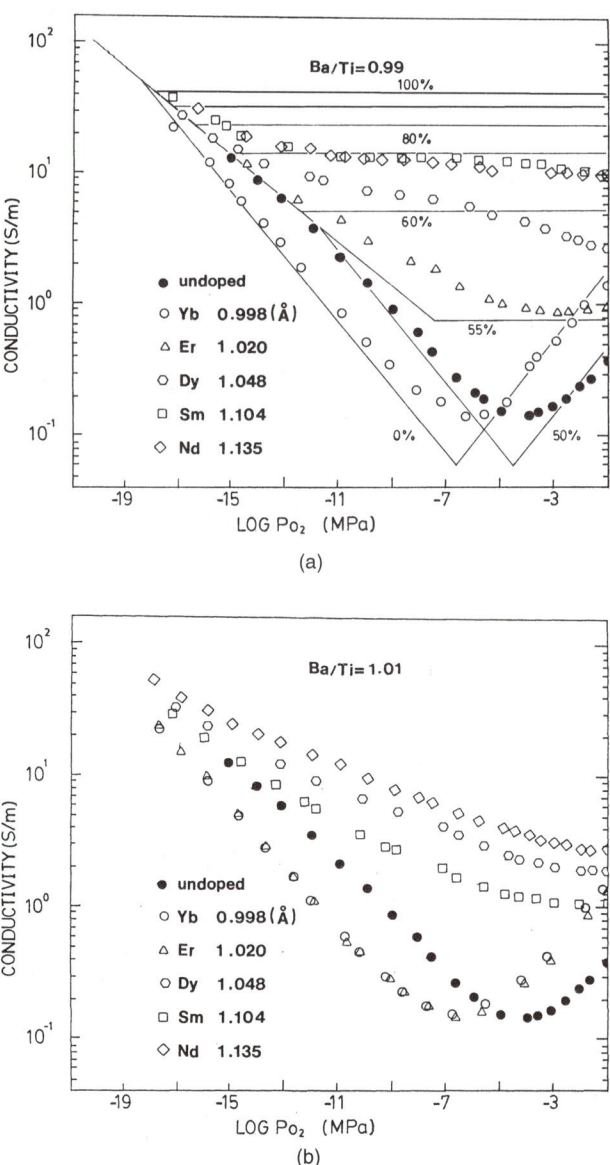

Figure 14.22 The equilibrium conductivity of BaTiO$_3$ with 1200 ppm of various trivalent rare earth impurities as a function of oxygen activity. The samples contain 1% excess TiO$_2$ (a) and 1% excess BaO (b). (Reproduced from Takada et al., 1987, by permission of the American Ceramic Society.)

of these two cations are approximately midway between those of Ba^{2+} (0.135 nm) and Ti^{4+} (0.0605 nm). The dashed lines in Fig. 14.22a represent predicted levels of conductivity for different site occupation ratios, based on an assumed mobility value for the electrons. It is clearly possible to determine the actual site occupation ratios by comparison. It is probable that the two largest dopant cations are pure donors in the samples with excess TiO_2, and a small error in the assumed mobility value accounts for the position of their conductivity plateau below the line calculated for 100% occupancy on the A sites.

The ionic radius can be a determining factor even when the resulting dopant center has a formal defect charge of 2. When $BaTiO_3$ is doped with the other alkaline earth cations, Sr^{2+} (0.118 nm) always substitutes for Ba^{2+}, while we have seen that Ca^{2+} (0.100 nm), while preferring the A site, can be induced to substitute for Ti^{4+} up to about the 2% level. The smaller Mg^{2+} (0.0720 nm) substitutes predominantly, or perhaps entirely, for Ti^{4+}. This might be anticipated by the earlier observation that $MgTiO_3$ has the ilmenite structure, where both cations are six-coordinate, rather than the perovskite structure, where the A site cation is twelve-coordinate.

SUMMARY

With the exception of the peculiar and abrupt transition from semiconducting to insulating properties with increasing donor content, it is seen that the defect chemistry of $BaTiO_3$ behaves quite ideally. Because the system shows such a wide variety of behavior (i.e., both n- and p-type conductivity in the impurity-controlled region and the resulting conductivity minima, and intrinsic behavior under the most reducing conditions), it is a particularly useful system for demonstrating the principles of defect chemistry. This is of more than academic interest given the usefulness of this material, as well as of other compounds of similar composition and structure.

REFERENCES

Chan, H. M., M. P. Harmer, and D. M. Smyth. Compensating defects in highly donor-doped BaTiO₃. *J. Am. Ceram. Soc.* 69(6):507–510, 1986.

Chan, N.-H., and D. M. Smyth. Defect chemistry of donor-doped BaTiO₃. *J. Am. Ceram. Soc.* 67(4):285–288, 1984.

Chan, N.-H., R. K. Sharma, and D. M. Smyth. Nonstoichiometry in undoped BaTiO₃. *J. Am. Ceram. Soc.* 64(9):556–562, 1981.

Chan, N.-H., R. K. Sharma, and D. M. Smyth. Nonstoichiometry in acceptor-doped BaTiO₃. *J. Am. Ceram. Soc.* 65(3):167–170, 1982.

Chang, E. K., A. Mehta, and D. M. Smyth. Ionic transport numbers from equilibrium conductivities. In *Proceedings of a Symposium on Electro-Ceramics and Solid-State Devices*, H. L. Tuller and D. M. Smyth, Eds. Pennington, NJ: The Electrochemical Society, 1988, pp. 35–45.

Choi, G. M., and H. L. Tuller. Defect structure and electrical properties of single-crystal Ba₀.₀₃Sr₀.₉₇TiO₃. *J. Am. Ceram. Soc.* 71(4):201–205, 1988.

Dong, C. C., M. V. Raymond, and D. M. Smyth. The insulator-semiconductor transition in perovskite oxides. In *Electroceramics IV*, Vol. I, R. Waser, S. Hoffmann, D. Bonnenberg, and C. Hoffmann, Eds. Aachen: Augustinus Buchhandlung, 1994, pp. 47–52.

Eror, N. G., and D. M. Smyth. Oxygen nonstoichiometry of donor-doped $BaTiO_3$ and TiO_2. In *The Chemistry of Extended Defects in Non-Metallic Solids*, L. Eyring and M. O'Keeffe, Eds. Amsterdam: North-Holland, 1970, pp. 62–74.

Eror, N. G., and D. M. Smyth. Nonstoichiometric disorder in single-crystalline $BaTiO_3$ at elevated temperatures. *J. Solid State Chem.* 24:235–244, 1978.

Han, Y. H., J. B. Appleby, and D. M. Smyth. Calcium as an acceptor impurity in $BaTiO_3$. *J. Am. Ceram. Soc.* 70(2):96–100, 1987.

Ihrig, H. On the polaron nature of the charge transport in $BaTiO_3$. *J. Phys. C.* 9(18):3469–3474, 1976.

Jonker, G. H. Some aspects of semiconducting barium titanate. *Solid-State Electron.* 7:895–903, 1964.

Lewis, G. V., and C. R. A. Catlow. Computer modeling of barium titanate. *Radiat. Eff.* 73:307–314, 1983.

Ma, F. Oxidation enthalpy measurements for acceptor-doped perovskite materials. M.S. thesis, Lehigh University, Bethlehem, PA, 1995.

Moulson, A. J., and J. M. Herbert. *Electroceramics, Materials, Properties, and Applications.* London: Chapman & Hall, 1990.

Raymond, M. V., and D. M. Smyth. Defect chemistry and transport properties of $Pb(Zr_{1/2}Ti_{1/2})O_3$. *Integrated Ferroelectr.* 4:145–154, 1994.

Sakabe, Y, T. Takagi, K. Wakino, and D. M. Smyth. Dielectric materials for base-metal multilayer ceramic capacitors. *Ad. Ceram.* 19:103–115, 1987.

Schmalzried, H. Point defects in ternary ionic compounds. *Prog. Solid State Chem.* 2:265–303, 1965.

Seuter, A. M. J. H. Defect chemistry and electrical transport properties of barium titanate. *Philips Res. Rep. Suppl..* 3:1–84, 1974.

Smyth, D. M. Thermodynamic characterization of ternary compounds. I. The case of negligible defect association. *J. Solid State Chem.* 16:73- 81, 1976.

Smyth, D. M. Thermodynamic characterization of ternary compounds. II. The case of extensive defect association. *J. Solid State Chem.* 20:359–364, 1977.

Smyth, D. M. Oxidative nonstoichiometry in perovskite oxides. In *Properties and Applications of Perovskite-Type Oxides*, L. G. Tejuca and J. L. G. Fierro, Eds. New York: Marcel Dekker, 1993.

Takada, K., E. Chang, and D. M. Smyth. Rare earth additions to $BaTiO_3$. *Adv. Ceram.* 19:147–152, 1987.

Waser, R. Bulk conductivity and defect chemistry of acceptor-doped strontium titanate in the quenched state. *J. Am. Ceram. Soc.* 74:1934–1940, 1991.

Order versus Disorder $\qquad$ 15

So far, we have dealt only with individual lattice defects or with small defect complexes that usually involve only two oppositely charged defects. There is persuasive evidence for more complex defect clusters in systems that have very high defect concentrations, such as nonstoichiometric FeO or UO_2. Moreover, as techniques for structural characterization become more sophisticated and sensitive, additional phenomena have been observed. In the 1950s, using improved x-ray diffraction techniques, investigators such as Arne Magnéli in Sweden and David Wadsley in Australia began to see systematic structural changes in nonstoichiometric TiO_2 (Andersson et al., 1957) and solid solutions such as Cr_2O_3–TiO_2 and Nb_2O_5–TiO_2 (Wadsley, 1955; Andersson et al., 1959). This type of structural change is shown in Fig. 15.1, which represents a two-dimensional view of an oxygen-deficient oxide having the ReO_3 structural network of corner-sharing octahedra. The left-hand side of the structure shows an ordered array of oxygen vacancies. After the blocks on either side of the line of vacancies move relative to one another by a process called "crystallographic shear," the vacancies are eliminated and are replaced by a cation-rich layer (shear planes) as shown on the right-hand side. The point defects have been replaced by periodic modification of the structure. Note that the cations in adjacent layers are interstitial relative to one another. In TiO_2, a shear plane is analogous to a single layer of the NiAs structure (see Chapter 2) extending through the rutile lattice. In the ideal case, the shear planes are equally spaced and show up on x-ray diffraction patterns as additional low angle lines. As the concentration of aliovalent impurity, or the amount of nonstoichiometry, increases, the spacing between the shear planes decreases. This results in homologous series of compositions and structures, such as $Ti_{n-2}Cr_2O_{2n-1}$ and Ti_nO_{2n-1} (Andersson et al., 1959). As n increases, the structures approach that of the parent reference structure, the rutile structure in this case. In the early work, these ordered structures were observed only for relatively small values of n (e.g., $n < 10$ or $TiO_{1.9}$). This corresponds to large concentrations of aliovalent dopants, or large deviations from stoichiometry. Thus these findings did not represent a serious threat to classical defect chemists, since there appeared to be plenty of room for randomly distributed point defects at small to modest defect concentrations. It

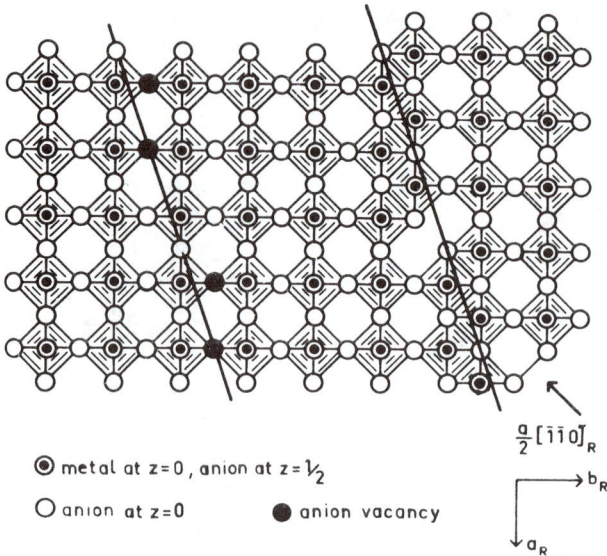

Figure 15.1 A single plane of the three-dimensional array of corner-shared octahedra that is the ReO_3 structure; note ordered array of oxygen vacancies (solid circles) on left. On the right, vacancies were eliminated by a shearing motion in the direction of the arrow and replaced by the metal-rich shear plane. (Reproduced from Hyde and Bursill, 1970, by permission of the author.)

was assumed that individual lattice defects were the dominant form of disorder up to some saturation value, after which the shear planes formed to separate blocks or slabs of defect-saturated material of the parent structure.

The situation became more threatening to defect chemists with the advent of lattice imaging techniques in transmission electron microscopy. Now the shear planes could be imaged directly as shown in Fig. 15.2 for $(Ti,Fe)O_{1.8}$ (Bursill, 1974). Of even greater significance, even for dilute dopant concentrations, or small amounts of nonstoichiometry, isolated clusters of ordered planes or even individual shear planes could be observed throughout the sample, as shown in Fig. 15.3 for "slightly reduced rutile" (Hyde and Bursill, 1970). Since these observations were made on samples that had been cooled from the equilibration temperature, it is possible that the ordered structures formed during cooling and do not represent the equilibrium situation at high temperatures. However, the extent of ordering seemed to increase with increasing time of anneal at the equilibration temperature, indicating that the order is probably stable at high temperatures. There appeared to be a competition between the accommodation of aliovalent impurities and nonstoichiometry by the distribution of individual point defects, or simple defect complexes, or by a change in structure that could be viewed as an ordering of defects along certain crystallographic planes. This competition extended to investigators using different experimental techniques. One leading scientist in the field even stated "Thus all the attempts to analyze the results in terms of point defect theory and a simple $P^{-1/n}$ law become meaningless"! Obviously, the author does not agree with this extreme view.

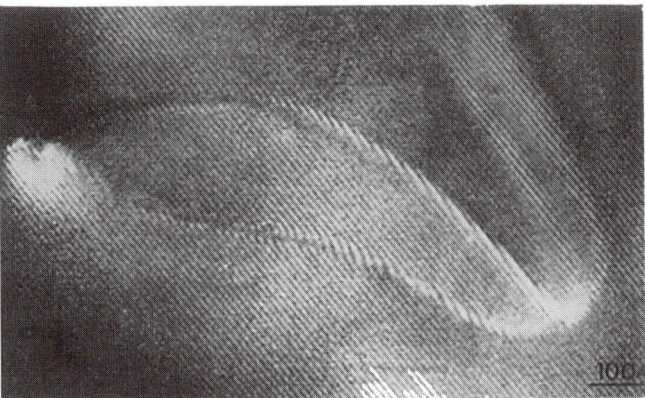

Figure 15.2 TEM image of $(Ti,Fe)O_{1.80}$ showing domains of ordered shear planes with a spacing of 0.98 nm. (Reproduced from Bursill, 1974.)

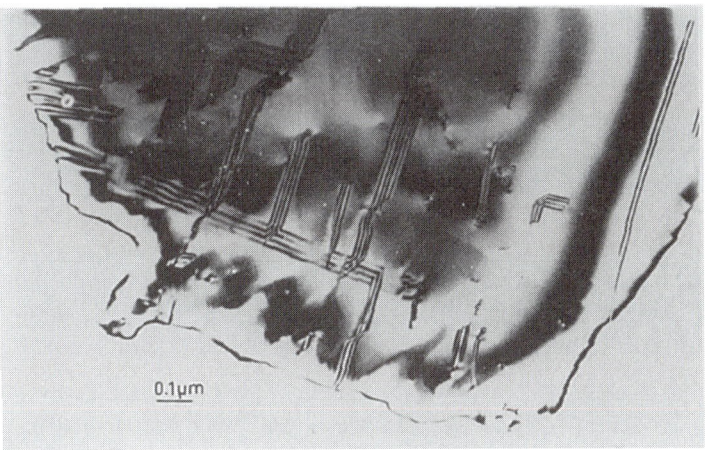

Figure 15.3 A low density of shear planes in very slightly reduced TiO_2. The sample was slow-cooled from 1343 K in a H_2–H_2O atmosphere. (Reproduced from Hyde and Bursill, 1970, by permission of the author.)

Theoretical calculations by Catlow served to clarify the competition (Catlow, 1979). When he compared the calculated energies of accommodation by a distribution of point defects with those of a superlattice ordering in the form of shear planes in a rigid lattice, he found that the random point defect model is always favored. However, when he allowed lattice relaxation to occur around the shear planes, the ordered structures became favored when the amount of relaxation was large. The ability of a lattice to relax is related to its polarizability, and the dielectric constant of the material is a measure of its polarizability. Thus in matrix materials that have large (low frequency) dielectric constants, we can expect superlattice ordering to prevail over randomly distributed point defects. This is in accord with observation of the ordered structures primarily in the titanates, niobates, tungstates, and so on. Of course these are also the systems for which there have been extensive studies of defect chemistry and for which there is very good agreement with the formalism of classical point defect chemistry, which also has a very sound thermodynamic base, as described in the earlier chapters of this book. Thus there still remains some philosophical competition: the microscopists know what they see, and the defect chemists know what they measure. Note that whatever the manner of incorporation, each oxygen atom that leaves the crystal leaves behind two electrons, and each oxygen atom that enters the crystal requires two electrons to become an oxygen ion, thus creating two holes. Thus the electronic conductivity may not be very sensitive to the precise mechanism of accommodation. It is somewhat ironic that while a high dielectric constant promotes organization of defects into periodic structural changes, it also reduces the electrostatic attraction between oppositely charged defects and thus reduces the tendency to form defect complexes or associates.

In a careful study of slightly reduced TiO_{2-x}, with $x < 0.003$, Blanchin et al. (1980), using transmission electron microscopy, found no shear planes in samples quenched from 1100°C but did observe them in slowly cooled samples and in samples annealed at lower temperatures. They collected data for reduced TiO_2 obtained by several investigators that indicates the phase field for material free of shear planes and that for material containing shear planes as a function of temperature and O/Ti ratio; these results are reproduced here as Fig. 15.4. All the compositional data shown in Fig. 12.1 and the equilibrium conductivity data shown in Fig. 12.2 were obtained at 1000°C and higher. According to Fig. 15.4, at that temperature the composition $TiO_{1.99}$ is just about on the boundary between the region of shear planes and the region of dispersed point defects. $TiO_{1.99}$ is the most oxygen-deficient composition shown in Fig. 12.1, and the conditions used to obtain the conductivity data in Fig. 12.2 do not extend further into the region at which crystallographic shear would be

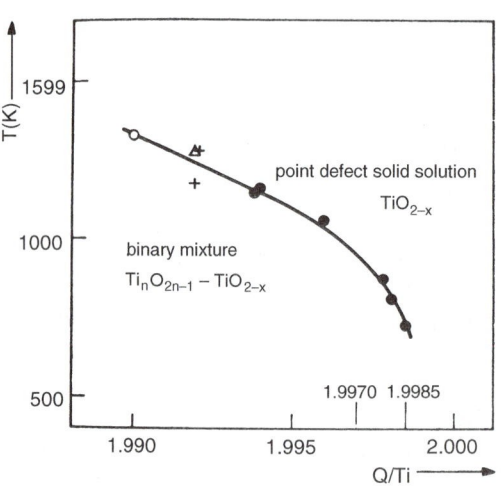

Figure 15.4 Phase diagram for TiO_{2-x} as determined from the data of several investigators showing the boundary between the point defect solid solution and material containing shear planes. (Reproduced from Blancin et al., 1980, by permission of Wiley VCH, SEM.)

expected. Only in the low temperature data shown in Fig. 12.5 would one expect the shear planes to be present in substantial amounts. It appears that there remains plenty of room for defect chemistry involving randomly distributed lattice defects.

BLOCK STRUCTURES

Another form of structural elimination of point defects, which might be called "block structures," involves columns of metal–oxygen octahedra linked by corner-sharing. In cross section the blocks have dimensions of a few octahedra (e.g., 3×4 octahedra or 3×3 octahedra). The blocks extend continuously in the third direction and are linked to one another most usually by edge-sharing. An idealized structure for $Ti_2Nb_{10}O_{29}$ is shown in Fig. 15.5 (Iijima, 1971) This corresponds to 16.7% replacement of Nb^{5+} by Ti^{4+} in Nb_2O_5. The blocks are $3 \times 4\times$ infinity and are linked to vertically offset adjacent blocks by edge-sharing. A lattice image of this material is shown in Fig. 15.6. The light areas are the empty columns between octahedra, as shown in Fig. 15.5. The spacings are not always perfectly uniform; another view of the same material (Fig. 15.7), shows some of the 3×4 columns replaced by layers of 3×3 columns. The structures are sometimes further complicated by the ordered appearance of cations in some of the interstices between the blocks.

Given the enormous variety of ways in which structures can eliminate defects by crystallographic shear and by arrangement of blocks, there is obviously a huge family of complex compositions, each of which corresponds to some specific ordering scheme. Thus we see mention of structures that correspond to compositions such as $MgNb_{14}O_{35}F_2$, $W_8Nb_{18}O_{69}$, and even $Ti_6Mo_{19}Nb_{144}O_{429}$! Oh, Dalton! Where have you gone?

Figure 15.5   Ideal block structure of $Ti_2Nb_{10}O_{29}$, consisting of 3 by 4 columns of corner-shared octahedra. The blocks are connected within each layer and between the layers by edge-sharing. The lighter octahedra lie half an octahedron height below the darker octahedra. (Reproduced from Iijima, 1971, by permission of the American Institute of Physics.)

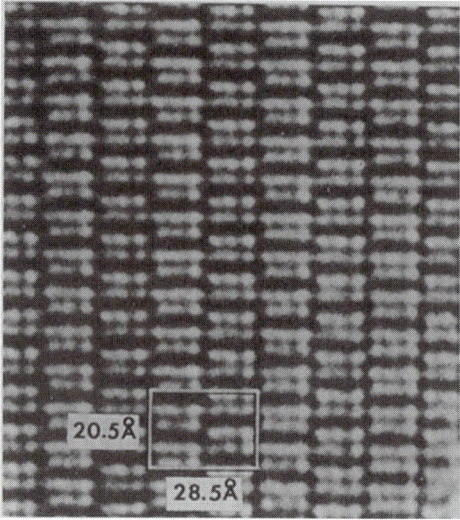

Figure 15.6 Two-dimensional lattice image of $Ti_2Nb_{10}O_{29}$. The unit cell is outlined. The light spots correspond to the unoccupied columns between the corner-shared octahedra shown in Fig. 15.5. (Reproduced from Iijima, 1971, by permission of the American Institute of Physics.)

Figure 15.7 Another lattice image of $Ti_2Nb_{10}O_{29}$ showing a disordered region in which layers of 3 by 3 octahedra are inserted randomly between the normal 3 by 4 regions. (Reproduced from Iijima, 1971, by permission of the American Institute of Physics.)

SUMMARY

It is clear that some compounds, especially those with high dielectric constants, begin to accommodate aliovalent dopants and deviations from stoichiometry at some level of concentration by periodic structural modifications rather than by individual point defects. The major question is, At what point does a given structure become saturated with point defects? Figure 15.4 gives some evidence for one system and indicates that the point defect field is still quite extensive, especially at high temperatures. Further support is given by some thoroughly studied systems for which a simple defect model is in excellent agreement with such diverse experimental techniques as precise compositional determinations, equilibrium electrical conductivities, Seebeck coefficients, and diffusion studies (e.g., see Chapter 13). In particular, one would not expect cation diffusion to correlate with electronic conductivity when the lattice defects reach a saturated limit and additional nonstoichiometry is accommodated by crystallographic shear or the formation of block structures. Moreover, the configurational entropy is a powerful weapon against the ordering of a disordered system. It is important to recognize that ordered structures are an important part of solid state chemistry, but they do not invalidate equilibrium point defect chemistry over a significant and important compositional range.

REFERENCES

Andersson, S., B. Collen, U. Kuylenstierna, and A. Magnéli. Phase analysis studies on the titanium–oxygen system. *Acta Chem. Scand.* 11:1641–1652, 1957.

Andersson, S., A. Sundholm, and A. Magnéli. A homologous series of mixed titanium chromium oxides $Ti_{n-2}Cr_2O_{2n-1}$ isomorphous with the series Ti_nO_{2n-1} and V_nO_{2n-1}. *Acta Chem. Scand.* 13:989–997, 1959.

Blanchin, M. G., P. Faisant, C. Picard, M. Ezzo, and G. Fontaine. Transmission electron microscope observations of slightly reduced rutile. *Phys. Stat. Sol.* A60:357–362, 1980.

Bursill, L. A. An electron microscope study of the FeO–Fe_2O_3–TiO_2 system and the nature of iron-doped rutile. *J. Solid State Chem.* 10:72–94, 1974.

Catlow, C. R. A. In *Modulated Structures, AIP Conference Proceedings*, No. 53, J. M. Cowley, J. B. Cohen, M. B. Salamon, and B. J. Wuench, Eds. New York: American Institute of Physics, 1979, pp. 149–161.

Hyde, B. G., and L. A. Bursill. Point, line and planar defects in some non-stoichiometric compounds. In *The Chemistry of Extended Defects in Non-Metallic Solids*, L. Eyring and M. O'Keeffe, Eds. Amsterdam, North-Holland, 1970.

Iijima, S. High-resolution electron microscopy of crystal lattice of titanium–niobium oxide. *J. Appl. Phys.* 42 (13):5891–5893, 1971.

Wadsley, A. D. The crystal chemistry of nonstoichiometric compounds. *Rev. Pure Appl. Chem.* 5(3):165–193, 1955.

Index